Environmental Impact and Remediation of Heavy Metals

Edited by Hosam M. Saleh and Amal I. Hassan

Published in London, United Kingdom

Environmental Impact and Remediation of Heavy Metals
http://dx.doi.org/10.5772/intechopen.97895
Edited by Hosam M. Saleh and Amal I. Hassan

Contributors
George Yaw Hadzi, Hoang Nguyen Van, Murugesan Manikkampatti Palanisamy, Akilamudhan Palaniappan, Venkata Ratnam Myneni, Padmapriya Veerappan, Minar Mohamed Lebba, Carlos Montalvo, Yunuen Canedo, Brenda Zermeño, Ruby S. Sheila Gines, Rosa M. Cerón, Claudia A. Alejandra Aguilar, Alejandro Ruiz, Gabriel Mustatea, Elena Loredana Ungureanu, Yıldırım İsmail Tosun, Mekonnen M. Tarekegn, Getaneh L. Lulseged Weldekidan, Garima Awasthi, Mahipal Singh Sankhla, Vanisree C. R., Prashant Singh, Kumud Kant Awasthi, Ekta B. Jadhav, Varad Nagar, Rohit Kumar Verma, Josefina Plaza Cazón, Matias Gonzalez, Marcela Ruscitti, Anna Bartkowiak, Lazare Etiégni, Elijah S. Omutange, Julius O. Manyala, Anne N. Wekesa, Odipo Osano, Thomas Munyao, Alfred A. Otieno, Everaldo G. Attard, Tamara Attard, Hosam M. Saleh, Riyam N. Khalef, Amal I. Hassan, Morufu Olalekan Raimi, Henry Olawale Sawyerr, Ifeanyichukwu Clinton Ezekwe, Salako Gabriel

First published in London, United Kingdom, 2022 by IntechOpen
IntechOpen is the global imprint of INTECHOPEN LIMITED, registered in England and Wales, registration number: 11086078, 5 Princes Gate Court, London, SW7 2QJ, United Kingdom

British Library Cataloguing-in-Publication Data
A catalogue record for this book is available from the British Library

Additional hard and PDF copies can be obtained from orders@intechopen.com

Environmental Impact and Remediation of Heavy Metals
Edited by Hosam M. Saleh and Amal I. Hassan
p. cm.
Print ISBN 978-1-80355-525-6
Online ISBN 978-1-80355-526-3
eBook (PDF) ISBN 978-1-80355-527-0

Meet the editors

Hosam Saleh is a Professor of Radioactive Waste Management at the Radioisotope Department, Atomic Energy Authority, Egypt. He was awarded an MSc and Ph.D. in Physical Chemistry from Cairo University. Saleh has more than 25 years of experience in hazardous waste management with an emphasis on treatment and developing new matrixes for the immobilization of these wastes. He is also interested in studying innovative economic and environment-friendly techniques for the management of hazardous and radioactive wastes. He authored many peer-reviewed scientific papers and chapters and served as an editor of several books. He has been selected among the top 2% of scientists in the world according to the Stanford University report for 2020 and 2021.

Amal I. Hassan is a Professor of Animal Physiology, Department of Radioisotopes, Nuclear Research Center, Atomic Energy Authority, Egypt. She has authored many peer-reviewed papers on chronic disease problems. She is a referee, reviewer, and editorial board member for several international scientific journals. She received a Certificate of Excellence in international scientific research arbitration from Publons. She was also selected for inclusion in Who's Who in 2014, 2015, and 2016.

Contents

Preface

Heavy metals are naturally occurring elements that accumulate in the environment primarily because of anthropogenic activities. Humans come into contact with these elements via occupational exposure and food consumption. These substances can have toxic effects on one's health and wellbeing.

The book contains thirteen chapters that provide basic information on heavy metals and their effects on the environment. The chapters are organized into three sections. The first section, "Heavy Metals and Their Effects on the Environment," includes seven chapters.

Chapter 1: "Heavy Metal's Environmental Impact"

Chapter 2: "Toxicity of Heavy Metals"

Chapter 3: "Influence of Heavy Metals on Quality of Raw Materials, Animal Products, and Human and Animal Health Status"

Chapter 4: "Heavy Metal Contamination of Food Crops: Transportation via Food Chain, Human Consumption, Toxicity and Management Strategies"

Chapter 5: "Soil-Skeleton and Soil-Water Heavy Metal Contamination by Finite Element Modelling with Freundlich Isotherm Adsorption Parameters"

Chapter 6: "Microwave Vitrification of Hazardous Sludge by Şırnak Asphaltite Slime, Shale and Sewage Sludge – Landfill Management by Vitrified Char Sand"

Chapter 7: "Heavy Metals in Cosmetics"

The second section, "Evaluation of Heavy Metals and Their Risks to Irrigation Water," includes five chapters.

Chapter 8: "Concentration Levels of Heavy Metals and Selected Ions in the Irrigation Water: The Case of Little Akaki River, Addis Ababa, Ethiopia"

Chapter 9: "Toxicants in Water: Hydrochemical Appraisal of Toxic Metals Concentration and Seasonal Variation in Drinking Water Quality in Oil and Gas Field Area of Rivers State, Nigeria"

Chapter 10: "Effect of Mining on Heavy Metals Toxicity and Health Risk in Selected Rivers of Ghana"

Chapter 11: "Distribution of Potentially Toxic Elements in Water, Sediment and Soils in the Riparian Zones around a Kraft Pulp and Paper Mill in Western Kenya"

Chapter 12: "Quantification of Heavy Metal Levels in Sediments of the "Palizada" River in a Protected Natural Area of Southeastern Mexico"

Finally, the third section "Remediation of Heavy Metals," includes two chapters.

Chapter 13: "Leaching Technology for Precious Heavy Metal Recapture through (HCI + HNO_3) and (HCI + $H_2 SO_4$) from E-Waste

Chapter 14: "Phytoextraction of Zn(II) and Cu(II) by *Canna indica*: Related Physiological Effects"

The editors wish to thank all the participants in this book for their valuable contributions. We also gratefully acknowledge the staff at IntechOpen, especially Author Service Manager Ms. Maja Bozicevic for her continuous assistance in finalizing this work.

Hosam M. Saleh and Amal I. Hassan
Egyptian Atomic Energy Authority,
Cairo, Egypt

Section 1

Heavy Metals and Their Effects on the Environment

Chapter 1

Heavy Metal's Environmental Impact

Riyam N. Khalef, Amal I. Hassan and Hosam M. Saleh

Abstract

Heavy metals are inorganic elements with something like a density of more than 5 g/cm^3. Essential and non-essential heavy metals were divided into two groups based on their toxicity. Heavy metals, unlike organic pollutants, are non-biodegradable and tend to accumulate in living things. Many heavy metal ions are hazardous or carcinogenic. The majority of heavy metals, such as cadmium, copper, and zinc, are linked to pollution and hazardous concerns. There are more than 50 elements categorized as heavy metals, with 17 of them being extremely hazardous and easily accessible. Metal pollutants are often non-degradable and have no recognized homeostasis mechanism. Their mere presence in aquatic habitats is enough to have a direct or indirect impact on living systems. The anthropogenic pollution of heavy metals in ancient mining regions refers to areas where the concentration of one or more heavy metals exceeds normal values. Heavy metals disrupt cellular organelles and components in biological systems. Nanoscale zero-valent iron is a promising alternative for heavy metal cleanup. Heavy metal ions are poisonous, non-degradable, and tend to bioaccumulate and biomagnify. The purpose of this chapter is to display some heavy metals and the environmental impact of these minerals, which includes soil, plants, and humans.

Keywords: heavy metals, chromium, lead, cadmium, mercury, copper, zinc, toxicity of heavy metals, remediation of heavy metals

1. Introduction

Heavy metals are metallic elements with a higher density than water [1]. Heavy metals also include metalloids, such as arsenic, that can cause toxicity at low levels of exposure, based on the notion that heaviness and toxicity are linked. Because of its physical and chemical features, this group comprises arsenic (As), cobalt (Co), Iron (Fe), and manganese (Mn) are less common heavy metals (Mn) [1]. Essential and non-essential heavy metals were divided into two groups based on their toxicity [1]. At low concentrations, heavy metals essential are either nontoxic or considerably less harmful (Zn, Cu, Co, and Fe). Even at low concentrations, non-essential metals are very hazardous (such as Cd, Hg, Cr, and As) [2]. Apart from a few emissions into the atmosphere in the form of dust particles or vapors, these heavy metals are mostly found in the planet's aquatic and soil phases [3]. Because of their toxicity, persistence, and non-degradability, arsenic (As), cadmium (Cd), and lead (Pb) are considered primary hazardous elements [4]. Researchers devised this study to detect the presence or absence of these fatal heavy metals in samples obtained of infant formula milk using Atomic Absorption Spectrophotometry (AAS) since lead, mercury, and arsenic were named as the top three most harmful chemicals on the priority list. The quantity of these heavy metals is measured in parts per million (ppm) [5]. Because, lead, mercury and arsenic were listed as the top three most dangerous compounds on the priority list, researchers devised this study to detect the presence or absence of these deadly heavy metals in chosen samples of infant formula milk using Atomic Absorption Spectrophotometry (AAS). These heavy metals' presence is measured in parts per million (ppm) [5]. Heavy metals are non-biodegradable, poisonous, and easy to accumulate in living creatures in general, and the human body in particular, in low quantities [6]. Heavy metal bioaccumulation in humans, such as cadmium and copper, can lead to cancer, nerve damage, failure of the liver and kidneys, as well as death [7]. Some heavy metals, such as Cd, Pb, and Cr, have no known biological purpose, whereas others, such as Cu, Zn, and Mn, are required in small amounts for normal plant growth and development but are severely poisonous to plants and animals when concentrations are somewhat higher than those required [8]. Heavy metal toxicity is still a hot topic in science, and more research is required to better to recognize the effects of the damaging mechanism and how to control them to lessen medical problems [9]. Due to rapid development, heavy metals wastewaters are increasingly discharged into the environment, mainly in developing nations, due to the rapid development of businesses such as metal plating facilities, mining operations, fertilizer industries, tanneries, batteries, paper industries, and pesticides, among others. Heavy metals, unlike organic pollutants, are non-biodegradable and tend to accumulate in living things. Many heavy metal ions are hazardous or carcinogenic [10]. The majority of heavy metals, such as cadmium, copper, and zinc, are linked to pollution and hazardous concerns, particularly when they are dissolved. Because of their toxicity and mobility, the presence of any of these heavy metals at high levels is dangerous to individuals and can interfere with a variety of environmental benefits [11]. Heavy metals' capacity to penetrate membranes until cells is a key role in many of their harmful effects. Furthermore, metals' transmembrane transfer may be implicated in their absorption, distribution in the body, and excretion, hence transmembrane transfer aids in the determination of metal toxic kinetics. Because cell membranes are so important in metal toxicity [12]. Toxic metals are commonly found in industrial, municipal, and urban runoff, and they can affect humans and other living things. Rising levels of trace metals, particularly heavy metals, in our rivers are due to increased urbanization and industry. Many hazardous chemical components accumulate in the soil and sediments of water bodies after being released into the environment. There are more than 50 elements categorized as heavy metals, with 17 of them being extremely hazardous and easily accessible. Anions play a significant role in drinking water, and the outcomes have been shown to have an impact on people's health [13]. China was close to the city streets. The residents are exposed to street dust regularly. Metal pollutants are often non-degradable and have no recognized homeostasis mechanism. As a result, any large concentration of heavy metals poses a threat to biological life [14]. An increase in anthropogenic activity has resulted in the release of numerous dangerous compounds into water resources, endangering aquatic ecosystems and the environment. Because heavy metal ions are very poisonous, non-degradable, and tend to bioaccumulate and biomagnify as a result of the food chain, they are the most serious contributors to water pollution. Their mere presence in aquatic habitats is enough to have a direct or indirect impact on living systems.

Heavy metal ions are extremely harmful to both plants and animals in the soil environment, as they are absorbed by plants and eventually reach animals and people [15]. A set of metals and metalloids having an atomic density larger than 4000 kg/m^3 is referred to as "heavy metal" (four times the density of water). Heavy metals are present in rocks, soil, plants, and animals and occur naturally in the environment. Metals can be found in a variety of forms, including dissolved ions in water or vapor, as well as minerals in rocks, sand, and soil. These materials can also form bonds with organic and inorganic molecules, as well as cling to airborne particles. Metals are released into the air and water by both natural and manmade mechanisms [16]. The vast range of issues surrounding the presence of HMs in the food chain and their effects on human health necessitates more research in this area as part of a holistic approach to the environment in which humans live. The anthropogenic pollution of heavy metals in ancient mining regions refers to areas where the concentration of one or more heavy metals exceeds normal values in most soils, as well as some agricultural products used as plant food, such as vegetables and fruit, and even animal products (meat, eggs, and milk) [17]. It's also crucial to identify the various sources of heavy metals in the environment and establish their total concentration. Element speciation, profile distribution, and spatial distribution are common methods for distinguishing between anthropogenic and geogenic sources of potentially toxic elements, but they are insufficiently reliable to distinguish between sources of element concentration on their own and should be combined with additional information such as parent rock composition or known anthropogenic sources [18]. Heavy metals have been found to disrupt cellular organelles and components in biological systems, including the cell membrane, mitochondria, lysosomes, endoplasmic reticulum, nuclei, and several enzymes involved in metabolism, detoxification, and damage repair [19]. Nanoscale

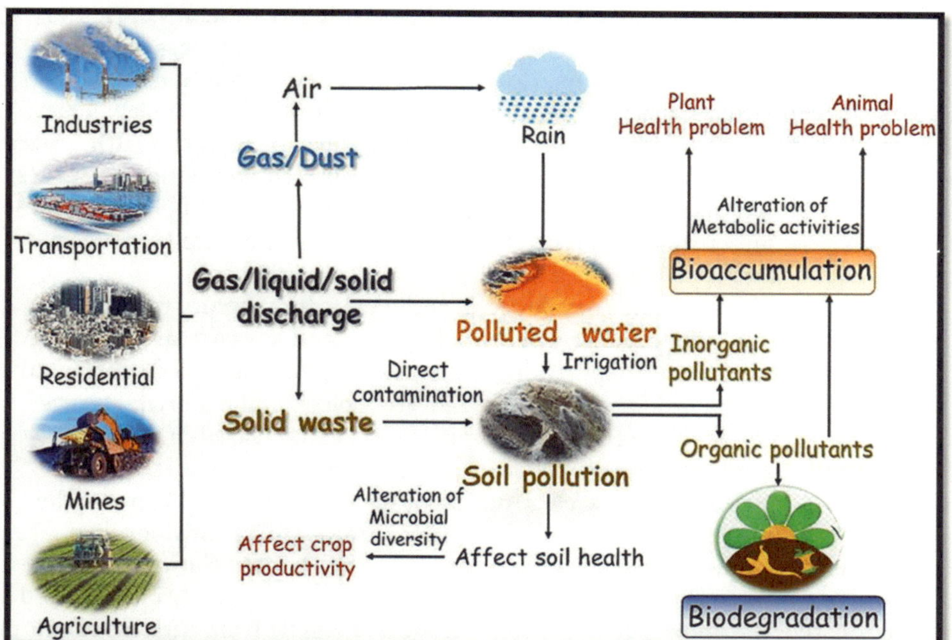

Figure 1.
Sources, migration and toxicity of heavy metals [23].

zero-valent iron is a promising alternative for heavy metal cleanup, with high efficiency and low economic costs [20]. It has a bigger specific surface area, reduction reactivity, and high surface energy. Such dangerous forms of heavy metals may survive in our environment for longer periods, and once they come into touch with the land, water, and soil, they may pose harm to living things [21].

Currently, the study region is endowed with significant mineral resources that span the whole state. Illegal mining of these minerals has resulted in the occurrence of HMs in soil and water supplies. Land application of fertilizers, animal manures, sewage sludge, pesticides, mining tailings, mechanic wastes, and disposal of heavy metal wastes are further sources of HMs [22]. Heavy metals are thought to arise from two basic sources: natural inputs (such as parent material weathering) and human inputs (such as metalliferous industries and mining, automobile emissions, agricultural practices, and so on) [23]. As a result, identifying metal sources is critical before implementing various pollution cleanup measures. The majority of current research is done on mid-to large-scale (for example, mining areas, industrial areas, large cities, etc.) as shown in **Figure 1** [23].

2. Some of the heavy metals

2.1 Mercury

Mercury is a naturally occurring metal that comes in a variety of shapes and sizes. Metallic mercury is a lustrous, silver-white liquid with no odor. Mercury forms inorganic mercury compounds or salts, which are usually white powders or crystals when it reacts with other elements like chlorine, sulfur, or oxygen [24]. Mercury enters the brain quickly, causing tremors, sadness, and behavioral issues [25]. Thousands of years have passed since the beginning of time. It's mostly used for precious metal extraction. Mercury (Hg) is used in thermometers, barometers, manometers, sphygmomanometers, float valves, mercury switches, mercury relays, and fluorescent lights, as well as in the paint industry [26]. In New Jersey, USA, levels of mercury above 0.5 ppm were found, a level of human health concern for those who consume fish regularly. 48.8% of the sampled population of 36,422 lakes in the USA had mercury tissue concentrations that exceeded 0.3 ppm [27]. For many years, mercury, which is on the US EPA's priority pollutants list, has received a lot of attention [28].

Among the various methods developed over the years for mercury removal, adsorption has substantial promise due to its simplicity and low cost, as well as the adsorption method's efficacy in purifying water [29]. We have observed substantial progress in Hg removal in adsorptive separation technology and materials science over the last 4 years [30]. The Hg(II) removal efficiencies increased in the order of ACCl impregnated ACBr-impregnated ACI-impregnated AC, indicating that the introduced functional groups' electron-donating ability may help increase the adsorption capacities of the adsorbents [30]. There are several promising advantages for the removal of aqueous Hg(II), such as rapid separation, easy elution of analytes, and reusable adsorbent [31]. Existing adsorbent materials have largely been limited in their effectiveness and efficiency for the removal of Hg(II) from aqueous solutions due to challenges such as low surface area and improper distribution of thio/thiol groups, resulting in low capacity and moderate affinity for Hg(II), sulfur leaching, and poor stability over a wide pH range [31].

2.2 Cadmium

Cadmium is a silver-white metal with a density of 8.7 g/mL and an atomic weight of 112.41 g/mol. Its main industrial applications are governed by its strong electrical conductivity, good chemical resistance, and low melting point [32]. Cadmium is released from a wide range of sources, including galvanized pipeline corrosion, erosion of natural deposits, discharge from metal refineries, runoff from waste batteries, and mining, smelting, and refining of nonferrous metals [33]. Cd impacts the brain by disrupting particular membrane function, primarily in the hippocampus, according to reports. Neurotoxicity is known to be higher in newborns than in adults, which could be owing to the absence of blood-brain barrier maturation in newborns. Cd accumulates greater in the choroid's plexus region of the brain in newborns. In prenatal exposure to Cd inhibits the acetylcholine esterase (AchE), Na^+/K^+-ATPase pump, which lowers neuronal activity in pups, according to in-vivo investigations [34]. Cadmium is one of the most toxic heavy metals, and this has piqued the attention of environmentalists. The primary sources of cadmium discharge into the environment through waste streams include electroplating, smelting, alloy production, pigments, plastic, battery, mining, and refining activities [35].

Compared to virgin biochar, biochar treated with MnO_2 has more hydroxyl groups, a bigger surface area, and a higher pore volume. They also verified that Cd(II) complexation with hydroxyl groups produces Cd–O or Cd–OH species, which is the major mechanism for Cd elimination [24]. The batch and column desorption experiments were carried out under the same conditions. One gram of adsorbent was used to treat 50 milliliters of sample water containing 20 mg of Cd(II). It was left in contact with the adsorbent for 24 hours. Then, as detailed in a previous paper [36] investigated the effects of pyrolysis temperature on biochar Pb removal ability [24].

2.3 Chromium

Chromium is an element that can be found as a liquid, solid, or gas in rocks, animals, plants, and soil. Chromium(VI) compounds are carcinogens and toxins, but chromium(III) is a required vitamin. Long-term exposure can affect the liver, kidneys, circulatory system, and nerves, as well as cause skin irritation. High levels of breathing can cause nasal irritation, nose ulcers, runny nose, and breathing problems such as asthma, cough, shortness of breath, or wheezing [37]. The use of cellulose ion exchangers for water and wastewater treatment was investigated. The ability of three cellulose ion exchangers to remove proteins, azo dyes, chromate, and heavy metal ions was investigated [38].

As a result, the elimination of Pb(II) and Cr(VI) has been recognized as a crucial environmental issue. To remediate Pb(II) and Cr(VI), mineral adsorption, coagulation, chemical precipitation, ion exchange, biosorption, chemical reduction, and membrane separation have all been developed [20].

2.4 Lead

Pb(II) ion has been discovered to be one of the most dangerous heavy metals, with the ability to cause harmful consequences in animals and plants. Drinking water is the primary route for Pb(II) ions to enter the food chain, and aquatic creatures' bioaccumulate them [39]. Lead poisoning and chelation therapy: The general public is exposed and its compounds through industries like automobile and battery manufacture, refining, and smelting. Lead disrupts several biological processes and is toxic to the nervous system, heart, kidneys, and gastrointestinal tract, with the neurological system being the most vulnerable. Lead also causes cognitive issues in children by interfering with brain growth [40]. Heavy metals, including Pb, are produced by the combustion of fossil fuels in automobiles. Pollution is also thought to be caused by the wear of motor tires and the corrosion of parts. As a result, heavy metal pollution caused by automobiles continues to be a severe problem around the world [41].

Many strategies have been used to remove heavy metals from contaminated liquid effluents to reduce the negative impact. Chemical precipitation is a popular industrial procedure that takes multiple phases before the water can be considered drinkable [42]. The majority of adsorption research was done in a batch setting. Various low-cost adsorbents, such as bagasse pith sulphurised activated carbon, blast furnace sludge, biogas residual slurry, olive mill products, and peanut shell carbon, have been used in the search for effective and cheap removal of Pb(II) from wastewater [43].

2.5 Copper

Copper is common metal contamination that, by definition, is essential for organism functioning, but it is also potentially dangerous copper may in the soil be either static or migratory. Immobile copper that is not bioavailable can be absorbed or precipitated into the soil matrix. Because copper is usually cationic, it forms complexes with negatively charged clay minerals, anionic salts, organic materials, hydroxides, phosphorus, and sulfate [44]. Copper is a versatile metal with many uses due to its excellent qualities. It's utilized in electronics, as well as the manufacture of wires, sheets, and tubes, as well as the formation of alloys. Copper is resistant to the effects of the atmosphere and many chemicals; yet, it is known that metal is prone to corrosion in aggressive media. Copper corrosion inhibitors are required in such situations since no protective passive layer may be envisaged. Copper corrosion is a possibility [45]. Even though copper is required for human metabolism, it is hardly used. Excessive copper use, on the other hand, can cause major health problems including high cholesterol, rapid breathing, kidney and liver damage, convulsions, cramps, vomiting, and even death [46].

The disposal of precipitated cupric hydroxide is a major issue with this form of treatment. Ion exchange treatment, which is the second most extensively used approach for copper removal, does not have a sludge disposal issue and has the advantage of Cu(II) reclamation [47].

2.6 Zinc

Zinc belongs to the periodic table's group IIB and is a beautiful bluish-white metal. When heated between 110°C and 150°C, it transforms from brittle and crystalline to ductile and pliable. It's a moderately reactive metal that reacts with oxygen and other non-metals to form hydrogen, as well as dilute acids. The majority of zinc is added through industrial processes such as mining, coal and waste combustion, and steel processing [48]. Zinc is an important trace element for human health. It regulates various metabolic processes and is necessary for the physiological functioning of living tissue. Too much zinc, on the other hand, can cause serious health problems such as stomach pains, rashes, vomiting, nausea, and anemia [10]. Vomiting, diarrhea, bloody urine, icterus (yellow mucus membrane), liver failure, renal failure, and anemia have all been documented as symptoms of zinc toxicosis [49]. Zinc is widely used in various industries such as galvanization, paint, batteries, smelting, fertilizers and pesticides, fossil fuel combustion, pigment, polymer stabilizers, and so on, and zinc is present in large amounts in effluent from these industries [50].

The process is affected by a variety of factors. Adsorption (chemisorption), complexation on the surface and pores, ion exchange, microprecipitation, heavy metal hydroxide condensation onto the biosurface, and surface adsorption are some of the processes involved in the biosorption process [51].

3. Effect of heavy metals on water

An increase in anthropogenic activity has resulted in the release of numerous dangerous compounds into water resources, endangering aquatic ecosystems and the environment. Heavy metal ions, which are very poisonous, non-degradable, and tend to bioaccumulate and biomagnify, are the most serious contributors to water pollution [15]. Toxic contaminants from anthropogenic businesses, such as mining or agricultural operations that do not use environmentally friendly procedures, or natural phenomena, such as volcanoes, earthquakes, or storms, are virtually always present in wastewaters [52]. Heavy metals are now one of the most important environmental hazards. To safeguard people and the environment, hazardous heavy metals should be eliminated from wastewater. Chemical precipitation, ion exchange, adsorption membrane filtration, electrochemical treatment procedures, and other ways are utilized to remove heavy metal ions [53]. It is regarded as one of the most hazardous to human health among heavy metals. Precipitation, coagulation/flocculation, ion exchange, reverse osmosis, complexation/sequestration, and electrochemical processes are all common ways of removing heavy metal ions from wastewater. These technologies are not commercially viable, and they may result in the production of harmful by-products [54]. As a result, removing unwanted metals from water systems effectively and thoroughly remains critical yet difficult work for environmental engineers. Several strategies for removing heavy metals from water have been presented in recent years [55].

3.1 Treatment of water contaminated with heavy metals

Because the composition of wastewater is exceedingly intricate, and the characterization of target species will be severely influenced by the many coexisting compounds, current technologies will have a difficult time recognizing the detailed composition. Physicochemical tests such as complexometric titration, ion exchange, and stripping voltammetry were used to evaluate the complexation features in early investigations, making it difficult to obtain the exact coordination condition of heavy metals [56]. Micellar-enhanced ultrafiltration (MEUF) is a newly discovered water treatment technique that has been proved to be an effective technology for removing heavy metals even at low concentrations [7].

4. Effect of heavy metals on plant

Soil heavy metal pollution would result in two major issues: loss of soil value and increased health risks for persons living near affected areas. Soil that has been poisoned by heavy metals will lose at least some of its function. When heavy metal concentrations are within legal limits, soils may be able to continue to function. However, more attention should be made to soil heavy metal intervention and goal values [57]. In light of the phytotoxicity and biological relevance of the metal species that control various plant processes, the term "heavy metal" is coined. Few metals, such as Zn, Fe, Cu, Mn, Ni, and Co, are important micronutrients for plants, but others, such as Hg, Al, Cd, Pb, As, Ga, Ag, and Cr, are non-essential for plants and have no recognized physiological function. The HMs' critical limit thresholds and reactions at the cellular and whole-plant levels are summarized [58]. The overall visual toxic reaction differs between heavy metals due to their varied locations of action inside the plant. The most common visual indication of heavy metal toxicity is a reduction in plant development, which includes leaf chlorosis, necrosis, turgor loss, a drop-in seed germination rate, and crippled photosynthetic machinery, which is commonly linked to senescence or plant mortality [59]. The concentration of this element in food varies depending on where it comes from, how it's stored, and how it's processed. These metals have several peculiar characteristics, including the fact that (1) they do not degrade over time (2) they can be necessary or beneficial to plants at certain levels but can be toxic when levels exceed specific thresholds, (3) they are always present at a background level of non-anthropogenic origin, with their input in soils being related to weathering of parent rocks and paedogenesis, and (4) heavy metals in soils can become mobile as a result of changing environmental conditions because they frequently appear as cations that interact strongly with the soil matrix [5]. Multiple studies have found that anthropogenic sources are the principal contributors of heavy metal contamination in the environment. Traffic emission (vehicle exhaust particles, tire wear particles, weathered street surface particles, brake lining wear particles, etc.), industrial emission (power plants, coal combustion, metallurgical industry, auto repair shop, chemical plant, etc.), domestic emission, building and pavement surface weathering, and atmospheric deposited heavy metals are all anthropogenic sources of heavy metals in urban soils and urban road dust [60]. Soil bacteria have been shown to alter heavy metal mobility and bioavailability by solubilizing metal phosphates, releasing chelating agents, producing redox changes, and acidification [61]. Due to the textural composition of distinct soil strata, heavy metals differ in different soil horizons. Different remediation strategies have been developed to avoid metal deposition and movement within the soil profile. Those based on the addition of materials capable of immobilizing mobile forms of metals, such as compost and biosolid, are sufficient. Another approach is phytoremediation, which is based on heavy metal absorption by various plant species [62]. pH, organic matter, and redox conditions are all factors that affect the chemistry of metals in soil and their intake by organisms; of these, pH is the most important and easiest controllable. Soil pH influences the availability and plant uptake of micronutrients [63].

4.1 Treatment of soil contaminated with heavy metals

Soil washing, which comprises pretreatment, separation, coarse-grained treatment, fine-grained treatment, water treatment, and residual management, can reduce heavy metal concentrations in soils by physical/chemical desorption, chelation, dissolution, and oxidation processes. The distribution of heavy metal compounds between soil and washing solution impacts soil cleaning performance. The efficacy of treatment varies depending on the washing technique and solution agents utilized [57]. Heavy metals and metalloids can accumulate in soils due to emissions from rapidly expanding industrial areas, mine tailings, disposal of high metal wastes, leaded gasoline and paints, fertilizer application, animal manures, sewage sludge, pesticides, wastewater irrigation, coal combustion residues, petrochemical spillage, and atmospheric deposition [64]. Minerals are dissolved in most cases by reacting with carbonic acid and water. Insoluble minerals are distributed into fine particles. Metals and metalloids from metal wastes, gasoline, animal feces, sludge, wastewater irrigation, and atmospheric deposition contaminate soils. Heavy metals can be removed from soil and water via phytobiological remediation, which is a cost-effective and environmentally benign method. Heavy metals are removed from soil and water through phytobial remediation, which incorporates plants and bacteria. Plants are used to ingest heavy metals, and microbes aid in the breakdown of those metallic elements in phytobial-based remediation [65]. Integrating an appropriate bacteria that can release numerous plant growth-promoting substances can improve these mechanisms [66]. Phytobial remediation, in contrast to other invasive technologies, is widely considered the safest and most cost-effective option. It's in situ treatment method has also been demonstrated to reduce heavy metal distribution in soil and aid in topsoil preservation. Phytoremediation is aided by the mobilization and volatilization of free-living microorganisms. Metals are mobilized through a variety of processes, including volatilization, redox transformation, leaching, and chelation. Endophytes are bacteria and fungus that dwell on the inside of plants. They spend at least part of their life cycle inside the plant without harming it. They are found in almost every plant, with certain of them having the ability to encourage plant development [67]. Secondary metabolites are produced by a few fungal endophytes. Heavy metal tolerance has been discovered in Methylobacterium strains from the *Pteris vittata* plant [68]. Algae are considered an essential constituent of the aquatic system, playing a significant role in the biogeochemical cycle. Because of its exceptional absorption and sequestration capability, it has piqued the interest of researchers all over the world [69]. Though several integrated techniques, recombinant genetic engineering of bacteria and plants has also proven to be worthwhile in terms of heavy metal removal applications. If microbes are genetically modified, they can perform better than the natural variety, which has enormous remedial potential. Similarly, genetic engineering can be used to stimulate phytoremediation to increase heavy metal accumulation and absorption [70].

5. Effect of heavy metals on human

These do not degrade and accumulate in live beings, resulting in a variety of illnesses and disorders of the neurological, immunological, reproductive, and gastrointestinal systems. Because these heavy metal ions (HMI) are non-biodegradable, they can last for decades or even centuries once released into the environment. Lead (Pb), mercury (Hg), cadmium (Cd), chromium (Cr), and arsenic (As) are among the most poisonous heavy metals [71]. Melanin may protect tissue by filtering or detoxicating heavy metals from the surrounding neuronal retina and photoreceptor cells. The choroid plexus of the brain, like the retinal pigment epithelium, sequesters lead and acts as a protective barrier against harmful materials entering the brain [72]. During pregnancy, potentially dangerous contaminants circulating in a pregnant woman's blood might reach the fetus, posing a risk to the child's health. Because of their ubiquitous exposure, cadmium (Cd), lead (Pb), manganese (Mn), and mercury (Hg) have gotten a lot of attention [73]. Exposure to heavy metals and other contaminants has resulted in a variety of problems in humans and wildlife, including carcinogenic, mutagenic, and teratogenic effects. Structure anomalies, nutritional imbalance, metabolic disruption, and low have all been observed in plants cultivated in contaminated areas [74]. Furthermore, exposure to these hazardous metals has been linked to several serious disorders, including Alzheimer's disease [75]. Due to the difficulty of completely avoiding heavy metal exposure, chemoprevention is a prominent technique for shielding humans and animals from the risk of major health problems caused by toxic metal exposure. The usefulness of many antioxidants, including vitamins taurine, in reducing heavy metal-induced oxidative DNA damage was investigated [75]. Liquid pollutants can hurt human health as well as the environment. Landfill leachate and mine drainage, among other sources of these toxins, cause serious health and environmental hazards [76]. Lead (Pb), cadmium (Cd), mercury (Hg), and arsenic are the most prevalent heavy metals that can cause health concerns when taken in contaminated foods (As). For thousands of years, heavy metals have been employed in a variety of situations throughout human culture. Even though the severe health consequences of heavy metals have long been known, heavy metal exposure persists and, in certain countries, is even rising. Unfortunately, food and food containers are one of the most prevalent causes of heavy metal contamination in the general population [77].

Cadmium-rich foods can significantly increase the amount of cadmium in people's bodies. Liver, mushrooms, prawns, mussels, cocoa powder, and dried seaweed are just a few examples. The circulatory system is a significant route of exposure, and blood vessels are thought to be the main organs of cadmium poisoning. Chronic inhalation exposure to cadmium particles is linked to pulmonary function abnormalities and chest radiographs that are suggestive of emphysema [78]. Cigarette smokers have greater blood and urine cadmium levels, while former smokers have intermediate levels and nonsmokers have lower levels [79]. High quantities of chromium(VI) in the air might irritate the lining of the nose and cause ulcers. Irritation and ulcers in the stomach and small intestine, anemia, and sperm impairment are the most common health concerns found in animals after consuming chromium(VI) compounds [80]. Lead is the most systemic toxin, affecting the kidneys, liver, central nervous system, hematopoietic system, endocrine system, and reproductive system among other organs [81]. Mercury has an extremely low excretion rate once absorbed. The kidneys, neurological tissue, and the liver store a large part of what is absorbed. Mercury is harmful in all forms, with gastrointestinal toxicity, neurotoxicity, and nephrotoxicity among the side effects [82]. Several additional critical elements, like copper, are required for biological function; nevertheless, excessive amounts of these metals cause cellular and tissue damage, resulting in a range of negative impacts and human diseases. There is a relatively limited range of concentrations between helpful and hazardous effects for several elements, such as chromium and copper [82].

6. Conclusion and recommendations

Heavy metals are found naturally throughout the earth's crust, anthropogenic activities such as mining and smelting, industrial production and use, and domestic and agricultural use of metals and metal-containing compounds cause environmental contamination and human exposure. Physical parameters such as temperature, phase association, adsorption, and sequestration influence their bioavailability. Metals such as Co, Cu, Cr, Fe, Mg, Mn, and Zn have been identified as crucial nutrients necessary for a variety of biochemical and physiological functions. A lack of certain micronutrients leads to several deficient illnesses or syndromes.

The most important steps taken to protect the environment are the establishment of natural reserves that ensure that one element is not tyrannized over another, by placing endangered animals and rare plants in an ideal environment suitable for them to grow and reproduce without imbalance or imbalance. It decreases because the environment is safe in our necks to maintain it well and use modern methods or environmentally friendly materials for the treatment process. Remediation of heavy metal-contaminated soil is required to eliminate the related dangers, make the land resource available for agricultural development, improve food security, and reduce land tenure issues. In the future, more emphasis should be placed on assessment methodologies for measuring remediation efficacy while creating new remediation technologies.

Author details

Riyam N. Khalef[1,2], Amal I. Hassan[3] and Hosam M. Saleh[3*]

1 Faculty of Science, Chemistry Department, Al Azhar University, Egypt

2 College of Applied Sciences, University of Fallujah, Al-Fallujah, Iraq

3 Radioisotopes Department, Nuclear Research Center, Egyptian Atomic Energy Authority (EAEA), Egypt

*Address all correspondence to: hosam.saleh@eaea.org.eg; hosamsaleh70@yahoo.com

References

[1] Kim J-J, Kim Y-S, Kumar V. Heavy metal toxicity: An update of chelating therapeutic strategies. Journal of Trace Elements in Medicine and Biology. 2019;**54**:226-231

[2] Anjulo TK, Mersso BT. Assessment of dairy feeds for heavy metals. American Academic Scientific Research Journal for Engineering, Technology, and Sciences. 2015;**11**(1):20-31

[3] Zenk MH. Heavy metal detoxification in higher plants-a review. Gene. 1996;**179**(1):21-30

[4] Zhang N et al. Simultaneous determination of arsenic, cadmium and lead in plant foods by ICP-MS combined with automated focused infrared ashing and cold trap. Food Chemistry. 2018;**264**:462-470

[5] Das SK, Grewal AS, Banerjee M. A brief review: Heavy metal and their analysis. Organization. 2011;**11**(1):3

[6] Al-Saydeh SA, El-Naas MH, Zaidi SJ. Copper removal from industrial wastewater: A comprehensive review. Journal of Industrial and Engineering Chemistry. 2017;**56**:35-44

[7] Häyrynen P, Landaburu-Aguirre J, Pongrácz E, Keiski RL. Study of permeate flux in micellar-enhanced ultrafiltration on a semi-pilot scale: Simultaneous removal of heavy metals from phosphorous rich real wastewaters. Separation and Purification Technology. 2012;**93**:59-66

[8] Adrees M et al. Mechanisms of silicon-mediated alleviation of heavy metal toxicity in plants: A review. Ecotoxicology and Environmental Safety. 2015;**119**:186-197

[9] Singh M, Kumar J, Singh S, Singh VP, Prasad SM, Singh M. Adaptation strategies of plants against heavy metal toxicity: A short review. Biochemistry and Pharmacology (Los Angeles). 2015;**4**(161):501-2167

[10] Fu F, Wang Q. Removal of heavy metal ions from wastewaters: A review. Journal of Environmental Management. 2011;**92**(3):407-418. DOI: 10.1016/j.jenvman.2010.11.011

[11] Vardhan KH, Kumar PS, Panda RC. A review on heavy metal pollution, toxicity and remedial measures: Current trends and future perspectives. Journal of Molecular Liquids. 2019;**290**:111197

[12] Foulkes EC. Transport of toxic heavy metals across cell membranes (44486). Proceedings of the Society for Experimental Biology and Medicine. 2000;**223**(3):234-240

[13] Mohod CV, Dhote J. Review of heavy metals in drinking water and their effect on human health. International Journal of Innovative Science Engineering and Technology. 2013;**2**(7):2992-2996

[14] Zheng N, Liu J, Wang Q, Liang Z. Health risk assessment of heavy metal exposure to street dust in the zinc smelting district, Northeast of China. Science of the Total Environment. 2010;**408**(4):726-733

[15] Malik LA, Bashir A, Qureashi A, Pandith AH. Detection and removal of heavy metal ions: A review. Environmental Chemistry Letters. 2019;**17**(4):1495-1521

[16] Sarubbo LA, Rocha RB Jr, Luna JM, Rufino RD, Santos VA, Banat IM. Some aspects of heavy metals

contamination remediation and role of biosurfactants. Chemistry and Ecology. 2015;**31**(8):707-723

[17] Gergen I, Harmanescu M. Application of principal component analysis in the pollution assessment with heavy metals of vegetable food chain in the old mining areas. Chemistry Central Journal. 2012;**6**(1):1-13

[18] Zhou J, Ma D, Pan J, Nie W, Wu K. Application of multivariate statistical approach to identify heavy metal sources in sediment and waters: A case study in Yangzhong, China. Environmental Geology. 2008;**54**(2):373-380

[19] Wang S, Shi X. Molecular mechanisms of metal toxicity and carcinogenesis. Molecular and Cellular Biochemistry. 2001;**222**(1):3-9

[20] Liu J, Mwamulima T, Wang Y, Fang Y, Song S, Peng C. Removal of Pb(II) and Cr(VI) from aqueous solutions using the fly ash-based adsorbent material-supported zero-valent iron. Journal of Molecular Liquids. 2017;**243**:205-211

[21] Yadav VK et al. Synthesis and characterization of amorphous iron oxide nanoparticles by the sonochemical method and their application for the remediation of heavy metals from wastewater. Nanomaterials. 2020;**10**(8):1551

[22] Eyankware MO, Obasi PN. A holistic review of heavy metals in water and soil in Ebonyi SE, Nigeria; with emphasis on its effects on human, plants and aquatic organisms. World News of Natural Sciences. 2021;**38**:1-19

[23] Chen T et al. Heavy metal sources identification and sampling uncertainty analysis in a field-scale vegetable soil of Hangzhou, China. Environmental Pollution. 2009;**157**(3):1003-1010

[24] Li Y, Yu H, Liu L, Yu H. Application of co-pyrolysis biochar for the adsorption and immobilization of heavy metals in contaminated environmental substrates. Journal of Hazardous Materials. 2021;**420**:126655

[25] Langford NJ, Ferner RE. Toxicity of mercury. Journal of Human Hypertension. 1999;**13**(10):651-656

[26] Pratush A, Kumar A, Hu Z. Adverse effect of heavy metals (As, Pb, Hg, and Cr) on health and their bioremediation strategies: A review. International Microbiology. 2018;**21**(3):97-106

[27] Kim K-H, Kabir E, Jahan SA. A review on the distribution of Hg in the environment and its human health impacts. Journal of Hazardous Materials. 2016;**306**:376-385

[28] Zhang F-S, Nriagu JO, Itoh H. Mercury removal from water using activated carbons derived from organic sewage sludge. Water Research. 2005;**39**(2-3):389-395

[29] Li B, Zhang Y, Ma D, Shi Z, Ma S. Mercury nano-trap for effective and efficient removal of mercury(II) from aqueous solution. Nature Communications. 2014;**5**(1):1-7

[30] Yu J-G et al. Removal of mercury by adsorption: A review. Environmental Science and Pollution Research. 2016;**23**(6):5056-5076

[31] Liu J, Du X. Fast removal of aqueous Hg(II) with quaternary ammonium-functionalized magnetic mesoporous silica and silica regeneration. Journal of Materials Chemistry. 2011;**21**(19):6981-6987

[32] Remelli M, Nurchi VM, Lachowicz JI, Medici S, Zoroddu MA, Peana M. Competition between Cd(II)

and other divalent transition metal ions during complex formation with amino acids, peptides, and chelating agents. Coordination Chemistry Reviews. 2016;**327**:55-69

[33] Purkayastha D, Mishra U, Biswas S. A comprehensive review on Cd(II) removal from aqueous solution. Journal of Water Process Engineering. 2014;**2**:105-128

[34] Karri V, Schuhmacher M, Kumar V. Heavy metals (Pb, Cd, As and MeHg) as risk factors for cognitive dysfunction: A general review of metal mixture mechanism in brain. Environmental Toxicology and Pharmacology. 2016;**48**:203-213

[35] Benguella B, Benaissa H. Cadmium removal from aqueous solutions by chitin: Kinetic and equilibrium studies. Water Research. 2002;**36**(10):2463-2474

[36] Ajmal M, Rao RAK, Anwar S, Ahmad J, Ahmad R. Adsorption studies on rice husk: Removal and recovery of Cd(II) from wastewater. Bioresource Technology. 2003;**86**(2):147-149

[37] Staniek H, Krejpcio Z. The effects of supplementary Cr3 (chromium(III) propionate complex) on the mineral status in healthy female rats. Biological Trace Element Research. 2017;**180**(1):90-99. DOI: 10.1007/s12011-017-0985-3

[38] Barbooti MM. Simultaneous removal of chromium and lead from water by sorption on iraqi montmorillonite. Journal of Environmental Protection (Irvine, California). 2015;**06**(03):237-249. DOI: 10.4236/jep.2015.63024

[39] Oladoye PO. Natural, low-cost adsorbents for toxic Pb(II) ion sequestration from (waste) water: A state-of-the-art review. Chemosphere. 2022;**287**:132130

[40] Yu D, Li L, Yuan W. Neonatal anesthetic neurotoxicity: Insight into the molecular mechanisms of long-term neurocognitive deficits. Biomedicine & Pharmacotherapy. 2017;**87**:196-199

[41] Ozaki H, Watanabe I, Kuno K. Investigation of the heavy metal sources in relation to automobiles. Water, Air, and Soil Pollution. 2004;**157**(1):209-223

[42] Awual MR et al. Offering an innovative composited material for effective lead(II) monitoring and removal from polluted water. Journal of Cleaner Production. 2019;**231**:214-223

[43] Arbabi M, Hemati S, Amiri M. Removal of lead ions from industrial wastewater: A review of removal methods. International Journal of Epidemiologic Research. 2015;**2**(2):105-109

[44] Mackie KA, Müller T, Kandeler E. Remediation of copper in vineyards–a mini review. Environmental Pollution. 2012;**167**:16-26

[45] Antonijevic MM, Petrovic MB. Copper corrosion inhibitors. A review. International Journal of Electrochemical Science. 2008;**3**(1):1-28

[46] Abbas A et al. Heavy metal removal from aqueous solution by advanced carbon nanotubes: Critical review of adsorption applications. Separation and Purification Technology. 2016;**157**:141-161

[47] Oncel MS, Muhcu A, Demirbas E, Kobya M. A comparative study of chemical precipitation and electrocoagulation for treatment of coal acid drainage wastewater. Journal of Environmental Chemical Engineering. 2013;**1**(4):989-995

[48] Gakwisiri C, Raut N, Al-Saadi A, Al-Aisri S, Al-Ajmi A. A critical review of removal of zinc from wastewater. In: Proceedings of the World Congress on Engineering. 2012;**4**(1):627-630

[49] Duruibe JO, Ogwuegbu MOC, Egwurugwu JN. Heavy metal pollution and human biotoxic effects. International Journal of Physical Sciences. 2007;**2**(5):112-118

[50] Parmar M, Thakur LS. Heavy metal Cu, Ni and Zn: Toxicity, health hazards and their removal techniques by low cost adsorbents: A short overview. International Journal of Plant, Animal and Environmental Sciences. 2013;**3**(3):143-157

[51] Elham A, Hossein T, Mahnoosh H. Removal of Zn(II) and Pb(II) ions using rice husk in food industrial wastewater. Journal of Applied Sciences and Environmental Management. 2010;**14**(4):159-162

[52] Azimi A, Azari A, Rezakazemi M, Ansarpour M. Removal of heavy metals from industrial wastewaters: A review. ChemBioEng Reviews. 2017;**4**(1):37-59. DOI: 10.1002/cben.201600010

[53] Tahir MT, Badamasi H, Suleiman AK. Methods of removing heavy metal ions from waste waters: A review. Elixir Environment and Forestry. 2017;**110**:48354-48357

[54] Khoramzadeh E, Nasernejad B, Halladj R. Mercury biosorption from aqueous solutions by sugarcane bagasse. Journal of the Taiwan Institute of Chemical Engineers. 2013;**44**(2):266-269

[55] Hua M, Zhang S, Pan B, Zhang W, Lv L, Zhang Q. Heavy metal removal from water/wastewater by nanosized metal oxides: A review. Journal of Hazardous Materials. 2012;**211**:317-331

[56] Xu Z, Zhang Q, Li X, Huang X. A critical review on chemical analysis of heavy metal complexes in water/wastewater and the mechanism of treatment methods. Chemical Engineering Journal. 2022;**429**:131688

[57] Chen W, Li H. Cost-effectiveness analysis for soil heavy metal contamination treatments. Water, Air, & Soil Pollution. 2018;**229**(4):1-13

[58] Thakur M et al. Metal tolerance in plants: Molecular and physicochemical interface determines the 'not so heavy effect' of heavy metals. Chemosphere. 2022;**287**:131957

[59] Gill M. Heavy metal stress in plants: A review. International Journal of Advanced Research. 2014;**2**(6):1043-1055

[60] Wei B, Yang L. A review of heavy metal contaminations in urban soils, urban road dusts and agricultural soils from China. Microchemical Journal. 2010;**94**(2):99-107

[61] Xie Y et al. Identification of Cd-resistant microorganisms from heavy metal-contaminated soil and its potential in promoting the growth and Cd accumulation of bermudagrass. Environmental Research. 2021;**200**:111730

[62] Zubillaga MS, Bressan E, Lavado RS. Heavy metal mobility in polluted soils: Effect of different treatments. American Journal of Environmental Sciences. 2008;**4**(6):620

[63] Bose S, Bhattacharyya AK. Heavy metal accumulation in wheat plant grown in soil amended with industrial sludge. Chemosphere. 2008;**70**(7):1264-1272

[64] Maurya A, Kesharwani L, Mishra MK. Analysis of heavy metal

in soil through atomic absorption spectroscopy for forensic consideration. International Journal for Research in Applied Science and Engineering Technology. 2018;**6**(6):1188-1192

[65] Dubois H, Verkasalo E, Claessens H. Potential of birch (*Betula pendula* Roth and *B. pubescens* Ehrh.) for forestry and forest-based industry sector within the changing climatic and socio-economic context of Western Europe. Forests. 2020;**11**(3):336

[66] Etesami H, Maheshwari DK. Use of plant growth promoting rhizobacteria (PGPRs) with multiple plant growth promoting traits in stress agriculture: Action mechanisms and future prospects. Ecotoxicology and Environmental Safety. 2018;**156**:225-246

[67] Roy M, Giri AK, Dutta S, Mukherjee P. Integrated phytobial remediation for sustainable management of arsenic in soil and water. Environment International. 2015;**75**:180-198

[68] Alcántara-Martínez N, Figueroa-Martínez F, Rivera-Cabrera F, Gutiérrez-Sánchez G, Volke-Sepúlveda T. An endophytic strain of Methylobacterium sp. increases arsenate tolerance in *Acacia farnesiana* (L.) Willd: A proteomic approach. Science of the Total Environment. 2018;**625**:762-774

[69] He M et al. Antimony speciation in the environment: Recent advances in understanding the biogeochemical processes and ecological effects. Journal of Environmental Sciences. 2019;**75**:14-39

[70] de Oliveira Carvalho A, Moreira Gomes V. Plant defensins and defensin-like peptides-biological activities and biotechnological applications. Current Pharmaceutical Design. 2011;**17**(38):4270-4293

[71] Bansod B, Kumar T, Thakur R, Rana S, Singh I. A review on various electrochemical techniques for heavy metal ions detection with different sensing platforms. Biosensors & Bioelectronics. 2017;**94**:443-455

[72] Erie JC, Butz JA, Good JA, Erie EA, Burritt MF, Cameron JD. Heavy metal concentrations in human eyes. American Journal of Ophthalmology. 2005;**139**(5):888-893

[73] Kabamba M, Tuakuila J. Toxic metal (Cd, Hg, Mn, Pb) partition in the maternal/foetal unit: A systematic mini—Review of recent epidemiological studies. Toxicology Letters. 2020;**332**:20-26

[74] Rahman Z. An overview on heavy metal resistant microorganisms for simultaneous treatment of multiple chemical pollutants at co-contaminated sites, and their multipurpose application. Journal of Hazardous Materials. 2020;**396**:122682

[75] Turkez H, Geyikoglu F, Tatar A, Keles MS, Kaplan I. The effects of some boron compounds against heavy metal toxicity in human blood. Experimental and Toxicologic Pathology. 2012;**64**(1-2):93-101

[76] Francisca FM, Glatstein DA. Environmental application of basic oxygen furnace slag for the removal of heavy metals from leachates. Journal of Hazardous Materials. 2020;**384**:121294

[77] Cheraghali AM, Kobarfard F, Faeizy N. Heavy metals contamination of table salt consumed in Iran. Iranian Journal of Pharmaceutical Research (IJPR). 2010;**9**(2):129

[78] Davison AG et al. Cadmium fume inhalation and emphysema. Lancet. 1988;**331**(8587):663-667

[79] Gharibshahi N, Javaherizadeh H, Khorasgani ZN, Mahdavinia M. Relationship between lead and cadmium levels in blood and refractory chronic constipation among Iranian children. Arquivos de Gastroenterologia. 2021;**58**:329-336

[80] Wilbur S, Abadin H, Fay M, Yu D, Tencza B, Ingerman L, et al. Health effects. In Toxicological Profile for Chromium. US: Agency for Toxic Substances and Disease Registry; Sep 2012

[81] ATSDR U. Toxicological Profile for Lead. Agency for Toxic Substances and Disease Registry; 2007

[82] Tchounwou PB, Ayensu WK, Ninashvili N, Sutton D. Environmental exposure to mercury and its toxicopathologic implications for public health. Environmental Toxicology. 2003;**18**(3):149-175

Chapter 2

Toxicity of Heavy Metals

Elena Loredana Ungureanu and Gabriel Mustatea

Abstract

Heavy metals are elements who exist naturally in the environment, but rapid urbanization and industrialization led to increased levels of these metals. These metals can reach the human body through food, water or air, where they have the property to accumulate in various tissues and organs for long periods of time and to produce serious effects on certain organs and the proper functioning of the body. Studies have also shown that heavy metals can have important effects, including on plants or animals. Their toxicity is dependent on factors such as dose, route of exposure, time of exposure, level of concentration, as well as age, gender, genetics, and nutritional status of exposed individuals. There is a growing interest from researchers to detect various physical, physical-chemical or microbiological methods to reduce or eliminate the presence of these metals, especially from surface or wastewater, which are mainly responsible for food contamination. This chapter present the main characteristics of heavy metals, the sources of contamination of exposure, as well as their toxicity on some environmental segments and especially on living organisms.

Keywords: contamination, heavy metals, human health, sources of exposure, toxicity

1. Introduction

Heavy metals (HM) represent a group of metallic elements and metalloids characterized by a relatively density higher than 5 g/cm^3, an atomic number greater than 20 and with properties like conductance of heat, current and luster surface [1–3].

Pollution or contamination of the environment with heavy metals is a major concern, due to their capacity to bioaccumulate and persistence in the environment, non-biodegradable nature, contaminate the food chains and their toxicity on the environment and living organisms (humans, animals and plants) [1–3]. Heavy metal toxicity is a concern of ecological, nutritional, evolutionary and environmental reasons [1].

Heavy metals are among the most investigated pollutants and received a higher attention by researchers, because of their toxicity [2, 4]. These elements are naturally present in the environment, but on which modern industrialization and urbanization, anthropogenic activities and use of fertilizers, led to increased levels of these metals in the environment and implicitly to a high exposure of living things to them [2, 5]. Among the heavy metals and the most toxic metalloids are chromium, mercury, arsenic, cadmium, lead, nickel, copper, zinc, but the most common heavy metals in the environment are chromium, manganese, nickel, lead, cadmium, copper and zinc [2].

Regarding their functions in biological systems, heavy metals can be essential and nonessential. The nonessential heavy metals do not possess biological functions in living organisms, being non-essential to metabolic system, both for plants and animals. Their category includes lead, cadmium, mercury, aluminum and arsenic [2, 6, 7], being able to exert toxic effects even at low concentrations [8]. The essential heavy metals are elements, which are indispensable for plant and animals, which play a vital role in biological processes and entire metabolism and may be required in living organism in different concentrations [2, 8]. These heavy metals are considered as trace elements because of their presence in trace concentrations (less than 10 ppm) in different environmental matrices [9]. The essentiality and toxicity of the trace metals depending on the dose of exposure [10]. This category includes 19 elements, among which the most important are manganese, iron, copper, zinc, nickel and chromium [2].

Trace elements or trace minerals are minerals necessary for the body, but in amounts between 1 and 100 mg/day for adults and represents less than 0.01–0.02% of the total body weight [10–12]. When they exceed these threshold concentrations, they become dangerous to the health of living organisms [1].

According to WHO classification, trace elements can be divided into three groups, such as essential elements (zinc, iodine, molybdenum, copper, selenium, chromium), probably essential elements (manganese, silicon, boron, vanadium, nickel) and potentially toxic elements (lead, cadmium, fluorine, mercury, aluminum, arsenic, barium, lithium, tin [13, 14].

Another classification of the trace elements was made by Frieden in 1981, based on their levels in biological tissues, being divided into 3 groups, namely essential trace elements (boron, cobalt, copper, iodine, manganese, molybdenum, zinc), probably essential trace elements (chromium, fluorine, nickel, selenium, vanadium) and physically promotive trace elements (bromine, lithium, silicone, tin) [13, 15].

The present chapter presents the characteristics of heavy metals, the main sources of heavy metal contamination of the environment, as well as human exposure sources. The impact of their toxicity on various environmental segments, such as water, air, soil, as well as on living organisms, animals, but especially humans, has also been described.

2. Sources of environmental contamination

Heavy metals contamination of environment can come both from natural sources and from anthropogenic processes. Natural emissions of heavy metals include volcanic eruptions, rock weathering, sea-salt sprays, forest fires, biogenic sources, windborne soil particles and can be found in the nature as oxides, hydroxides, silicates, sulphates, sulphides, phosphates, organic compounds [4].

Anthropogenic processes which can release heavy metals in different environmental compartments, are industries, agriculture (insecticides, pesticides which can release As), fossil fuels combustion (Ni, V, Hg, Se, Sn), wastewater, mining, smelting (As, Cu, Zn), corrosion, metallurgical processes, residual organic matter, transportation (Pb) [4, 7, 16].

3. Toxicity

Heavy metals can produce side effects on soil, on water, on air, but also on plants, animals and humans [3, 4, 17]. In soil, high levels of heavy metals can produce alteration of soil quality through modification of pH, color, porosity and natural composition [4, 18], but also low crop production, loss of many types of normal flora and habitat [19]. Their accumulation into the water imposes serious problems on humans and ecosystems [4], due to decreasing of drinking water quality and purity, decreasing water supplies for all living organisms [19]. High levels of heavy metals in air can lead to harmful health problems, including respiratory infections, cardiovascular disease, premature mortality, eyes and skin irritation, but also can cause infrastructure deterioration, acid rain increasing, corrosion, eutrophication and haze [4], low yields of the crop, not enough oxygen [19]. In plants, they can produce damage of roots or leaves, interfere in important biochemical process, such as photosynthesis, alteration of minerals absorption, damage of chlorophyll, reduce the growth and development of the roots, which leading to reduction pf overall growth of the plant [3, 20, 21].

The toxicity of heavy metals in animal is manifested through decreased body weight, kidney damage, liver affections, shortened life span, increased oxidative stress, modifications of cells composition, DNA damage [17]. In humans they can produce kidney damage, liver affections, pulmonary effects, several types of cancer [3].

Heavy metals became toxic when are not metabolized by the body and accumulates in organs and soft tissues [4]. They reach the human body by ingesting contaminated water or food, inhalation of absorption through the skin. Among the pathways, ingestion in the common route that helps the heavy metals to enter to the animal bodies [3, 4]. The effect of this metals can be inhibitory, stimulatory and toxic for some biochemical processes [3], being able to produce various health problems on nervous system (Alzheimer, Parkinsoma, depression, dementia), on bone system (bone mineralization) an on reproductive system. Also, can produce DNA damage, RNA affection, or cancer of lungs, skin, bladder, due to production of ROS [3]. Their toxicity depends the dose of exposure, time of exposure, pollutant concentration, organism which are exposed to it, nature and oxidation state of the metal [3, 4].

3.1 Toxicity of lead (Pb)

Lead is the most important toxic heavy metal in the environment because can cause serious environmental contamination and health problems [1, 10]. The main sources of environmental contamination including industrial processes, such as fossil fuel burning, mining, smelting, manufacturing, recycling activities. It is also used for leaded pipes, lead-glazed or lead-soldered containers, leaded paint, leaded gasoline, leaded aviation fuel [10, 22].

The inorganic lead can enter into the human body by inhalation (pulmonary absorption) of contaminated air or by smoking (15%), or by ingestion (gastrointestinal absorption) of food (65%) and water (20%) [1, 3, 22, 23]. Although organic compounds are absorbed through skin, inorganic compounds cannot be absorbed [10].

According to the WHO guidelines, the international level of concern of poisoning with lead is 25 μg/dl of blood for adults and for children, it must be less than 5 μg/dl of blood [23]. Their absorption is influenced by the age and physiological status of the exposed person [22].

However, the nervous system is most affected by exposure to high concentrations of lead, in both children and adults. Because children absorb 4–5 times more ingested lead, it can cause impaired neurobehavioral development, learning disabilities, speech and language handicaps, poor attention span, lower IQ, diminished intelligence, anti-social behavior [10, 22]. At high concentration, lead can produce coma, convulsions and even death on children and may be left with mental retardation and behavioral disorders [10]. In adults it can be manifested headache, poor attention, irritability, loss of memory, dullness [9, 22]. Increased absorption rate was observed when other nutrients such as calcium or iron are lacking. Even at lower concentrations, known as safe levels, children face learning or behavioral problems, decreased intelligence in children [10]. Although it mainly affects the nervous system, the largest amount of lead is found in the kidneys [9, 22].

Research has shown that this heavy metal can cross the placental barrier in pregnant women who have high levels of it in the blood, causing fetal abnormalities such as low IQ level, encephalopathy, neurological disorders, disruption of calcium levels in nerve cells [3]. Pregnant women exposed to lead, can manifest miscarriage, premature birth, reduced birth weight, stillbirth [10, 22].

After absorption, 99% of lead is bound to the hemoglobin, being circulated through the vascular system to soft tissues, bones, liver, kidneys (organs of lead excretion), hair [3, 10, 19], being stored especially in teeth and bones (where in incorporated into the mineral in place of calcium) [10, 22]. The stored Pb can be reintroduced into the bloodstream, especially during pregnancy, exposing the fetus [10].

Lead can produce lungs disorders, reduced pulmonary function, anemia, liver damage, cardiovascular dysfunction, renal impairment, immunotoxicity, disturbance of the balance free radicals-antioxidant system, cognitive impairments [1, 5, 10, 17]. Anemia occurs as a result of the interaction that this metal has with the important enzymes involved in the synthesis of hemoglobin, enzymes that are responsible and transport oxygen. Thus, by retardation of these enzymes, the hemoglobin concentration is reduced [3]. At high concentration, it can produce high risk of hypertension, gastrointestinal disorders, Alzheimer's disease, kidneys damage, interfere in vitamin D metabolism and thyrotoxicity, by affecting the normal function of thyroid gland, [3, 19, 22].

In people with high levels of lead in the blood, there was an impairment of sexual function, manifested by decreased libido, decreased sperm count and their mobility, changes in sperm composition [3, 22].

Also, this metal can cause changes at cellular level, such as decreased cell viability, cell distortion, reduced cohesion, lipid peroxidation, damage of protein folding, stop structural protein synthesis, intra- and inter-cellular signaling, apoptosis, ionic transportation, especially of calcium, cell adhesion, release of neurotransmitters, inhibiting enzymes activity, inhibits mineral absorption, affecting the activities of mitochondria and endoplasmic reticulum, decreases level of glutathione, generation of reactive oxygen species or reducing antioxidants [1, 3, 17, 22]. Lead has ability to inhibit or mimic the activity of calcium and perturbs their intracellular cycling, may interfere with proteins, can be bound to biological molecules and interfering with their function by various mechanisms [22].

Studies demonstrated that lead can produce genetic damage by mechanisms which include inhibition of DNA synthesis and repair, oxidative damage, being considered by the International Agency for Research on Cancer (IARC) as a probable human carcinogen [22].

Studies performed on animal models have shown altered homeostasis, induced kidney damage, decreases of antioxidant levels, decreased body weight, shortened life span, increases of total protein, albumin, histamine, creatinine, decreased red blood cells count [5, 17].

DOI: http://dx.doi.org/10.5772/intechopen.102441

3.2 Toxicity of cadmium (Cd)

Cadmium is an industrial compound, used in plastic industry, for obtaining plastic stabilizer, but also for production of color pigments, alloys (being a by-product of zinc production), glass production, electroplating industries, welders, rechargeable batteries (about three-fourths of cadmium production). Others important sources include emissions from industrial activities, such as mining or smelting [1, 5, 9, 19, 22, 24].

Exposure to cadmium is achieved by ingestion of food or water, inhalation of contaminated dust, especially for employers which work in primary metal industries or in cadmium-contaminated places, or by smoking cigarettes [3, 5, 10, 19, 22]. Because this metal could not penetrate the skin barrier, dermal exposure not represent a health concern [10].

The main way of exposure for smokers is the smoking, while, for non-smokers, the primary source of exposure is food, such as peanuts, crustaceans and mollusks, leafy vegetables, sunflower seeds, cocoa powder, rice, grains, soybeans, mushrooms, potatoes [3, 10, 22, 25]. Biomonitoring studies have shown that in the case of cigarette smokers, blood and urine levels were generally high, moderate in former smokers and in non-smokers they were reduced [22]. This is related the capacity of this metal to accumulate in high concentrations in tabaco leaves [5, 26]. Their toxicity depends both, the dose of exposure and the exposure time [3]. The percentage of cadmium, absorbed after ingestion is 5–10%, but in diets with a low intake of iron, calcium or protein, the percentage absorbed is higher [10].

In case of occupational workers, in industries which uses this metal, inhalation is the primary way of exposure, so that a percentage of 5–35% of inhaled cadmium is absorbed into the blood, depending the form, particle size, or site of deposition. If this metal reaches the level of the alveoli, its absorption into the blood could be 100% [10]. Their chronic exposure has been associated with changes in pulmonary function, emphysema, decreases in olfactory function [22].

The most toxic form is divalent cadmium ion (Cd^{2+}), which is the most common form and may disturb the basic cellular functions and can cause various side effects [3, 22]. This element can cause side effects even at low concentrations, due to its low excretion rate [17, 27].

Also, it has the capability to replace iron and copper in different cytoplasmic and membrane proteins, and these unbounded substituted metals participate in oxidative stress processes, due to their increased levels [17].

When it binds to cysteine-rich proteins, its concentration inside the body increases 3000 times, forming compounds, such as metallothionein, which can produce hepatotoxicity, nephrotoxicity [1, 3]. If attached to compounds such as histidine, glutamate or cysteine, it can cause iron deficiencies. As a result of exposure, the immune system and endocrine system is affected, even at a young age [3].

Studies have shown that women have higher levels of cadmium than men, and pregnant women have more levels than non-pregnant women. Cadmium does not cross the placental barrier, and remains trapped in it, preventing it from affecting the prenatal exposure of the fetus [3].

The target organs for cadmium are the liver, bones, vascular system, nerve tissues, but especially the kidneys, leading to their damage or malfunction [3, 17, 19, 28]. As their concentration inside the kidneys increases, the rate of calcium excretion from the body is high, which means an increased risk of kidney stones [3, 17, 29]. Also, its renal excretion causes damage to the renal tubules and tubular disfunction by promoting oxidative stress in proximal tubular cells [3, 17].

In case of acute ingestion, symptoms such as vomiting, vertigo, abdominal pain, burning sensation, muscle cramps, shock, loss of consciousness, nausea, convulsions appear in 15–30 min. Because this heavy metal is a severe pulmonary and gastrointestinal irritant, erosion of the intestinal tract, diseases of pulmonary, hepatic or renal or coma could appear, depending the route of poisoning [22].

The exposure to low levels, may affect the prostatic lipid metabolism and the increasing of the fatty acids used to synthesis of phospholipids, with effects on the composition and functions of the plasma membrane [3].

High levels of cadmium in the blood cause a decrease in bone density, especially in pregnant women. Also, it can produce Itai-itai bone disease, which is characterized by painful degenerative bone disease (such as osteomalacia and osteoporosis), renal tubular abnormalities, calcium and phosphate excretion, lung cancer [5, 10, 30].

Chronic exposure can cause effects such as anemia, emphysema, osteoporosis, renal disorders, anosmia, chronic rhinitis, but also have a depressant effect, by changing the levels of serotonin, norepinephrine or acetylcholine [3, 22].

By accumulating in the pancreas and blood, the both exocrine and endocrine function of the pancreas is affected, resulting in a reduction in serum insulin. It may also affect the pancreas to resisting the secretion of insulin, and producing diabetes type 2. Research has shown that it can affect adipose tissue and can lead to obesity. Research has shown that exposure to this element can alter the balance of pituitary hormones. On reproductive system, Cd can affect the synthesis of testosterone and progesterone, spontaneous abortion, low birth weight, changes and apoptosis of germ cells, reducing of semen quality, damage of DNA of sperm cells, apoptosis of Sertoli cells [3].

Long term exposure to cell, it could transform normal cell into malignant cells. Because it contributes to the development of certain types of cancer, such as lung, prostate, pancreatic or kidney cancer, especially in case of occupational exposure, it has been classified as no. 1 human carcinogen by the International Agency for Research on Cancer USA [3, 5, 17, 22, 31]. Rodent studies have demonstrated the capacity of this metal to causes pulmonary adenocarcinomas or prostatic proliferative lesions, leading to adenocarcinomas [22].

At the cellular level, Cd disrupts the respiratory chain of the mitochondria, involved in transport across cell membranes and cell damage through production of reactive oxygen species (ROS), blocking calcium channels, hinders sulfhydryl enzymes, interacts with some cell ligands, promote lipid peroxidation and protein carbonylation. It also affects oxidative phosphorylation pathways, mitochondrial genes involved in cell apoptosis, reducing the ATP level and the energy production. This heavy metal affects the activity of some antioxidant enzymes, such as glutathione reductase, catalase, glutathione peroxidase. Also, cadmium could interact with DNA and may reduce its binding capacity or repair, DNA damage or disruption of synthesis of nucleic acid or proteins [3, 17, 22, 24].

Animal studies have shown that it can produce disorders in the metabolism of zinc, copper and calcium, being able to decrease their absorption and resulting in low dietary intake [5, 32, 33]. The hepatotoxicity and nephrotoxicity of Cd was also observed, after administration of certain doses of cadmium [5, 33]. At cellular level, changes in cell-cell adhesion, autophagic response, changes in cellular signaling pathways, cell death [5], mitochondrial swelling, decrease in antioxidant levels, increases in urinary proteins, more vacuoles and lysosomes in proximal tubule cells were observed [17].

3.3 Toxicity of arsenic (As)

Arsenic is one of the most important heavy metals, with property of a semi metallic, is found in nature in the form of metalloid (As^0) inorganic and organic form, and arsine (AsH_3) [1, 17, 22, 34]. The main inorganic forms include the trivalent form, arsenite (As^{3+}), and the pentavalent form, arsenate (As^{5+}). Among the organic compounds of arsenic are the methylated metabolites, such as monomethylarsonic acid (MMA), dimethylarsinic acid (DMA) and trimethylarsine oxide [9, 22]. Inorganic arsenic compounds, found in water is more toxic than organic compounds, found in seafood, which is less harmful [1, 10, 17, 23, 35]. Studies demonstrated thar trivalent arsenite is 2–10 times more toxic than pentavalent arsenate [22]. The order of increasing toxicity of arsenic compounds is the following, organic arsenicals < metalloid (As^0) < inorganic forms (As^{5+} < As^{3+}) < arsine [5, 36, 37].

Arsenite, which is prevalent and more mobile, has the capability to bind to thiol or sulfhydryl groups of proteins and inactivate more than 200 enzymes, with effects on different organ systems, but also to inhibits the uptake of glucose into cells, fatty acid oxidation, production of acetyl coenzyme A, gluconeogenesis, synthesis of glutathione reductase and thioredoxin reductase. Arsenate can replace phosphate, involved in biological processes, including the transport system [3, 17, 22, 23, 38]. Environmental pollution with this heavy metal, occur as a result of volcanic eruptions, soil erosion or some anthropogenic activities [9, 22]. It is used to obtain industrially products, such as, insecticides, herbicides, fungicides, algicides, smelting, mining, sheep dips, ceramics and glass making, wood preservatives, refining of metallic ores, paints, dye stuffs or for some medicinal treatments for syphilis, yaws, amoebic dysentery, trypanosomiasis [1, 22].

The exposure to elevated levels of inorganic arsenic occurs through ingestion (oral route) of food and water contaminated, inhalation of smoking tobacco, dust or burning smoke from arsenic-treated wood, working in a place where this metal is made or used, dermal contact and parenteral route [5, 10, 22]. Diet, and especially water, is the most important source of exposure, with an intake of about 12–50 μg/day, but the dietary requirement has been suggested to be between 12 and 25 12–50 μg/day [22, 23, 39]. Food sources of arsenic are seafood, poultry, grains (especially rice), bread, cereal products, mushrooms, dairy products [23, 40].

Exposure from air and soil is much smaller, but in areas with a high contamination, the intake through these ways may become significant [22]. Inorganic and organic compounds leave the body through renal excretion. Most of inorganic compounds are eliminated within several days, but some will remain stored for several months or even longer. Organic compounds are eliminated by the body much faster than inorganic arsenic, so most of them will leave the body in a few days [10]. After the absorption in the body, the target organs are lungs, spleen, kidneys, liver, but also, hair, skin and nails, but the last three for long-term accumulation [5].

Researcher showed a strong association between arsenic exposure and increased risks of carcinogenic and systemic health effects, including cardiovascular, dermatologic, nervous, hepatobiliary, renal, gastrointestinal and respiratory diseases [3, 9, 22]. So, in the case of poisoning, the symptoms manifested are abdominal pain, hemolysis, keratosis and hyperkeratosis, edema, gangrene and finally skin cancer [3, 23, 35]. The severity of symptoms varies depending upon the oxidation state and chemical species of arsenic, the solubility, frequency and exposure time, exposure dose, individual susceptibilities, age, gender, genetic and nutritional factors of exposed person [3, 9, 22].

It has been observed that in the case of persons exposed to high concentrations, symptoms such as developmental abnormalities, diabetes, cardiovascular and peripheral vascular disease, pulmonary disease, hearing loss, liver fibrosis, cirrhosis, melanosis, hematologic disorders (anemia, leukopenia, eosinophilia), neurologic and neurobehavioral disorders and different carcinoma have occurred [1, 9, 17, 22, 41, 42].

Long term exposure influences the promotion of carcinogenesis in various tissues or organs, so in areas with higher pollution, was observed a higher mortality rate for different types of cancers, such as kidney, skin, liver, lungs and bladder [3, 9, 10, 22]. For this reason, arsenic and arsenic compounds has been classified as carcinogenic to humans by International Agency for Research on Cancer (IARC) [3, 10]. Also, symptoms like, pigmentation changes, skin lesions, hyperkeratosis, was observed, which may be a precursor to skin cancer. Even at low concentration for a long time, it could change the color of the skin [1, 10]. Chronic arsenic toxicity is termed arsenicosis [1].

At lower concentration, for shorter exposure, arsenic and its compounds may cause nausea and vomiting, reduced production of erythrocytes and leukocytes, abnormal heart beat, damage of blood vessels [1].

This heavy metal could cross the placenta, particularly during early gestation, and affect the fetus, leading to adverse pregnancy outcomes, such as spontaneous abortion, stillbirth, preterm birth, low birth weight), higher infant mortality [5, 10, 43]. Numerous studies demonstrated that in utero or in childhood exposure to this metal, can lead to increases mortality in young adults due to multiple cancers, cardiovascular diseases, kidney failure, lung damage [10, 44], but also negative impact on cognitive developments, intelligence and memory [10, 45].

Their genotoxicity was demonstrated through its capacity to inhibit DNA repair, induce some chromosomal anomalies and DNA damage, sister-chromatid exchanges, arrest cells in mitosis, induce expression of some genes and gene amplification, interfere with formation of micronuclei in different cells, promote oxidative stress, altered growth factors, interfere with cell signaling pathways, inhibition of cell proliferation, promote apoptotic mechanism in various cell (monocytes, T-cells, cancer cells, melanocytes, dermal cells, keratinocytes), mitochondrial disfunctions [5, 17, 22, 46].

In addition to the ability to bind certain structures or to replace some compounds, at cellular level, arsenic compounds could inhibit the mitochondrial enzymes involved in cellular respiration, inactivate some enzymes, such as thiolase and dihydrolipoyl dehydrogenase and affects the oxidative phosphorylation [22].

Animal studies released that arsenic could produce deficits of growth, altered liver and breast milk triglyceride levels [17, 47], decrease of cell viability, induced apoptosis in some cells, increased oxidative stress, increased phosphorylation [17, 48], lower levels of corticosterone receptor, reduced learning and memory [17, 49].

3.4 Toxicity of mercury (Hg)

Mercury or hydrargyrum is a heavy metal which belong to the transition elements series of periodic table [9, 22] and exist in the nature in three chemical forms, such as elemental or metallic or elementary mercury (Hg^0), inorganic mercurous (Hg^{+1}) and mercuric (Hg^{+2}) and organic mercury compound, methylmercury (MeHg or CH3-Hg) and ethylmercury (EtHg or CH3CH2-Hg), the last two being obtained through methylation of inorganic mercuric form by microorganisms found in water and soil [5, 9, 17, 22, 50, 51]. Each chemical form has its own toxicity and chemical properties [9, 22]. Organic Hg compounds are more harmful than inorganic Hg, the order of increasing toxicity being following: metallic mercury (Hg^0) < inorganic mercuric (Hg^{2+}) < inorganic mercurous (Hg^{1+}) < organic compounds [5]. At room temperature, elementary mercury is a liquid with high vapor pressure and released into nature as Hg vapor, which are more hazardous than liquid form [5, 9, 23].

It is used in numerous industrial processes, including mining (for extraction of gold), electrical industry (switches, thermostats, batteries), in lamp production factories (for fluorescent light bulbs), caustic soda production, measurement instruments (thermometers, manometers, barometers, mercury switches), nuclear reactors, paint industries, antifungal agents for wood processing, fungicides in agriculture (methylmercury and ethylmercury), soaps and some skin lightening creams (as mercury chloride) [1, 5, 22, 23, 52].

This metal can reach into the body through inhalation and ingestion of food contamination, especially of fish and seafood, but also by dental amalgams (which contain over 50% elemental mercury), preventive medical practices, industrial and agricultural operations, occupational operations [17, 22].

The most absorbed chemical species are elementary and methyl mercury (Me-Hg) [5, 22]. Metallic mercury, which is highly lipophilic, is absorbed by lungs (80%) and tissues lining the mouth and then passed into the cell through cell membranes when in oxidized and became inorganic mercuric (Hg^{2+}), highly reactive. The elementary mercury has the capacity to cross the blood-brain barrier and the placental barrier [5, 22], having a higher neurotoxicity compared to inorganic mercury, which passes the cell membrane in a slower rate, but cannot cross the blood brain barrier and placenta [5]. Metallic mercury is slightly absorbed in the gastrointestinal tract, the toxicity in this case being reduced [5].

Methyl mercury is easily absorbed in gastrointestinal tract (95%) and circulated in the body, where bound to thiol groups, such as cysteine, with which it can form compounds able to pass the blood brain barrier [5, 17, 53]. Toxicokinetic of ethylmercury is similar with that of methylmercury [5, 53].

Methyl mercury entered in organism through the consumption of fish [5, 54], is absorbed in the gastrointestinal tract and due to its lipophilicity can pass the blood-brain barrier and placental barrier [22]. Cooking of fish does not diminish or eliminate mercury content [5]. Exposure to methyl mercury can produce mental retardation, cerebral palsy, deafness, blindness, dysarthria (especially at children exposed in utero) [17]. Instead, at higher concentration for short time, this could produce lung damage, nausea, vomiting, skin rashes, increased heart rate and blood pressure. Symptoms of organic mercury poisoning are depression, fatigue, memory problems, headache, tremors, hair loss [1].

Mercury and its compounds excretion rate depends on its oxidation state [10]. Elemental and inorganic mercury is eliminated by the kidney (urine) and minimally through gastrointestinal tract (feces), having a half-life of 30–60 days [10, 55, 56]. Organic compounds are excreted by feces, but are recirculated enterohepatic, in this case the half-life being 70 days.

Major of absorbed mercury accumulates into kidneys (where produce adverse effects on proximal tubules), hair, neurological tissues and liver [5, 22]. Because it accumulates in hair, it represents an index of exposure to methylmercury [5].

Elemental mercury exposure is associated with cough, dyspnea, fever, tremors, polyneuropathy of axonal sensor motor, malaise, gingivitis, delusions, hallucinations, mercurial erythrism, while exposure to inorganic mercury produce insomnia, renal tubular damage, wight loss, erythema, pruritus, hypersalivation, excessive perspiration [17].

Chronic mercury exposure produces neurological disorders, such as ataxia, shyness, tremors, numb limbs, memory problems, inability to speak, irritability, chewing, swallowing, muscle weakness, but also renal system disorders [1, 5, 23, 57]. Patients exposed to higher levels of methylmercury present increased tendon reflex [5, 57]. Low dose mercury can produce effects on neuronal systems, both on developing fetus and adolescent stage [17, 58], but also cell cytotoxicity, oxidative stress, which are associated with neurodegenerative disorders like Alzheimer and Parkinson [17, 51, 59]. At low concentration, it can affect the human endocrine system, through reduced production of thyroid gland hormone, affecting physiological functions of endocrine glands, reduced binding capacity of hormone to receptor, the most affected hormones being adrenaline, estrogen, testosterone and insulin [3].

On reproductive system, studied demonstrated their capacity to produce infertility in both, men and women. In male the spermatogenesis is affected, while in women could affect the levels of progesterone and estrogens, which produce disfunctions in ovaries, irregular menstruation and sloped uterus [5].

Because mercury can pass the placenta during pregnancy, it can affect fetus and can cause various abnormalities of the baby, such as developmental disabilities, dysplasia of the cerebral and cerebral cortexes and neuronal ectopia, especially after exposure to methylmercury [3, 5, 17, 57].

Into the cell, inorganic compounds and methylmercury interact with cysteine residues of proteins, product oxidative stress through generation of reactive oxygen species (ROS), which can produce enzymes, nucleic acid and lipids damage and may proceed to cell death [17]. They can affect the calcium homeostasis, by increasing intracellular calcium through acceleration the influx from extracellular medium and mobilizing intracellular stores [22]. Methylmercury also interact with sulfhydryl (–SH) and selenohydryl (–SeH) groups of the proteins and could produce damage of nucleophilic groups involved in catalytic, binding and transport functions [17]. Inorganic mercury also produces reactive oxygen species (ROS) through affecting oxidative phosphorylation and electron transport [22].

A number of compounds, such as vitamin C, vitamin E, selenium, melatonin and enzymes, including, glutathione reductase, glutathione peroxidase, catalase, superoxide dismutase, can have a protective effect on the body through antioxidant mechanisms to reduce or avoid the formation of reactive oxygen species. Mercury genotoxicity was associated with DNA damage, conformational changes in proteins responsible for DNA repair, genetic mutations, mitotic spindle, chromosomal segregation, action on nucleic acids [22].

3.5 Toxicity of aluminum (Al)

Aluminum, the third most common metal of the earth crust, exist in the environment in only one oxidation state (Al^{3+}). It is naturally present in food, but also in the environment, as silicates, oxides and hydroxides. Aluminum and its compounds are poorly absorbed through ingestion and inhalation, but the rates of absorption are not yet known [1, 10].

The ways in which this metal can reach the body are ingestion, inhalation, dermal contact or drugs [3, 10, 60]. Human exposure takes place through the consumption of drinking water, food and beverages that are high in aluminum content, working in environment with high levels of this metal, hemodialysis, long term intravenous nutrition, cosmetic products, utensils and medicines which contains it, dusty environments [1, 3, 10]. Patients with kidney dialysis are more exposed to this metal, through contaminated dialysates and phosphate binders [1]. The bioavailability of aluminum from diet is influenced by its form, as well as the presence of other food constituents which help him to form complexes [10].

The primary way of excretion is through urine. Due to the its natural presence and intake from food, all people have some levels in the body, and also in the urine [10]. People suffering from kidney disease has a low rate of elimination from the body, which involves its accumulation in the body, affecting the bones and brain [1, 3]. Also, their accumulation in the body, leading to changes in proximal tubules, such as increases in number and size of lysosomes, damage of mitochondria [3].

After entry to body, aluminum accumulates in soft tissues where interact with proteins and lipids and may produce changes in their structure [3].

In case of poisoning, the principal symptoms are nausea, ulcer of mouth and skin, skin rashes pain, vomiting, diarrhea and arthritic pain [1, 3].

On nervous system, aluminum may produce loss of memory and coordination, problems with balance, neurodegenerative disorders, such as Alzheimer, dementia, Parkinson, sclerosis. The studies demonstrated that higher concentration of aluminum found in different parts of brain could initiate the development of Alzheimer disease in humans [1, 3]. This metal could form a complex with adenosine triphosphate (ATP) from neuronal cells, which can affect their signaling and cause excitotoxicity [3].

Dialysis patients treated with dialysis fluids which contain aluminum, showed neurotoxic effects, while humans exposed to high aluminum dust in the workplace, manifested aluminosis [10, 61].

Humans exposed to higher levels could manifest changes of secondary hyperparathyroidism, adynamic bone disease, osteomalacia, the last two being characterized by low bone remodeling. Their toxicity is associated with lung disorders, anemia, nervous system problems, impaired iron absorption [1]. The accumulation of aluminum in bones impaired the bone formation process, known as osteodystrophy and put antiproliferative effects on osteoblasts [3]. Workers chronically exposed to aluminum, developed contact dermatitis and irritant dermatitis [1].

At cellular level, studies conducted demonstrated that it can disturbs the homeostasis of magnesium, calcium and iron, lower cholinergic elevations, apoptotic death of neuronal cells, inhibition of enzymes involved in DNA repair, inhibition of activity of antioxidant enzymes, cross linking of DNA, affecting cell viability, plasma membrane, microvilli and cell function in cells kidney [3, 62]. This increases the peroxidation of lipids from plasma membrane, by enhancement of lipid hydroperoxides, which can reduce the molecular arrangement of lipoprotein at the surface of membrane, but, also physical and chemical properties change in high density lipid (HDL). Also, aluminum is involved in high production of reactive oxygen species (ROS), which may obstruct normal process of mitochondria, initiation of inflammatory events and accumulation of iron, which induces genotoxicity in neuronal cells and death cells, affects the gene expression through interaction between aluminum and nucleic acid and monophosphate nucleotides [3].

3.6 Toxicity of chromium (Cr)

Chromium exists in environment in oxidation states and from Cr^{+2} to Cr^{+6} [1, 3, 5, 22, 63]. It does not exist in elementary state (Cr^{0}) [3, 22]. Trivalent oxidation state of Cr is considered more stable, followed by Cr^{+4}. The most commonly forms are Cr^{+3} and Cr^{+6}, both oxidation states being toxic to animals, humans and plants [5, 63].

Cr^{+3} is immobile and insoluble in water, while Cr^{+6} is mobile and highly soluble in water [1, 13]. The solubility of chromium depends on its pH, Cr^{+3} is soluble only in acidic pH, while in neutral and alkaline pH, Cr^{+3} gets precipitated [23].

Environmental contamination with it, occurs by oil burning, catalyst, pigments production, chromium steel, tannery facilities, but also fertilizers and sewage, because is extensively used in several industries, like metallurgy, refractory, tannins, production of paints and pigments, pulp and paper production, wood preservation [1, 9, 22]. Chromium released by the anthropogenic activities in the environment occurs mainly in the hexavalent form [22].

Human exposure occurs through ingestion of food and water which contain, inhalation, especially in case of occupational workers or by dermal contact [5, 64]. Through their bioaccumulation in the body, a variety of affections can appear, such as, dermal, renal, neurological and gastrointestinal diseases, but also development of several types of cancer, on lung, larynx, kidney, testicles, bones, bladder, thyroid [5, 65]. Chromium can affect the reproductive function in men, due to sperm count decline [19]. Ingestion of drinking water containing high level of chromium may cause tumor in stomach [3]. The target organs are lungs, but significant chromium exposure can take place through skin [3, 22].

Occupational exposure to chromium increases the risk of cancer of lung, liver, gastrointestinal tract and central nervous system, while in female workers cause abortion [3, 13, 38]. Excess of chromium can produce thyroid cancer through reduction of requirement level of thyroid hormone in the body, disrupting hormones synthesis and secretion, interfering in its metabolism or interaction with their receptors [3, 66].

Some humans are sensitive to Cr^{3+} and after exposure allergic reactions, including redness and swelling of the skin, can appear. This oxidation state is poorly absorbed by any way, the toxicity being attributable to Cr^{+6} oxidation form [22].

Ingestion of Cr^{+4} can cause irritation and ulcer of stomach and small intestine, anemia, disfunctions of male reproductive system and at high dose produces sever problems on nervous, respiratory and cardiovascular systems, digestive organs, excretory function [3]. Researcher studies demonstrated that high levels in water were associated with cancers of liver, lung and genitourinary system [5, 67].

Cr^{+6} can produce adverse effects on excretory system, reproductive system, asthma, allergy, irritation and ulcers in the stomach and small intestine, anemia, increased mortality due the development of cancer of lung, larynx, kidney, testicular, thyroid, bones [3, 5, 22, 68], and in case of excess inhalation appear irritation and ulcer of nose [3, 22]. Also, it can reduce the DNA replication, damage DNA transcription, chromosome aberrations and affection of RNA [3, 5, 69]. Inside the cell, Cr^{+6} is converted into Cr^{+5}, as intermediate, and then in Cr^{+3}, which can form complexes with proteins and DNA [1, 3]. Cr^{+5} and other intermediate compounds, including reactive species of carbon and oxygen, that form during the reduction of Cr^{+6} to Cr^{+3}, can react with DNA [3]. When hexavalent cation reacts with cellular reductants, Cr^{+4} and Cr^{+3} can also be obtained. Cr^{+6} was classified as group I occupational carcinogen [5, 70].

In cell, mechanism of chromium toxicity generates reactive oxygen species (ROS), which bring cell apoptosis, damage of DNA, genomic instability [3, 5, 71], suppression of DNA synthesis and genes expression [3], but also induces hyperexpression of some antioxidant enzymes, such as, peroxidase, catalase, superoxide dismutase [23].

Their carcinogenicity and toxicity depend of concentration, time of exposure, tissue and cell type [5, 72], route of exposure (ingestion, inhalation or dermal) [10], generation of free radicals [5, 73], oxidation state and its reactivity [5, 10, 22],

3.7 Toxicity of copper (Cu)

Copper is a trace element, component of many enzymes, including ceruloplasmin and cytochrome C oxidase, tyrosinase and dopamine beta-hydrolase, zinc-copper superoxide dismutase (antioxidant defense) and others, having function in transport functions, detoxification, antioxidant defense, immune function, pigmentation and melanin production [10, 74]. When it is present in high levels in the body, it may become toxic [3].

Human exposure take place through its release from water carrying pipes, fungicides, cooking utensils, birth control tablets, food. Copper has the highest redox activity, which lead to production of reactive oxygen species. Also, it binds to thiol groups of proteins and cause changes in liver enzymes involved in biotransformation processes [3].

At cell level, it can change the activity of natrium (Na^+)/potassium (K^+) ATP-ase and change of plasma membrane permeability, due the affection of the natrium/potassium pumps and increases of level of natrium in cytoplasm [3]. Large amounts of copper are stored in the liver [74], while the target organs are nervous system organs, including ganglia, neurons, cerebellum and hippocampus [3].

Excess of copper in the body or hypercupremia, occurs naturally during pregnancy, but also by chronic exposure to it, being associated with a number of diseases including Wilson's disease, hepatic disorders (cirrhosis, hepatitis, gastroenteritis), neurodisorders, hyperceruplasmin [3, 10, 74]. Neurodisorders produced by chronic exposure to copper include neurodegenerative disorders, like Alzheimer and Parkinson, but also Huntington disease, amyotrophic lateral sclerosis [3], cognitive impairment, personality and behavioral changes [74].

Cells studies demonstrated that copper is accumulated in some cancer cell, such as colon cancer cell, ovarian cancer cells, breast cancer cell, more than in normal cells. Also, at cellular level, it can cause oxidative damage of DNA, their reduction can be made by use of Cu specific chelating agents [3].

Hypocupremia or copper deficiency are represented by serum level less than normal value of 0.64–1.56 μg/mL. Extreme hypocupremia could produce Menkes disease, known as Menkes kinky hair syndrome, a genetic disorder, characterized by steely hair, due to a mutation of the transport protein mediating the copper uptake from the intestine, but also by progressive neurological deterioration and early childhood death [10].

3.8 Toxicity of zinc (Zn)

Another trace element, zinc, is involved in over 200 enzymes, with action in immune system, catalytic and structural structures, but also, in processes like synthesis and degradation of some components, including lipids, proteins, carbohydrates, nucleic acids, transcription and translation of polynucleotide, genetic expression, cell proliferation and differentiation, normal growth and development during pregnancy, childhood, adolescence, reduced growth rate and impaired resistance to infection [10].

Exposure to zinc of human is made by inhalation of zinc vapors and ingestion of a large overdose of zinc supplements, which contain zinc sulfate, overusing denture cream, but also by consumption of contaminated food and water [75, 76].

Zinc poisoning, at intakes higher than 100 mg/day, has been associated with abdominal pain, vomiting, diarrhea, nausea.

Long term exposure can cause malabsorption of copper and in case of diabetics, it can affect immune function associated with diabetes mellitus [10]. Severe toxicity present symptoms like kidney injury, pancreatic function damage, liver failure, dehydration and acute gastrointestinal bleed, septic shock, lethargy, sideroblastic anemia and dizziness [74, 76]. Zinc inhalation could produce dyspnea, airway inflammation and acute respiratory distress symptom, especially in case of occupational exposure [76].

Because this metal could interfere in copper absorption in the gastrointestinal tract, leading to copper deficiency [10], chronic exposure can cause polyneuropathy and can affect bone marrow [76].

3.9 Toxicity of nickel (Ni)

Nickel is an essential trace element for plant, animals and human, but also a chemical pollutant which exist in several oxidation states, but most common is Ni^{2+}. In the body, it is involved in activation of some enzymes, in protein structure and function, in prolactin production [3, 10].

Environmental contamination with nickel comes from natural sources, like volcanic emissions, weathering of soils, but also from industry, being used in catalysts for automobile, electroplating, electroforming, jewelry production, medical prostheses, production of nickel-cadmium batteries, cast coins [19].

This metal can reach the body through ingestion of contaminated water and food, inhalation of dust or smoking cigarettes and dermal contact, leading to increases level of Ni in blood, urine and body tissues. However, less than 10% of ingested nickel is absorbed by gastrointestinal tract [3, 10].

It can pass through plasma membrane through diffusion, calcium transport channels and phagocytosis, is circulated to various tissues, where bind with albumin, histidine and macroglobulin. In case of nickel, the target organs are kidneys, bons, lungs, liver, brain and glands of endocrine system, but it is not accumulated in those, being excreted outside [3].

Nickel exposure can produce disorders of liver, kidney, spleen, brain and tissues, but also vesicular eczema, nasal and lung cancer. Also, it interferes in iron resorption, which lead to anemia, disturb the incorporation of calcium into skeleton, causing parakeratosis damage [10]. On reproductive system, this metal affects the quality of semen and cause abnormalities in it, including the tail of sperms [3].

Occupational exposure can cause allergic dermatitis, known as "nickel allergy". In case of dermal contact, skin rash or allergic dermatitis appear, due to wearing of nickel-plated jewelry. Women are more sensible to nickel than men, especially in pregnant women which work in metallurgic industry and their babies hence structure abnormalities [3, 10].

At cellular level, it can produce breaking of DNA strands, cross linking of DNA protection, DNA oxidation, nucleotides removal, genes mutations, modifications of chromatids, binding to enzymes involved in DNA repair and degradation of protein, generation of ROS, enhances lipid peroxidation, affecting calcium and sulfhydryl homeostasis, degradation of glutathione [3].

4. Conclusions

Heavy metal pollution is global treat and increasing day by day, due to many natural and anthropogenic activities, which disturb natural composition of soil, water and air, but also of living organisms [3, 23].

DOI: http://dx.doi.org/10.5772/intechopen.102441

These metals can enter the body from sources of contamination by ingestion, inhalation or dermal contact, where they are absorbed, then bioaccumulated in various organs or target tissues, for different periods of time [5, 22]. The most important is the occupational exposure for those working in industries where these metals are produced or used, which can be reduced by various engineering solutions [1].

Heavy metals can affect organs and their functions, causing adverse effects in humans like, cardiovascular, neurologic, gastrointestinal, immunologic, endocrine, reproductively disorders, but also various types of cancer, including lungs, bladder, skin. But, the severity of those side effects depends on chemical state, time and dose of exposure, solubility [22].

In order to prevent exposure to these metals, as well as the occurrence of health problems, it is important to establish safety limits for different matrices [19].

Acknowledgements

This work was achieved through Core Program, with the support of the Ministry of Research, Innovation and Digitization, contract 22 N/2019, project PN 19 02 03 02 and CNCS/CCCDI—UEFISCDI, project number PN-III-P3-3.6-H2020-2020-0011/ Ctr. 1/2020.

Author details

Elena Loredana Ungureanu* and Gabriel Mustatea
National Research and Development Institute for Food Bioresources, Bucharest, Romania

*Address all correspondence to: elena_ungureanu93@yahoo.com

References

[1] Jaishankar M, Tseten T, Anbalagan N, Mathew BB, Beeregowda K. Toxicity, mechanism and health effects of some heavy metals. Interdisciplinary Toxicology. 2014;**7**(2):60-72. DOI: 10.2478/intox-2014-0009

[2] Ali H, Khan E, Ilahi I. Environmental chemistry and ecotoxicology of hazardous heavy metals: Environmental persistence, toxicity and bioaccumulation. Journal of Chemistry. 2019;**2019**:6730305. DOI: 10.1155/2019/6730305

[3] Rehman AU, Nazir S, Irshad R, Tahir K, Ur Rehman K, Ul Islam R, et al. Toxicity of heavy metals in plants and animals and their uptake by magnetic iron oxide nanoparticles. Journal of Molecular Liquids. 2021;**321**:114455. DOI: 10.1016/j.molliq.2020.114455

[4] Masindi V, Muedi KL. Environmental contamination by heavy metals. In: Saleh HEDM, Aglan RF, editors. Heavy Metals. Chichester: IntechOpen; 2018. pp. 115-133. DOI: 10.5772/intechopen.76082

[5] Balali-Mood M, Naseri K, Tahergorabi Z, Khazdair MR, Sadeghi M. Toxic mechanisms of five heavy metals: Mercury, lead, chromium, cadmium and arsenic. Frontiers in Pharmacology. 2021;**12**:643972. DOI: 10.3389/fphar.2021.643972

[6] Rai PK, Lee SS, Zhang M, Tsang YF, Kim KH. Heavy metals in food crops: Health risks, fate, mechanisms and management. Environment International. 2019;**125**:365-385. DOI: 10.1016/j.envint.2019.01.067

[7] Edelstein M, Ben-Hur M. Heavy metals and metalloids: Sources, risks and strategies to reduce their accumulation in horticultural crops. Scientia Horticulturae. 2018;**234**:431-444. DOI: 10.1016/j.scienta.2017.12.039

[8] Koller M, Saleh HM. Introductory chapter: Introducing heavy metals. In: Saleh HEDM, Aglan RF, editors. Heavy Metals. Vol. 2018. Chichester: IntechOpen; 2018. pp. 3-11. DOI: 10.5772/intechopen.74783

[9] Kaur R, Sharma S, Kaur N. Heavy metals toxicity and the environment. Journal of Pharmacognosy and Phytochemistry. 2019;**SP1**:247-249

[10] Aliasgharpour M. Trace elements in human nutrition (II)—An update. International Journal of Preventive Medicine. 2020;**11**(2):1-17. DOI: 10.4103/ijpvm.IJPVM_48_19

[11] Frieden E. New perspective on the essential trace element. Journal of Chemical Education. 1985;**62**(11):917-923

[12] Kulkarni N, Kalele K, Kulkarni M, Kathariya R. Trace elements in oral health and disease: An updated review. JDRR. 2014;**1**(2):100-104. DOI: 10.4103/2348-2915.133959

[13] Bhattacharya PT, Misra SR, Hussain M. Nutritional aspects of essential trace elements in oral health and disease: An extensive review. Scientifica. 2016;**2016**:5464373. DOI: 10.1155/2016/5464373

[14] Aliasgharpour M, Marjan RF. Trace elements in human nutrition: A review. The Journal of Medical Investigation. 2013;**2**(3):115-128. DOI: 10.4103/ijpvm.IJPVM_48_19

[15] Prashanth L, Kattapagari KK, Chitturi RT, Baddam VRR, Prasad LK. A review on role of essential trace elements in health and disease. Journal Dr.NTR University of Health Sciences. 2015;**4**(2):75-85. DOI: 10.4103/2277-8632.158577

[16] He Z, Shentu J, Yang X, Baligar VV, Zhang T, Stoffella PJ. Heavy metal contamination of soils: Sources, indicators, and assessment. Journal of Environmental Indicators. 2015;**9**:17-18

[17] Wu X, Cobbina SJ, Mao G, Xu H, Zhang Z, Yang L. A review of toxicity and mechanism of individual and mixtures of heavy metals in the environment. Environmental Science and Pollution Research. 2016;**23**:8244-8259. DOI: 0.1007/s11356-016-6333-x

[18] Musilova J, Arvay J, Vollmannova A, Toth T, Tomas J. Environmental contamination by heavy metals in region with previous mining activity. Bulletin of Environmental Contamination and Toxicology. 2016;**97**:569-575. DOI: 0.1007/s00128-016-1907-3

[19] Al-Lami AMA, Khudhaier SR, Aswad OA. Effects of heavy metals pollution on human health. Annals of Tropical Medicine and Public Health. 2020;**23**:1-4. DOI: 10.36295/ASRO.2020.231125

[20] Zeng F, Qiu B, Wu X, Niu S, Wu F, Zhang G. Glutathione-mediated alleviation of chromium toxicity in rice plants. Biological Trace Element Research. 2012;**148**:255-263. DOI: 10.1007/s12011-012-9362-4

[21] Najeeb U, Ahmad W, Zia MH, Malik Z, Zhou W. Enhancing the lead phytostabilization in wetland plant *Juncus effusus* L. through somaclonal manipulation and EDTA enrichment. Arabian Journal of Chemistry. 2014;**14**(2):3310-3317. DOI: 10.1016/j.arabjc.2014.01.009

[22] Tchounwou PB, Yedjou CG, Patlolla AK, Sutton DJ. Heavy metals toxicity and the environment. EXS. 2012;**101**:133-164. DOI: 10.1007/978-3-7643-8340-4_6

[23] Pratush A, Kumar A, Hu Z. Adverse effect of heavy metals (As, Pb, Hg, and Cr) on health and their bioremediation strategies: A review. International Microbiology. 2018;**21**:97-106. DOI: 10.1007/s10123-018-0012-3

[24] Rashid K, Sinha K, Sil PC. An uppdate on oxidative stress-mediated organ pathophysiology. Food and Chemical Toxicology. 2013;**62**:584-600. DOI: 10.1016/j.fct.2013.09.026

[25] Chunhabundit R. Cadmium exposure and potential health risk from foods in contaminated area, Thailand. Toxicology Research. 2016;**32**(1):65-72. DOI: 10.5487/tr.2016.32.1.065

[26] Proshad R, Zang D, Uddin M, Wu Y. Presence of cadmium and lead in tobacco and soil with ecological and human health risks in Sichuan province China. Environmental Science and Pollution Research. 2020;**27**:18355-18370. DOI: 10.1007/s11356-020-08160-1

[27] Pretto A, Loro VL, Morsch VM, Moraes BS, Menezes C, Santi A, et al. Alterations in carbohydrate and protein metabolism in silver catfish (*Rhamdia quelen*) exposed to cadmium. Ecotoxicology and Environmental Safety. 2014;**100**:188-192. DOI: 10.1016/j.ecoenv.2013.11.004

[28] Tsutsumi T, Ishihara A, Yamamoto A, Asaji H, Yamakawa S, Tokumura A. The potential protective role of lysophospholipid mediators in nephrotoxicity induced by chronically

exposed cadmium. Food and Chemical Toxicology. 2014;**65**:52-62. DOI: 10.1016/j.fct.2013.12.019

[29] Hambach R, Lison D, D'Haese PC, Weyler J, De Graef E, De Schryver A, et al. Co-exposure to lead increases the renal response to low levels of cadmium in metallurgy workers. Toxicology Letters. 2013;**222**:213-218. DOI: 10.1016/j.toxlet.2013.06.218

[30] Nishijo M, Nakagawa H, Suwazono Y, Nogawa K, Kido T. Causes of death in patients with Itai-itai diseases suffering from severe chronic cadmium poisoning: A nested case-control analysis of follow-up study in Japan. BMJ Open. 2017;**7**(7):e015694. DOI: 10.1136/bmjopen-2016-015694

[31] Kim TH, Kim JH, Le Kim MD, Suh WD, Kim JE, Yeon HJ. Exposure assessment and safe intake guidelines for heavy metals in consumed fishery products in the Republic of Korea. Environmental Science and Pollution Research International. 2020;**27**:33042-33051. DOI: 10.1007/s11356-020-09624-0

[32] Borowska S, Brzóska M, Galażyn-Sidorczuk M, Rogalska J. Effect of an extract from *Aronia melanocarpa* L. berries on boy status of zinc and cooper under chronic exposure to cadmium: An in vivo experimental study. Nutrients. 2017;**9**(12):1374. DOI: 10.3390/nu9121374

[33] Fay MJ, Alt LAC, Ryba D, Salamah R, Peach R, Papaeliou A. Cadmium nephrotoxicity is associated with altered microRNA expression in the rat renal cortex. Toxics. 2018;**6**(1):16

[34] Mulware SJ. Trace elements and carcinogenicity: A subject in review. 3 Biotech. 2013;**3**(2):85-96. DOI: 10.1007/s13205-012-0072-6

[35] Parmar M, Thakur LS. Heavy metal Cu, Ni and Zn: Toxicity, health hazard and their removal techniques. International Journal of Plant, Animal and Environmental Sciences. 2013;**3**(3):143-157

[36] Sattar A, Xie S, Hafeez MA, Wang X, Hussain HI, Iqbal Z. Metabolism and toxicity of arsenical in mammals. Environmental Toxicology and Pharmacology. 2016;**48**:214-224. DOI: 10.1016/j.etap.2016.10.020

[37] Kuivenhoven M, Mason K. Arsenic (Arsine) Toxicity. Treasure Island. FL: StatPearls Publishing;

[38] Yang L, Li X, Chu Z, Ren Y, Zhang J. Distribution and genetic diversity of the microorganisms in the biofilter for the simultaneous removal of arsenic, iron and manganese from simulated groundwater. Bioresource Technology. 2014;**156**:384-388. DOI: 10.1016/j.biortech.2014.01.067

[39] Mahmood A, Malik RN. Human health risk assessment of heavy metals via consumption of contaminated vegetables collected from different irrigation sources in Lahore, Pakistan. Arabian Journal of Chemistry. 2014;**7**(1):91-99. DOI: 10.1016/j.arabjc.2013.07.002

[40] Cheyns K, Waegeneers N, De Wiele TV, Ruttens A. Arsenic relies from foodstuff upon food preparation. Journal of Agricultural and Food Chemistry. 2017;**65**(11):2443-2453. DOI: 10.1021/acs.jafc.6b05721

[41] Ahmed MK, Habibullah-Al-Mamun M, Parvin E, Akter MS, Khan MS. Arsenic induced toxicity and histopathological changes in gill and liver tissue of freshwater fish, tilapia (*Oreochromis mossambicus*). Experimental and Toxicologic Pathology. 2013;**65**:903-909. DOI: 10.1016/j.etp.2013.01.003

[42] He J, Wang M, Jiang Y, Chen Q, Xu S, Xu Q, et al. Chronic arsenic exposure and angiogenesis in human bronchial epithelial cells via the ROS/miR-199a-5p/HIF-1α/COCS-2 pathway. EHP. 2014;**122**(3):255-261. DOI: 10.1289/ehp.1307545

[43] Quansah R, Armah FA, Essumang DK, Luginaah I, Clarke E, Marfoh K, et al. Association of arsenic with adverse pregnancy outcomes/infant mortality: A systematic review and meta-analysis. Environmental Health Perspectives. 2015;**123**:412-421. DOI: 10.1289/ehp.1307894

[44] Farzan SF, Karagas MR, Chen Y. In utero and early life arsenic exposure in relation to long-term health and disease. Toxicology and Applied Pharmacology. 2013;**272**:384-390. DOI: 10.1016/j.taap.2013.06.030

[45] Tolins M, Ruchirawat M, Landrigan P. The developmental neurotoxicity of arsenic: Cognitive and behavioral consequences of early life exposure. Annals of Global Health. 2014;**80**:303-314. DOI: 10.1016/j.aogh.2014.09.005

[46] Mochizuki H. Arsenic neurotoxicity in humans. International Journal of Molecular Sciences. 2019;**20**(14):3418. DOI: 10.3390/ijms20143418

[47] Rahman MK, Choudhary MI, Arif M, Morshed MM. Dopamine-β-hydroxylase activity and levels of its cofactors and other biochemical parameters in the serum of arsenicosis patients of Bangladesh. International Journal of Biomedical Sciences. 2014;**10**:52-60

[48] Rai NK, Ashok A, Rai A, Tripathi S, Nagar GK, Mitra K, et al. Exposure to As, Cd and Pb-mixture impairs myelin and axon development in rat brain, optic nerve and retina. Toxicology and Applied Pharmacology. 2013;**273**:424-258. DOI: 10.1016/J.TAAP.2013.05.003

[49] Ramsey KA, Foong RE, Sly PD, Larcombe AN, Zosky GR. Early life arsenic exposure and acute and long-term responses to influenza A infection in mice. Environmental Health Perspectives. 2013;**121**:1187-1193. DOI: 10.1289/ehp.1306748

[50] Li R, Wu H, Ding J, Fu W, Gan L, Li Y. Mercury pollution in vegetables, grains and soils from areas surrounding coal-fired power plants. Scientific Reports. 2017;**7**:1-9. DOI: 10.1038/srep46545

[51] Carocci A, Rovito N, Sinicropi MS, Genchi G. Mercury toxicity and neurodegenerative effects. In: Whitacre DM, editor. Reviews of Environmental Contamination and Toxicology. 1st ed. Vol. 229. Switzerland: Springer; 2014. 1-18 p. DOI: 10.1007/978-3-319-03777-6_1

[52] Morais S, Costa FG, Pereira ML. Heavy metals and human health. In: Oosthuisen J, editor. Environmental Health—Emergencies Issues and Practice. Chichester: IntechOpen; 2012. pp. 227-246

[53] Bridges CC, Zalups RK. Mechanisms involved in the transport of mercuric ions in target tissues. Archives of Toxicology. 2017;**91**(1):63-81. DOI: 10.1007/s00204-016-1803-y

[54] Moriarity RJ, Liberda EN, Tsuji LJS. Subsistence fishing in the Eeyou Istchee (James Bay, Quebec, Canada): A regional investigation of fish consumption as a route of exposure to methylmercury. Chemosphere. 2020;**258**:127413. DOI: 10.1016/j.chemosphere.2020.127413

[55] Caito SW, Jackson BP, Punshon T, Scrimale T, Grier A, Gill SR. Variation

in methylmercury metabolism and elimination status in humans following fish consumption. Toxicological Sciences. 2018;**161**:443-453. DOI: 10.1093/toxsci/kfx226

[56] Bjørklund G, Dadar M, Mutter J, Aaseth J. The toxicology of mercury: Current research and emerging trends. Environmental Research. 2017;**159**:545-554. DOI: 10.1016/j.envres.2017.08.051

[57] Dos Santos AA, Chang LW, Liejun Guo G, Aschner M. Chapter 35—Fetal Minamata disease: A human episode of congenital methylmercury poisoining. In: Slikker W, Paule MG, Whang C, editors. Handbook of Developmental Neurotoxicology. 2nd ed. Cambridge, MA: Academic Press; 2018. pp. 399-406

[58] Kovacic P, Somanathan R. Recent developments in the mechanism of teratogenesis-electron transfer, reactive oxygen species, and antioxidants. In: Laher I, editor. Systems Biology of Free Radicals and Antioxidants. 1th ed. Chichester: Springer; 2014. pp. 567-580. DOI: 10.1007/978-3-642-30018-9_39

[59] Goldman SM. Environmental toxins and Parkinson's disease. Annual Review of Pharmacology and Toxicology. 2014;**54**:141-164. DOI: 10.1146/annurev-pharmtox-011613-135937

[60] Nampoothiri M, John J, Kumar N, Mudgal J, Nampurath GK, Chamallamudi MR. Modulatory role of simvastatin against aluminum chloride-induced behavioral and biochemical changes in rats. Behavioural Neurology. 2015;**2015**:210169. DOI: 10.1155/2015/210169

[61] Klotz K, Weistenhöfer W, Neff F, Hartwig A, Thriel CA, Drexler H. The health effects of aluminum exposure. Deutsches Ärzteblatt International. 2017;**114**:653-659. DOI: 10.3238/arztebl.2017.0653

[62] Maya S, Prakash T, Madhu KD. Multifaced effects of aluminum in neurodegenerative diseases: A review. Biomedicine & Pharmacotherapy. 2016;**83**:746-754

[63] Shekhawat K, Chatterjee S, Joshi B. Chromium toxicity and its health hazard. International Journal of Advanced Research. 2015;**3**(7):167-172

[64] Nickens KP, Patierno SR, Ceryak S. Chromium genotoxicity: A double-edged sword. Chemico-Biological Interactions. 2010;**188**(2):276-288. DOI: 10.1016/j.cbi.2010.04.018

[65] Fang Z, Zhao M, Zhen H, Chen L, Shi P, Huang Z. Genotoxicity of tri-and hexavalent chromium compounds in vivo and their modes of action an DNA damage in vitro. PLoS One. 2014;**9**(8):e103194. DOI: 10.1371/journal.pone.0103194

[66] Vigneri R, Malandrino P, Giani F. Heavy metals in the volcanic environment and thyroid cancer. Molecular and Cellular Endocrinology. 2017;**457**:73-80

[67] Linos A, Petralias A, Christophi CA, Christoforidou E, Kouroutou P, Stoltidis M. Oral ingestion of hexavalent chromium through drinking water and cancer mortality in an industrial area of Greece—An ecological study. Environmental Health. 2011;**10**(1):50. DOI: 10.1186/1476-069x-10-50

[68] Deng Y, Wang M, Tian T, Lin S, Xu P, Zhou L. The effect of hexavalent chromium on the incidence and mortality of human cancers: A meta-analysis based on published epidemiological cohort studies. Frontiers in Oncology. 2019;**9**:24. DOI: 10.3389/fonc.2019.00024

[69] Wise SS, Wise JP. Chromium and genomic stability. Mutation Research. 2012;**733**(1-2):78-82. DOI: 10.1016/j.mrfmmm.2011.12.002

[70] Loomis D, Guha N, Hall AL, Straif K. Identifying occupational carcinogens: An update from the IARC monographs. Occupational and Environmental Medicine. 2018;**75**(8):593-603. DOI: 10.1136/oemed-2017-104944

[71] Pavesi T, Moreira JC. Mechanisms and individuality in chromium toxicity in humans. Journal of Applied Toxicology. 2020;**40**:1183-1197. DOI: 10.1002/jat.3965

[72] Ferreiera LMR, Cunha-Oliviera T, Sobral MC, Abreu PL, Alpoim MC, Urbano AM. Impact of carcinogenic chromium on the cellular response to proteotoxic stress. International Journal of Molecular Sciences. 2019;**20**(19):4901. DOI: 10.339/ijms20194901

[73] De Loughery Z, Luczak MW, Zhitkovich A. Monitoring Cr intermediates and reactive oxygen species with fluorescent probes during chromate reduction. Chemical Research in Toxicology. 2014;**27**(5):843-851. DOI: 10.1021/tx500028x

[74] Strachan S. Trace elements. Current Anaesthesia & Critical Care. 2010;**21**:44-48

[75] Rahimzadeh MR, Kazemi S, Moghadadamnia AA. Zinc poisoning—Symptoms, causes, treatments. Medicinal Chemistry. 2020;**20**(15):1489-1498. DOI: 10.2174/1389557520666200414161944

[76] Agnew UM, Slesinger TL. Zinc Toxicity. Florida, USA: StatPearls Publishing; 2021

Chapter 3

Influence of Heavy Metals on Quality of Raw Materials, Animal Products, and Human and Animal Health Status

Anna Bartkowiak

Abstract

Heavy metals constitute one of the threats to the natural environment and the health of living organisms. The sources of contamination of the environment with heavy metals are mainly industry, thermal and chemical processing of mineral resources, burning of coal, gases and liquid fuels, municipal economy (rubbish dumps, sewage), and agriculture consuming mineral fertilizers, plant protection agents, utilizing huge loads of pollution accompanying animal production. Accumulation of toxic elements in plant tissues leads to disturbances in plant reproduction and thus to lowering of their nutritional value. In humans and animals, in turn, it may cause poisoning and the occurrence of various disorders and diseases, including cancer. There are different ways to reduce the penetration of heavy metals into crops and deactivation in animal organisms—by using the tolerance of plants to heavy metals, cleaning the environment through phytoremediation, the use of antagonistic type interactions to reduce bioaccumulation in animal tissues, as well as the properties of compounds of organic and mineral origin. The aim of the chapter is to present the problems of environmental pollution and accumulation of heavy metals (mainly cadmium, mercury, and lead) in tissues of farm animals, their impact on human and animal health, as well as the possibility of inactivation of heavy metals in animal organisms.

Keywords: heavy metals, soil and plants, quality of raw materials, animal products, health status

1. Introduction

Problems concerning heavy metals are studied all over the world, including in Europe. This is because they constitute one of the threats to the natural environment and the health of living organisms, and thus pose a problem in the implementation of the concept of sustainable development [1, 2].

In the accepted classification of elements participating in life processes, heavy metals, such as mercury, lead, and cadmium, do not play any significant role in metabolism. They are considered to belong to the group of extreme toxicants with known embryotoxic, teratogenic, mutagenic, and carcinogenic effects. Even in trace amounts, they pose a real threat to living organisms. In the natural environment, there are also other heavy metals, such as arsenic, zinc, nickel, copper, which are also toxic and may cause poisoning and cancer. The effects of heavy metals entering the human body can be revealed even after many years [3].

In the past, the danger resulting from the presence of these metals in the environment concerned only certain groups of people employed in specialized branches of industry. Currently, as a result of development and civilization changes, toxic elements may appear in high concentrations far beyond the sources of contamination. This violates the biological balance of ecosystems, and their presence in the trophic chain creates exposure conditions for wide groups of the population [3].

The sources of contamination of the environment with heavy metals are the mining industry and mechanical, thermal, and chemical processing of mineral raw materials. Other important sources of emission of metallic elements are large-scale burning of coal, gases, and liquid fuels (motorization), municipal economy (waste dumps, sewage), and agriculture consuming mineral fertilizers, plant protection agents, utilizing huge loads of pollution accompanying animal production and processing industry. Monitoring studies of soils, drinking water, plants, as well as tissues of farm animals and products of animal origin (milk, eggs, honey) indicate considerable variability of heavy metal concentrations in this material, from trace to many times exceeding permissible values. In the case of animals, it depends on the animal species, life stage, husbandry system, and location of the breeding facility [3–8].

It has been repeatedly shown that mineral fertilizers have heavy metals in their composition, which lead to soil pollution. Fertilizers can be ranked in ascending order of heavy metal contamination—nitrogenous, potassic, calcareous, and phosphatic. The accumulation of these elements in the fertilizer material depends primarily on the technology used and the material from which they are produced (**Table 1**) [7, 9, 10].

Agriculture is the dominant land-use industry where large amounts of agrochemicals are applied. Unfortunately, metals and pesticides in the soil can reach aquatic ecosystems through leaching, soil erosion, and surface runoff [11]. The combination of heavy metals with pesticides is very dangerous because they can cause very serious health consequences for humans and animals. They contribute to neurodegenerative disorders, musculoskeletal diseases, and hormonal imbalances, are carcinogenic, cause genetic damage [12].

Source	Ace	Cd	Cu	Hg	Ni	Pb	Zn
Sewage sludges	2–26	2–1500	50–3300	0.1–55	16–5300	50–3000	700–49,000
Nitrogen fertilizers	2–120	0.05–8.5	1–15	0.3–3	7–38	2–1450	1–42
Phosphate fertilizers	2–1200	0.1–170	1–300	0.01–1.2	7–38	7–225	50–1450
Limestones	0.1–24.0	0.04–0.1	2–125	0.05	10–20	20–1250	10–450
Manure	3–25	0.3–0.8	2–60	0.09–0.2	7.8–30	6.6–15	15–250
Pesticides (%)	22–60	—	12–50	0.8–42	—	60	1.3–25

Source: own study based on [7].

Table 1.
Agricultural sources of heavy metals contamination in soils (ppm DW).

Therefore, the aim of the chapter is to present the problems of environmental pollution and accumulation of heavy metals (mainly cadmium, mercury, and lead) in tissues of farm animals, their impact on human and animal health, as well as the possibility of inactivation of heavy metals in animal organisms.

2. Heavy metal contamination of soils, plants, and drinking water

The excess of heavy metals in the soil inhibits the development of microorganisms and disrupts processes related to the transformation of organic matter. It also causes the accumulation of toxic elements in plant tissues, leading to disturbances in plant reproduction and thus lowering their nutritional value [13–15]. Excessive accumulation of the mentioned elements in the soil, however, is harmful to plants in particular [3, 7, 16, 17].

In soil, heavy metals can occur in different forms—dissolved in soil solution, exchangeable in organic and inorganic components, being structural components of soil grids, and as insoluble sediments with other soil components. The first two forms are the most available to plants. The concentration of elements in the soil depends on the pH of the soil—the higher (to slightly alkaline) the higher the immobilization of elements. The mobility of heavy metals in the soil varies. In acidic soils, Cd, Ni, and Zn are particularly mobile, Cr is moderately mobile and Cu and Pb are immobile. In neutral and alkaline soils, Cr is highly mobile, Cd and Zn are moderately mobile, and Ni is immobile. Other factors, such as cation exchange capacity, redox potential, organic matter content, type and amount of clay minerals, and oxide content of antagonistic elements Fe, Al, and Mn, also determine the increase of heavy metals in the soil and thus their availability to plants [18].

The natural cadmium content in soils is 0.2–1.05 mg kg^{-1} d.m. It is a highly mobile, active element, and easily assimilated by plants because of the available form of Cd^{2+} ion. The environmental hazard is related to the fact that Cd is one of the most toxic metals that show adverse effects on soil biological activity, plant metabolism, human health, and the animal kingdom. Excess cadmium in the plant manifests itself by twisting of leaves and the appearance of brown spots on the leaves [7].

Another toxic element is lead. Its natural content in soil is strongly related to the composition of the rock substrate. It is characterized by the lowest mobility among heavy metals. The highest Pb content in soil is found in highly industrialized areas. Lead can enter the body from two sources—the food chain and through inhalation of soil dust. It is a very dangerous metal with negative effects on humans, animals, and plants. Excess lead leads to reduced yields and dark green or red spots on leaves. Lead content in soil exceeding 500 mg kg^{-1} is a toxic value. A characteristic feature of this heavy metal is its accumulation in the human body, as it does not disintegrate in this environment. Getting into the human body a dose of about 20–50 g leads to death [7, 19–23].

In the case of copper, its excess causes tissue damage and elongation of root cells, alteration of membrane permeability and leakage of ions (e.g., K) and solutes from roots, peroxidation of chloroplast membrane lipids and inhibition of photosynthetic electron transport, immobilization of Cu in cell walls, in cell vacuoles and nondispersive Cu-protein complexes, as well as DNA damage and consequently inhibition of photosynthetic processes. Manganese, on the other hand, has a significant effect on some soil properties, particularly raising pH. Mn compounds are known for their rapid oxidation and reduction under varying soil conditions, thus oxidizing [7].

Nickel has become a major pollutant that is released during emissions from metal processing and increasing coal and oil burning, sludge application. Some phosphate fertilizers may also be an important source of Ni [3, 7].

In Poland, the permissible content of heavy metals (so-called risk-causing substances) on agricultural land is defined by the Regulation of the Minister of Environment of September 1, 2016, on the manner of conducting the assessment of land surface pollution (Annex No. 1 to the Regulation) [24]. According to the Ordinance, risk-causing substances that are particularly important for the protection of the earth surface and the permissible contents of these substances in the soil and the permissible contents of these substances in the soil [mg kg^{-1} dry mass of the earthy parts of the soil (<2 mm)] are defined (**Table 2**)—for depths 0–0.25 m ppt and more than 0.25 m ppt, with the division taking into account the soil groups and separated based on their use, the soil subgroups separated based on the soil properties (defined for the soil group II) and the soil and groundwater permeability.

Koncewicz-Baran and Gondek [25] investigated the content of general forms and bioavailable elements (Cd, Cr, Ni, Cu, Pb, Zn, and Mn) in agriculturally used soils showed that among the investigated soils, the natural content of Ni, Cu, and Pb was the highest. In other soils, increased content of Cd and Zn was determined.

Li et al. [26] pointed out the problem of soil contamination in China due to increased industrial development and urbanization. In Hunan Province, Central China, a study was conducted on the content of heavy metals in soil (Pb, Zn, Cu, Cd, As, Hg, Cr, and Ni). The results showed that the content of each heavy metal in the soil varied spatially. The highest accumulation was shown for Cd, followed by Pb, Zn, As, and Hg.

Plants can accumulate toxic metals from the soil over a very wide range. This depends on its temperature, reaction, water capacity, and potential. Under conditions of high immission, plants take up heavy metals from the air through the leaf blades. Strong accumulation of heavy metals in the root system and the aboveground parts of plants is a result of a poorly developed mechanism of chemical homeostasis in plants, which leads to non-selective absorption of elements and creates a high risk of including heavy metals in the food chain system. Changes in toxic metal concentrations in fodder plants may occur as a result of drying, ensiling, and granulation processes. In

Element	Permissible content
Arsenic (As)	10–50
Chromium (Cr)	150–500
Cadmium (Cd)	2–5
Cobalt (Co)	20–60
Copper (Cu)	100–300
Molybdenum (Mo)	10–50
Nickel (Ni)	100–300
Lead (Pb)	100–500
Mercury (Hg)	2–5

Source: own study based on [24].

Table 2.
Permissible content of selected elements causing the risk.

addition, heavy metals can be introduced into animal feed rations through enrichment with yeast, meat, bone, and fish meals and inorganic mineral additives, such as phosphate, dolomite, and chalk [3].

Karimi et al. [27] investigated the levels of toxic heavy metals (As, Cd, Hg, Pb) in agricultural products, such as legumes, wheat, and potatoes, in Markazi Province, Iran. Markazi Province is the most industrialized region in the country. Lead mines and other industrial activities are located there, which carries the presence of heavy metals in the soil. The results showed that among the samples analyzed, the carcinogenic risk index was within the acceptable level. However, in the case of wheat, it was found to be the most important source of toxic metal exposure due to its high consumption compared to the other crops, i.e., earthlings and legumes.

Studies conducted in Nigeria, where soils are contaminated with oil, showed that among the heavy metals tested (Pb, Cd, Cr, Mn, Fe, and Zn), for crops in the test samples, the elements Pb and Cr exceeded the limits set by WHO [28]. In Turkey of 12 districts of Sakarya city where cucurbit crops are grown, organochlorine pesticides have been applied to fields for more than 30 years. Studies of heavy metals (As, Cd, Cu, Cr, Ni, Pb, Zn) showed that the concentrations of Cu, Ni, and Cr were at 108.2 mg kg^{-1}, 219.9 mg kg^{-1}, and 173.1 mg kg^{-1}, respectively, and were the highest, i.e., 2–7 times higher than the limits given in the Turkish Soil Pollution Control Regulation [29].

Exposure of animals to heavy metals from drinking water is a very big problem. It is difficult to estimate, especially when animals use random intakes and when they are kept in an extensive (poultry) or grazing (sheep, cattle) way. Naveedullah et al. [11] conducted a study on the distribution of selected metals (Zn, Cu, Mn, Fe, Cr, Cd, and Pb) in soils in the Siling reservoir watershed in China, the various ecological and health risks associated with selected metals to the inhabitants. Protection of soil quality in the reservoir watershed is of great importance to preserve water quality, which is a source of drinking water. The study revealed seasonal variations of selected heavy metal content in soil samples. In addition, the multivariate analysis conducted showed significant anthropogenic, point, and non-point pollution of selected metals in the Siling reservoir watershed. Through the use of enrichment factor, geoaccumulation index and contamination factor, moderate to high contamination was found in soil samples during the summer and winter seasons. Low soil pH and high organic matter content increase the leaching of some elements from the soil into the aquifer formations and increase the toxic metal content in the water sources. According to the authors, the quality and quantity of fertilizers used were important causes leading to the accumulation of heavy metals in soils depending on land use.

3. Heavy metal content of animal tissues and raw materials of animal origin

Among animal products, cow's milk generally contains a number of trace elements in its composition, and its value as an environmental bioindicator is quite low. The mammary gland of cows forms a natural biological barrier that limits the passage of toxic elements from the mother's body to the food. However, milk may contain higher amounts of heavy metals, as a result of their breeding in industrial areas or secondary contamination in technological processes. The conducted studies on the content of heavy metals in animal tissues and raw materials of animal origin mostly confirm bioaccumulation of these elements in living organisms and products of animal origin.

For example, in Poland, on farms located in the direct vicinity of the Turów power plant, high concentrations of Pb, reaching even 1.865 mg kg^{-1},were detected in milk. Also, in milk collected directly from cows on smallholder farms in Lower Silesia, Pb content in 22.6% of samples exceeded the permissible level, Cd in 29.4%, and Hg in 50% of samples [30].

In pigs, the main source of possible contamination of pig tissues is feed, which due to the use of various additives may increase the concentration of heavy metals. The sources may be also dust and gas emissions from the industry caused by intensive air exchange in the livestock building. The average Pb content in the muscle tissue of fattening pigs is 0.05–0.58 mg kg^{-1} d.m. The concentration of Cd in muscles ranges from 0.02 to 0.04 mg kg^{-1} [30].

Sheep belong to good bioindicators of the environment because they are kept in extensive and pasture systems. Research shows that in industrial areas, cadmium accumulated in kidneys and udder. On the other hand, lead was found in the ribs, liver, and long bones. Zinc was most abundant in ribs and long bones, and Cu in kidneys, bones, and udder [30].

Conventional farms in central Greece were analyzed for the heavy metal content of copper (Cu), zinc (Zn), cadmium (Cd), lead (Pb), nickel (Ni), and chromium (Cr) in muscle tissues, livers, kidneys, feces and staple mixtures for livestock, such as cows and sheep. The study showed that the transfer of heavy metals from feed to animal products varied below acceptable risk levels. In experiment 2, feed for animals kept under different feeding systems and seasons on different farms (sheep, dairy cows, and pigs) was studied. The analyses conducted showed relatively high concentrations of Cu in pig feces (155 ± 9.13) and Zn in sheep feces (144.56 ± 5.78) [31].

Excessive accumulation of heavy metals can occur in poultry, particularly from backyard rearing. Poultry is exposed to landfills (farm waste), animal feces, and contaminated roadside ditches. It may eat contaminated soil, plants, and geohelminths. The study by Kołacz et al. [30] shows that in hens kept in the Lower Silesia region exceedances of Pb in chicken muscles occurred in 20% of samples, and chicken eggs in I4.3% of samples. In the case of Cd, exceedances concerned I4.8% and 7.9%, whereas Hg exceedances concerned 45.5% and 64.5%, respectively. Relatively high accumulation of mercury occurred in hen eggs, with averages often exceeding the nationally permissible limit of 0.02 mg Hg kg^{-1}. Kołacz et al. [32] performed an assessment of the degree of bioaccumulation of heavy metals (Cd, Cu, Hg, Pb, and Zn) in the muscles and livers of free-range hens in the copper Belt region. The results showed that the organ accumulating the higher amounts of heavy metals was chicken liver, in which the permissible content of Pb and Cd was exceeded. In the muscles, however, the Pb content was exceeded.

Duck and goose eggs are a good indicator of environmental contamination with heavy metals in the rural environment. Studies performed in industrialized areas exposed to metal-bearing dust emissions and other sources have shown many times higher concentrations of As, Cd, and Hg, slightly less Pb and Cu compared to sites not exposed to industrial emissions [33].

Bees are a very good bioindicator of the heavy metal content of micronutrients and toxic metals in the environment. Roman [34] conducted a study in two regions—industrial and agroforestry, with the aim of indicating whether worker bees and drones accumulate toxic metals. Toxic elements (Ni, Cr, Pb, Cd, and Se) were found in all samples, with higher concentrations of Ni, Cr, Cd, and Pb detected in the organisms of worker bees, while higher concentrations of Se were found in the organisms of drones. The organism of worker bees is a filter that retains from 20.45 (Pb) to 36.36%

(Cd) of toxic metals from the honey raw material in the process of processing it into honey. In a study conducted in Wroclaw, Poland, concerning the content of heavy metals in multiflower honey and propolis, it was found that the highest concentration of elements was in propolis: Zn—48.1, Cu—6.95, Pb—5.74, As—0.66, and Cd 0.19 mg kg^{-1}. Statistically, a significantly lower concentration of zinc, copper, arsenic, and cadmium was in honey, except for lead, whose content was higher than permissible values in 85% of samples [35].

Horses due to their use in different environmental conditions belong to a group of animals potentially exposed to bioaccumulation of heavy metals, especially in industrialized and urbanized areas. Few studies show that horse muscles do not contain excess As, Cd, Hg, and Pb, while kidneys almost always accumulate large amounts of Cd, and liver Zn and Cd [30].

Szkoda et al. [36] also conducted studies on the content of lead, cadmium, and mercury in tissues of game animals from selected main industrial areas in Poland—roe deer (*Capreolus capreolus*), red deer (*Cervus elaphus*), and wild boar (*Sus scrofa*). Studies are a useful source of information about the quality of ecosystems in which they live. The highest acceptable level of lead in meat was exceeded in 21% of analyzed samples. High cadmium concentration was determined in kidneys (above the maximum level of 1 mg kg^{-1} in 88% of samples). The highest concentration of toxic elements was found in roe deer and wild boar from Upper Silesia, which indicates high environmental contamination in comparison with other areas.

Heavy metal levels of wildlife biotopes from two different industrially exploited areas in Slovakia were studied in the Zemplín region of Slovakia. Various tissues (lungs, liver, kidneys, spleen, heart, and muscles) were sampled from animals, such as red deer, roe deer, mouflon, chamois, wild boar, European hare, among others. The content of elements exceeded the legal limits allowed for human consumption—mercury in 29%, cadmium—28%, and lead—23%. Chromium concentration did not exceed the limit in any sample. Of the wildlife living there, wild boar was the most heavily burdened species [37].

4. Effects of heavy metals on human and animal health

Heavy metals enter the human and animal body through the respiratory tract, sometimes even through the skin in the case of chemicals capable of crossing the skin barrier, and the digestive tract. Food is the main source of toxic elements. The deficiency of macro- and micro-nutrients in the body causes heavy metals to be absorbed in their place. When there is poor nutrition and a lack of essential nutrients in the body, the liver that performs detoxification cannot perform this important task [8, 38].

Copper is one of the elements with important functions in the human and animal body. It takes part in oxidation-reduction processes, regulates the metabolism, transport of iron. In the form of complexes with amino acids and albumin, it is transported to the liver, kidneys, intestines, and other tissues. Excess copper mainly causes decreased hemoglobin concentration, liver, and kidney damage [8, 14, 39].

Zinc accumulates in vertebrates in the liver, forming complexes with various proteins, and also in the kidneys and sex glands. Symptoms of zinc intoxication are—impaired immune response, reduction of HDL cholesterol fraction, decreased level of copper in the blood. On the other hand, acute zinc poisoning is manifested by vomiting, epigastric pain, fatigue, and sluggishness. Zinc chloride ($ZnCl_2$) is irritating to the skin, mucous membranes, and conjunctiva, causing burns in higher concentrations [8, 14, 39].

Cadmium is easily absorbed in living organisms. It enters the human body primarily through food, water, and inhalation as a result of tobacco smoking. Cadmium accumulates primarily in the kidneys and liver. Cadmium poisoning causes nausea, vomiting, salivation, abdominal pain, kidney damage, and circulatory failure. In addition, exposure to cadmium oxide fumes, exceeding the concentration of 0.5 mg Cd kg^{-3} causes, among others—emphysema, kidney function damage, changes in the skeletal system, pain in the limbs and spine [8, 14, 39].

Mercury poses a very high risk to living organisms. Mercury enters the body primarily from food via the digestive system and inhalation. Mercury compounds can interfere with most enzymatic reactions because they react with proteins containing sulfhydryl groups. The highest concentrations of mercury are found in the kidneys, but the brain is the most susceptible. Mercury vapor toxicly affects the lungs, eventually causing respiratory failure and death. In addition, chronic exposure to low concentrations of mercury vapor causes damage to the central nervous system with symptoms of weakness, memory impairment, mood swings, headache and limb pain, mucositis and gingivitis, and others [8, 14, 39].

Lead is also a highly toxic element. It enters the human body through the digestive system with water and food, and inhalation. It accumulates in the liver, heart, kidneys, as well as in the skin and muscles. Symptoms of lead poisoning of the human organism include disorders of the blood-forming system and the central nervous system. Lead is a mutagenic, carcinogenic, and embryotoxic element [8, 14, 39].

Kołacz et al. [40] conducted a study on the blood of cows from the copper industry region (LGOM) showed a significantly higher Cu content (1.67 mg l^{-1}) compared to the agricultural region (0.72 mg l^{-1}). In both regions, the content of Cd did not exceed 1.65 μg l^{-1}, Pb in the industrial region was 0.021 mg l^{-1}, and in the agricultural region 0.031 mg l^{-1}. On the other hand, the level of Hg in the blood of cows, in both regions was similar and was about 0.03 μg l^{-1}. It was found that the copper industry does not adversely affect the blood parameters of dairy cows. In the whole blood of dairy cows, a lead level of 0.1 mg l^{-1} is considered normal, 0.1–0.3 mg l^{-1} is considered elevated, and above 0.35 mg l^{-1} is considered toxic.

5. Ways to reduce heavy metal penetration into crops and inactivation in animals

Accumulation of heavy metals by crop plants varies genotypically. The transfer coefficient of elements from soil to plant is expressed as the ratio of the concentration in the plant to the total concentration in the soil. For example, low transfer coefficients for cadmium are found in maize, pea, oat, and wheat grains, while high Pb and Cd are found in wheat grains. Low coefficients are found for Zn transfer in spinach and lettuce leaves and roots of various plants. Therefore, it is better to grow crops with low metal uptake (some cereals, legumes, vegetables) on polluted soils, while leafy vegetables on unpolluted soils [41].

The mechanisms of plant tolerance to metal are important in plant breeding. It depends on the species, plant growth phase, tissue or organ, type of metal, time of action, and applied dose. In zinc tolerance, mainly processes related to its detoxification by organic acids and storage in vacuoles are involved. In tolerance to lead mainly processes in the cell walls are immobilized, while for cadmium mainly detoxification by phytochelatins or accumulation in the cell walls are involved. Due to the existing network of mechanisms in plants that protect against toxic effects of heavy metals, in the future, it may be a way to restore the biocenotic balance in ecosystems destroyed by industrial human activities [42].

Some plants have a natural ability to accumulate heavy metals, which is used in the process of environmental cleanup (so-called phytoremediation). Known species accumulate 1–2% of metals in tissues (so-called hyperaccumulators), for example, boll weed (Thlaspi sp.). However, due to the low biomass yields of these species, the practical usefulness of the plants is limited. Plants that are to be effective in the uptake of heavy metals should be characterized by the following features—fast growth, high biomass yield and easy harvesting, deep root system, and accumulation of large amounts of heavy metals in the aboveground parts. Several phytoremediation technologies can be distinguished—phytoextraction, i.e., removal of heavy metals by accumulation in the above-ground parts of plants, phytostabilization, i.e., immobilization of metals in soil and reduction of their availability in the environment, phytostimulation, i.e., support by plants of naturally occurring microbial degradation processes in the rhizosphere, phytodegradation, i.e., decomposition of organic substances by plants and related microorganisms and phytovolatilization, i.e., transformation of contaminants into a volatile state. The plants most commonly used for bioaccumulation belong to many families, of which the crucifers (Cruciferae), grasses (Poaceae), butterflies (Papilionacae), composite plants (Asteraceae), willow plants (Salicaceae), and clove plants (Caryophyllaceae) deserve special attention [17].

Barrero-Moreno et al. [43] conducted biofilter modeling using rice husk as filter material to remove heavy metals from water. The use of bioadsorption represents great potential because lignocellulosic materials can be obtained in large quantities, are inexpensive, and can selectively remove Cd (II), Cu (II), and Cr (VI) from aqueous solutions. Based on the results, rice husk was found to be a good alternative for making filters with the ability to remove Cd (II), Cu (II), and Cr (VI) with 83.21%, 67.11%, and 92.18% efficiency, respectively, for specific values of filter height, temperature, and pH.

In sustainable agricultural production, one of the ways to reduce environmental and human, and animal health risks is to use fertilizer from agricultural biogas plants. It can be used in liquid or solid form as fresh matter, granulate, or compost. The introduced organic matter can prevent the leaching of toxic elements on the one hand, and their uptake by plants on the other [41]. Studies on the content of selected heavy metals (Fe, Zn, Mn, Pb, Cd, Ni, and Cu) in soils fertilized with mineral fertilizers (NPK and CaNPK) and with digestate and granulate did not show exceeding of permissible standards. The content of elements was compared to unfertilized objects. It was found that the applied post-fermentation masses are safe for fertilizer use. Statistically lower contents of Zn, Cu, and Mn were found after fertilization with fresh digestate compared to control objects, while lower amounts of Cu, Fe, and Cd were found after the application of granules [44].

Compost from digestate is also a product used in fertilization. The content of selected heavy metals (Fe, Zn, Mn, Pb, Cd, Ni, and Cu) in soils fertilized with mineral fertilizers (NPK and CaNPK) and with compost was analyzed to demonstrate its environmental safety and, indirectly, animal and human health. The content of elements was compared to objects not fertilized with mineral fertilizers (object 0) and with compost (K—control). The study showed that the long-term application of nitrogen, phosphorus, or potassium fertilizers increased the content of available forms of heavy metals in the soil. On the other hand, application of compost from post-fermentation mass caused a statistically significant decrease in bioavailable forms of metals, especially in object CaNPK—Ni, Pb, and Fe and object 0—Zn and Ni. In the case of the NPK object, a significant reduction in the content of all the metals studied except Cu occurred [45].

Detoxification of contaminated natural environments is based on different solutions—the use of antagonistic type interactions in reducing bioaccumulation in animal tissues and the use of properties of compounds of organic and mineral origin.

Vitamins play a special role in reducing the bioaccumulation of heavy metals in the animal organism. They actively participate in body protection by increasing the absorption and cellular bioavailability of elements that are antagonists, maintaining the physiological concentration of ions of these elements, and participating in free radical reactions. Vitamin C in animal feed rations reduces cadmium retention and increases the effectiveness of elements antagonistic to this element, such as Fe, Cu, and Ca. Vitamin C can reduce cadmium concentration in the kidney and liver by 35–40%. Vitamin D, with Ca and Zn, and vitamin A with calcitriol and fluorine can also significantly reduce Cd concentration in tissues. On the other hand, vitamin E can counteract the activity of dehydrogenases, lowered by cadmium [46, 47]. Antagonists of cadmium are elements, such as Zn, Cu, Se, Fe, Mn, Mg, and Ca. In the case of lead, the antagonists are: Fe, Cu, Zn, Mg, Se, Ca, P, and K, while mercury may be selenium. Interactions involve competition in absorption for common transport sites and mutual displacement from metalloproteins, enzymes, DNA, RNA, and cell receptors [46].

Aluminosilicates (e.g., zeolites bentonites, kaolin), humic acids, or flavonoids play a special role in the process of heavy metals reduction in the animal organism. Aluminosilicates, due to their complex-forming, sorption, and ion-exchange properties, counteract the bioaccumulation of heavy metals in animal tissues. Studies using kaolin and zeolite as litter additives for broiler chickens showed a reduction of Hg by over 93%, Pb by almost 31%, and Cd by over 31% in birds' livers. The use of aluminosilicates in animal feed rations resulted in a reduction of Cu, Pb, Cd, Cr, and Ni content in animal tissues. The study with the application of a mixture of humic acids (brown coal and peat) and aluminosilicates (bentonite) showed more than twofold decreased Pb accumulation in animal livers. From the group of flavonoids, quercetin is an organic compound showing the ability to attach metal ions and form complex compounds. Flavonoids that are present in propolis can form chelate compounds with heavy metals, which contributes to the detoxifying effect of this bee product [46].

6. Conclusions

Industry, motorization, agriculture, and other sources are the cause of heavy metal contamination of the natural environment (soil, plants, water). Toxic metals easily pass into the organism, sometimes they excessively accumulate in tissues and organs, for example, in muscles, blood, liver, or animal hair, as well as in milk of cows, eggs of poultry, and honey, creating certain dangers for human health. There are various ways to reduce the accumulation of these elements in crop plants and deactivation in living organisms. For this purpose, the mechanisms of tolerance of plants to metal, plants having a natural ability to accumulate heavy metals in the process of cleaning the environment (phytoremediation), or the use of fertilizers which are a byproduct of agricultural biogas plants are used. For the deactivation of heavy metals in the body of humans and animals, antagonistic type interactions and the properties of compounds of organic and mineral origin are used.

Author details

Anna Bartkowiak
Institute of Technology and Life Sciences—National Research Institute, Falenty, Poland

*Address all correspondence to: a.bartkowiak@itp.edu.pl

References

[1] Bartkowiak A, Bartkowiak P. The determinants of sustainable development of agriculture sector in Poland. Transformations in Business & Economics. 2012;**11**(2A(26A)):42-59

[2] Tóth G, Hermann T, Da Silva MR, Montanarella L. Heavy metals in agricultural soils of the European Union with implications for food safety. Environment International. 2016;**88**:299-309

[3] Dobrzański Z, Kołacz R, Bodak E. Heavy metals in the animal environment. Medycyna Weterynaryjna. 1996;**52**(9):570-574. In Polish

[4] Dobrzański Z, Kołacz R, Czaban S, Bubel F, Malczewski M, Kupczyński R, et al. Assessing mercury content in plant and animal raw materials in an area impacted by the copper industry. Polish Journal of Environmental Studies. 2017;**26**(2):577-583. DOI: 10.15244/pjoes/66709

[5] Hejna M, Gottardo D, Baldi A, Dell'Orto V, Cheli F, Zaninelli M, et al. Review: Nutritional ecology of heavy metals. Animal. 2018;**12**(10):2156-2170

[6] Hejna M, Moscatelli A, Onelli E, Baldi A, Pilu S, Rossi L. Evaluation of concentration of heavy metals in animal rearing system. Italian Journal of Animal Science. 2019;**18**(1):1372-1384. DOI: 10.1080/1828051X.2019.1642806

[7] Kabata-Pendias A, Pendias H. Trace Elements in Soils and Plants. 3rd ed. Boca Raton London New York Washington, D.C: CRC Press LLC; 2001. p. 331. ISBN 0-8493-1575-1

[8] Rostański A, Cabała J, Słota M. Metal-bearing areas as a source of danger for the natural environment. In: Wierzbicka M, editor. Ecotoxicology, Plants, Soils, Metals. 1st ed. Warsaw: Warsaw University Publishers; 2015. Chapter: 5.5. pp. 522-543. In Polish

[9] Sady W, Smoleń S. Influence of soil-fertilizing factors on the accumulation of heavy metals in plants. In: Effectiveness of Using Fertilizers in Horticultural Crops. Poznan: Agricultural University Publisher August Cieszkowski in Poznan; 2004. pp. 269-277. Available from: https://www.researchgate.net/publication/263445462. In Polish

[10] Sommer SG, Schjoerring JK, Denmead OT. Ammonia emission from mineral fertilizers and fertilized crops. Advances in Agronomy. 2004;**82**(557622):557-622

[11] Naveedullah HMZ, Yu C, Shen H, Duan D, Shen C, Lou L, et al. Risk assessment of heavy metals pollution in agricultural soils of siling reservoir watershed in Zhejiang Province, China. Hindawi Publishing Corporation BioMed Research International. 2013;**2013**:590306; 10 pages. DOI: 10.1155/2013/590306

[12] Alengebawy A, Abdelkhalek ST, Qureshi SR, Wang MQ. Heavy metals and pesticides toxicity in agricultural soil and plants: Ecological risks and human health implications. Toxics. 2021;**9**(42): 1-33. DOI: 10.3390/toxics9030042

[13] Dobrzański Z, Górecka H, Opaliński S, Chojnacka K, Kołacz R. Contents of trace and ultratrace elements in milk and blood of cows. Veterinary Medicine. 2005;**61**(3):301-304. In Polish

[14] Ociepa-Kubicka A, Ociepa E. Toxic effects of heavy metals on plants,

animals and humans. Engineering and Environmental Protection. 2012; **15**(2):169-180. In Polish

[15] Srivastava V, Sarkar A, Singh S, Singh P, de Araujo ASF, Singh RP. Agroecological responses of heavy metal pollution with special emphasis on soil health and plant performances. Frontiers in Environmental Science. 2017;**5**:64. DOI: 10.3389/fenvs.2017.00064

[16] Bodak E, Kołacz R, Dobrzanski Z. Heavy metals—exposure conditions and defence mechanisms in animals. Medycyna Weterynaryjna. 1996;**52**(10):619-626. In Polish

[17] Żurek G. Alternative plants in phytoextraction of heavy metals from contaminated areas. Problemy Inżynierii Rolniczej. 2009;**17**(3):83-89. In Polish

[18] Aydinalp C, Marinova S. Distribution and forms of heavy metals in some agricultural soils. Polish Journal of Environmental Studies. 2003;**12**(5):629-633

[19] Górka A, Kogut A, Krzystyniak A. Content of selected heavy metals—lead and cadmium in soils and their toxic effect on plant development. Analit. 2017;**3**:32-39. In Polish

[20] Mendoza EO, Custodio M, Ascensión J, Bastos MC. Heavy metals in soils from high Andean zones and potential ecological risk assessment in Peru's Central Andes. Journal of Ecological Engineering. 2020;**21**(8):108-119. DOI: 10.12911/22998993/127094

[21] Ociepa E, Pachura P, Ociepa-Kubicka A. Impact of unconventional fertilization on migration of heavy metals in the soil-plant system. Engineering and Environmental Protection. 2014;**17**(2):325-338. In Polish

[22] Tchounwou PB, Yedjou CG, Patlolla AK, Sutton DJ. Heavy metal toxicity and the environment. In: Molecular, Clinical and Environmental Toxicology. Basel: Springer; 2012. pp. 133-164

[23] Wikarek-Paluch E, Rosik-Dulewska C. In: Rosik-Dulewska C, editor. Bioavailability of Selected Contaminants in Chemically Degraded Soils. Zabrze: Institute of Fundamentals of Environmental Engineering, Polish Academy of Sciences; 2020. ISBN 978-83-60877-14-2. In Polish

[24] Regulation of the Minister of Environment of 1 September 2016 on the manner of conducting the assessment of pollution of the earth surface (Dz. 2016, item 1395). In Polish

[25] Koncewicz-Baran M, Gondek K. The content of trace elements in agricultural soils. Infrastructure and Ecology of Rural Areas, Branch in Krakow. Polish Academy of Sciences. 2010;**14**:65-74. In Polish

[26] Li X, Zhao Z, Yuan Y, Wanga X, Li X. Heavy metal accumulation and its spatial distribution in agricultural soils: Evidence from Hunan province. China. RSC Advances. 2018;**8**:10665-10672

[27] Karimi F, Shariatifar N, Rezaei M, Alikord M, Arabameri M. Quantitative measurement of toxic metals and assessment of health risk in agricultural products food from Markazi Province of Iran. International Journal of Food Contamination. 2021;**8**:2. DOI: 10.1186/s40550-021-00083-0

[28] Nwaichi EO, Wegwu MO, Nwosu UL. Distribution of selected carcinogenic hydrocarbon and heavy metals in an oil-polluted agriculture zone. Environmental Monitoring and

Assessment. 2014;**186**:8697-8706. DOI: 10.1007/s10661-014-4037-6

[29] Isleyen M, Akpinar A, Eren B, Ok G. Heavy metal profiles of agricultural soils in Sakarya, Turkey. Environmental Engineering Research. 2019;**24**(3):427-433. DOI: 10.4491/eer.2018.274

[30] Kołacz R, Dobrzański Z, Bodak E. Bioaccumulation of Cd, Pb and Hg in animal tissues. Medycyna Weterynaryjna. 1996;**52**(11):686-691. In Polish

[31] Makridis C, Svarnas C, Rigas N, Gougoulias N, Roka L, Leontopoulos S. Transfer of heavy metal contaminants from animal feed to animal products. Journal of Agricultural Science and Technology. 2012;**A 2**:149-154

[32] Kołacz R, Dobrzański Z, Górecka H, Moryl A, Grudnik T. Contents of heavy metals in tissues of hens kept in the Copper Belt area. Acta Agrophysica. 2003;**1**(2):263-269. In Polish

[33] Dobrzański Z, Chojnacka K, Górecka H, Chojnacki A, Wiśniewski J. Water poultry eggs as an indicator of contamination of the rural environment. Acta Agrophysica. 2003;**1**(3):395-401. In Polish

[34] Roman A. Bioaccumulation of selected trace elements in the organism of worker bees and drones. Medycyna Weterynaryjna. 2006;**62**(12):1439-1442. In Polish

[35] Roman A, Madras-Majewska B, Popiela-Pleban E. Comparative study of selected toxic elements in propolis and honey. Journal of Apicultural Science. 2011;**33**(2):97-106

[36] Szkoda J, Durkalec M, Kołacz R, Opaliński S, Żmudzki J. Contents of lead, cadmium and mercury in tissues of wild game. Medycyna Weterynaryjna. 2012;**68**(11):689-692. In Polish

[37] Maľová J, Ciberej J, Maľa P, Zigo F, Semjon B. Heavy metal levels in the tissues of wild living animals from two distinct industrially exploited areas in Slovakia. Slovak Journal of Animal Science. 2019;**52**(3):100-110. ISSN 1337-9984

[38] Mudgal V, Madaan N, Mudgal A, Singh RB, Mishra S. Effect of toxic metals on human health. The Open Nutraceuticals Journal. 2010;**3**:94-99

[39] Piontek M, Fedyczak Z, Łuszczyńska K, Lechów H. Toxicity of copper, zinc and cadmium, mercury and lead to humans, vertebrates and aquatic organisms. Environmental Engineering, University of Zielona Góra, Zeszyty Naukowe. 2014;**155**(35):70-83. In Polish

[40] Kołacz R, Dobrzański Z, Kupczyński R, Cwynar P, Opaliński S, Pogoda-Sewerniak K. Impact of copper industry on the formation of selected heavy metals and biochemical indices in blood of dairy cows. Medycyna Weterynaryjna. 2017;**73**(3):171-175. DOI: 10.21521/mw.565. In Polish

[41] Puschenreiter M, Horak O, Friesl W, Hartl W. Low-cost agricultural measures to reduce heavy metal transfer into the food chain—a review. Plant, Soil and Environment. 2005;**51**(1):1-11

[42] Baranowska-Morek A. Plant mechanisms of tolerance to toxic effects of heavy metals. Cosmos, Problemy nauk Biologicznych. 2003;**52**(2-3):259-260; 283-298. In Polish

[43] Barrero-Moreno MC, Diaz-Vargas CA, Restrepo-Parra E.

Computational simulation of filters used in the removal of heavy metals using rice husks. Agriculture. 2021;**11**:146. DOI: 10.3390/agriculture11020146

[44] Bartkowiak A, Piątek M. Analysis of heavy metal content in soil fertilised with fresh and granulated digestate. Polish Journal of Environmental Studies. 2020;**29**(5):1-8. DOI: 10.15244/pjoes/117692

[45] Piątek M, Bartkowiak A. Analysis of heavy metals content in soil fertilized with compost from post-fermentation mass. In: Romaniuk W, Jankowska-Huflejt H, editors. Improving Technologies in Animal Production and Renewable Energy Taking into Account the Requirements of Sustainable Development. Falenty-Warsaw: ITP; 2019. pp. 222-228. In Polish

[46] Dobrzański Z, Kołacz R, Bodak E. Methods of preventing bioaccumulation of heavy metals in animals. Medycyna Weterynaryjna. 1996;**52**(12):763-768. In Polish

[47] Tutic A, Novakovic S, Lutovac M, Biocanin R, Ketin S, Omerovic N. The heavy metals in agrosystems and impact on health and quality of life. Open Access Macedonian Journal of Medical Sciences. 2015;**3**(2):345-355. DOI: 10.3889/oamjms.2015.048

Chapter 4

Heavy Metal Contamination of Food Crops: Transportation via Food Chain, Human Consumption, Toxicity and Management Strategies

Vanisree C.R., Mahipal Singh Sankhla, Prashant Singh, Ekta B. Jadhav, Rohit Kumar Verma, Kumud Kant Awasthi, Garima Awasthi and Varad Nagar

Abstract

Food security is a major concern that requires sustained advancement both statistically and on the basis of Qualitative assessment. In recent years, antagonistic impacts of unforeseen toxins have impacted the quality of crops and have created a burden on human lives. Heavy metals (e.g., Hg, As, Pb, Cd, and Cr) can affect humans, adding to dreariness and in severe cases even death. It additionally investigates the conceivable geological routes of heavy metals in the surrounding subsystems. The top-to-the-bottom conversation is additionally offered on physiological/atomic movement systems engaged with the take-up of metallic foreign substances inside food crops. At long last, the board procedures are proposed to recapture maintainability in soil–food subsystems. This paper reflects the contamination of the food crops with heavy metals, the way of transport of heavy metal to food crops, degree of toxicity after consumption and the strategies to maintain the problem.

Keywords: assimilation, crops, health, heavy metals exposure, soil, toxicity

1. Introduction

The contamination of heavy metals has spread widely across the environment, wreaking havoc on humans and the environment causing risks to the human race and hazard for the environment. A few hazardous Heavy metals (such as As, Pb, Cd, and Hg) relates to metabolic and other organic capacities as inconsequential. Certain metals, like Zn, Fe, Cr (III), etc., are important to carry metabolic cycles, including hemeproteins (like cytochrome p450) and catalysts. They are related to biota metabolism [1]. Despite the fact that metals like Nickel is an essential part of urease, it poses a significant health hazard for the human populations [1, 2] Accordingly, food and soil, crop/vegetable frameworks have given an exemplary illustration of the biodiverse ecosystem communications in the climate. For food the basic source of yield is the soil, but it is prone to contamination by heavy metals from various origins (e.g., energy-serious enterprises, for example, nuclear energy stations and coal mine shafts, and chloro-soluble base synthetic ventures, like goldmines, purifying, electroplating, materials, calfskin, and e-squander handling) and nonpoint sources (e.g., soil/residue disintegration, farming spillover, and open cargo stockpiling). Notwithstanding their human wellbeing suggestions, Heavy metals unfavorably influences soil biota through microbial cycles [3]. For instance, restorative plants utilized for customary human medical care ought to be analyzed for Heavy metals tainting to forestall unfavorable impacts. When exposed to near refining or other modern environments, a variety of restorative species of plants have been found to cause bioaccumulation of various heavy metals (e.g., Cd, As Cr, Cd, Cu, Pb, and Fe) [4, 5]. Both anthropogenic and natural sources release heavy metals into the environment. They can penetrate soils and groundwater, bioaccumulate in food webs, and harm biota because they are highly reactive and typically toxic at low quantities. The biotic effects of necessary and non-essential heavy metals are often highly distinct, and vary depending on the exact metal involved. This paper aims to provide a global overview of significant metal sources in agro-environments that are equivalent to the various anthropogenic activities and cycles [6]. This also includes details about heavy metals contamination in soil environments where food crops are grown in fully occupied main-lands. Moreover, ecological and lives of humans in these subsystems are needed to be discussed to aid in the understanding of the physiological/atomic systems which constitutes to the uptake of Heavy metals in the crops [7]. By suffocating or reducing the passage of metallic poisons from soil to food crops, management strategies are inextricably linked to human well-being government aid (soil-crop frameworks).

2. Heavy metal contamination

Natural contamination of these heavy metals is a significant environmental issue due to its negative impacts all over the world. Because of rapidly increasing horticultural and metal enterprises, improper rubbish removal, manures, and pesticides, organic wastes that are being disposed of in water bodies, land soil, and the environment. A few metals have an effect on natural capacities and development, while others collect in at least one organ, creating a variety of real illnesses such as cancer. Each metal's toxicological and pharmacokinetics cycles in humans are depicted. In general, environmental auditing can tell us the effects of heavy metal bioaccumulation in the human body with context to biochemical and physiological changes, as well as the severe illnesses occurring from their exposure.

Climate is an environmental factor in which people, plants, beings, minute living beings. It comprises land, the atmosphere and the hydrosphere component. Earth is a framework that is characterized by the four pillars, the living organisms (biosphere), the atmosphere (air), the lithosphere (land) and water (hydrosphere) which are all employed in a combination. Natural toxins, just like contaminations, are synthetic substances that are available at more elevated levels than in any segment of the climate [8–10]. During the most recent years, the process of industrialization has developed at a rapid phase. This manner has expanded the interest for abuse of the Earth's regular assets which is a growing concern ecological contamination [11]. A few poisons, such as inorganic particles, natural toxins, organometallic substances, vaporous poisons, radioactive substances and nano-based materials, have actually poisoned the climate [12, 13].

Heavy metals have existed on earth since the existence of earth. Because of an amazing increment in the usage of substantial metals which brought about a fast-approaching flood of metallic substances in both the earthbound climate and the oceanic climate [12]. The Heavy metal contamination has occurred due to anthropogenic activities, primarily due to metal mining, purification, foundries, and other metal-based ventures, as well as metal draining from various sources such as landfills, squander dumps, discharge, animals and chicken fertilizer, overflows, cars, and roadworks [7]. The usage of Heavy Metals (Heavy metals) in the industries like agriculture, pesticides, insect poisons, manures etc., has been an optional wellspring of heavy metal contamination. Eruptions from volcanoes, consumption of metals, from the environment are all examples of natural events that might increase heavy metal contamination. This contamination can travel from soil and water causing soil breakdown, and land degradation, which are all examples of natural factors that might increase heavy metal contamination [14].

3. Source of heavy metal in irrigation water

Heavy metals are commonly found in the environment as a result of substrate erosion. These naturally occurring heavy metals are generally in forms that are inaccessible to plant roots [15]. However, due to a rise in human activities that release more biologically accessible forms of heavy metals into the environment, this has altered in recent years [16] Agriculture is one of the most major human activity that contributes to heavy metals being released into the environment [17]. Using wastewater to irrigate agricultural soils resulted in considerably greater amounts of heavy metals in the edible parts of the crops produced on these soils, according to [18]. Heavy metals may be present in factory wastewater, which build over time in soil deposits along waste water routes as well as in creatures that live near them. Human exposure to polluted wastewater is common, especially in densely populated metropolitan areas or when wastewater is utilized for agricultural purposes.

Multiple businesses (textile, dyeing, garment, pharmaceutical, ceramic, paint, packing, etc.) dump untreated wastewater into neighboring canals, causing heavy metal pollution in irrigation water in Bangladesh's Gazipur District. Due to wastewater discharge from businesses, it was found that the surface water and soil of the industrial region in Dhaka and Gazipur District are significantly contaminated with Zn, Cr, Cu, Pb, and Cd. Farmers irrigate their crops with polluted water, resulting in heavy metal pollution in the district's veggies. Vegetables are polluted with high amounts of Cr, Zn, Cu, Fe, Pb, Ni, and Cd in the industrial region of the Dhaka Export Processing Zone (DEPZ). The majority of these veggies are sold at Dhaka's wholesale market and are consumed by a large number of people [19–21]. (**Table 1**) demonstrate the concentrations of heavy metals (mg L^{-1}) in effluent-contaminated water for irrigation within the DEPZ in Dhaka, Bangladesh [21].

3.1 Source of heavy metal in agricultural soil

Heavy metals and metalloids may be accumulated in soils as a result of discharges from rapidly expanding modern regions, mine tailings, removal of high metal

Values	Lead	Chromium	Cadmium
Safe limit	0.5	0.1	0.01
Mean	0.21	0.43	0.06
Median	0.19	0.43	0.04
Min	0.14	0.29	0.02
Max	0.30	0.53	0.08
SD	0.05	0.08	0.04

Table 1.
Concentrations of heavy metals (mg L^{-1}) in effluent-contaminated water [21].

squanders, leaded gas and paints, land use of composts, animal excrement, sewage slop, pesticides, wastewater water system, coal burning buildups, petrochemical spillage, and environmental statements [22, 23]. Frequently recognized heavy metals like Lead (Pb), chromium (Cr), arsenic (As), zinc (Zn), cadmium (Cd), copper (Cu), mercury (Hg), and nickel (Ni) are the pollutants. Soils are a significant sink for heavy metals delivered into the climate by recently referenced anthropogenic exercises, and dissimilar to regular unfamiliar substances that are oxidized to carbon (IV) oxide by microbial movement, most metals do not go through microbial or compound degradation, and their complete obsession in soils goes on for quite a while after discharge [24, 25]. Regardless, changes in their substance structure (speciation) and bioavailability are conceivable. The presence of harmful metals in soil can keep regular pollutants from biodegrading appropriately [26]. Significant metal contamination of soil can jeopardize individuals and the organic framework through: direct ingestion or contact with spoiled soil, the normal lifestyle (soil–plant-human or soil–plant–animal-human), drinking defiled ground water, decline in food quality (security and appeal) because of phytotoxicity, and reduction in land usability for agrarian creation because of phytotoxicity [27]. The adequate security and rebuilding of heavy metal-tainted soil biological systems need their depiction and repair. At both the public and global levels, current legislation on natural insurance and general well-being is based on data that describes the compound qualities of ecological wonders, particularly those that exist in our developed way of life [28]. Risk assessment is a powerful logical tool that enables leaders to manage contaminated areas in a cost-effective manner while protecting public and biological system health [29]. Heavy metals happen spontaneously in the soil environment as a result of pedogenetic cycles of parent material lasting at levels that are considered safe (1000 mg/kg-1) and occasionally hazardous (1000 mg/kg-1) [29]. In view of man's exacerbation and acceleration of nature's progressively happening geochemical example of metals, most soils in country and metropolitan regions may collect at any rate one of the heavy metals above described establishment regards adequately high to represent a danger to human wellbeing, plants, animals, organic frameworks, or other media [30]. Significant metals basically become toxins in soil conditions since (I) their paces of maturing through counterfeit cycles are quicker than normal ones, (ii) they are moved from mines to self-assertive regular regions with higher odds of direct receptiveness, and (iii) the metal unions in discarded things are tolerably high contrasted with those in get items. Critical metals in soil from anthropogenic sources will be more adaptable and henceforth bioavailable than those from pedogenic or lithogenic sources [30]. Metal-bearing solids in spoiled regions can emerge out of an assortment of anthropogenic sources, including metal mine tailings, expulsion of high metal wastes from inappropriately got landfills, leaded gas and poisonous paints, land utilization of fertilizer, creature feces, biosolids (sewage sludge), compost, pesticides, coal start developments, petrochemicals, and air contamination [31]. Fertilizer. Horticulture was the most important human effect on the land in general [32]. Plants require not just macronutrients but also essential micronutrients to build up and complete the lifecycle. Some soils are deficient in heavy metals which are necessary for solid plant development, and harvests might be supplemented with them as a soil development or foliar spray. Grain crops grown on Cu-deficient soils are treated with Cu on a regular basis as a soil expansion, and cereal and root crops may also be given Mn. In major cultivating frameworks, large volumes of compost are frequently applied to soils to provide adequate N, P, and K for crop development. Following measurements of heavy metals (e.g., Cd and Pb) as contaminants are present in the mixes utilized to supply these components [33] which may fundamentally extend their substance in the soil after being applied with compost. Cd and Pb, for example, have no recognized physiological effects. The use of some phosphatic composts accidentally introduces Cd and other potentially toxic components to the soil, such as F, Hg, and Pb [34]. Pesticides. In the past, a few common insecticides used widely in agribusiness and agriculture included significant metal centralizations. For example, in the recent past, around 10% of the synthetics used as insecticides and fungicides in the United Kingdom were based on intensities containing Cu, Hg, Mn, Pb, or Zn. Copper-containing fungicidal splashes like Bordeaux mixture (copper sulphate) and copper oxychloride are examples of such pesticides [35]. For a long time, lead arsenate was used in natural product plantings to suppress parasitic microorganisms. In New Zealand and Australia, arsenic-containing compounds were additionally broadly used to control dairy cows ticks and banana bugs, woods have been saved with Cu, Cr, and As (CCA) plans, and there are currently various disregarded regions where soil groupings of these parts endlessly outperform establishment obsessions. The narrative of arsenic pollution, infiltration, and its impact on a variety of crops, foods, mushrooms, fruits, vegetables, sea animals, and fish and animal products is being revealed via ongoing research [36]. Such contamination may cause issues, particularly if areas are revamped for non-horticultural or agrarian reasons. Rather than excrements, the utilization of such materials has been more compelled, in light of indicated objections or yields. Biosolids and Manures. The inadvertent social affair of heavy metals and others in the soil is brought about by the utilization of different biosolids (e.g., creature composts, fertilizers, and civil sewage ooze). In agribusiness, some living thing squanders like chickens, dairy cattle, and pig waste products are commonly applied to harvests and fields as solids or slurries [37]. In spite of the way that most fertilizers are viewed as significant excrements, the Cu and Zn added to things down as improvement advertisers and advertisers in the pig and poultry businesses. Metal spoiling of the dirt may happen because of the fixings in poultry wellbeing items. Fertilizers made by creatures on such weight reduction regimens incorporate high centralizations of As, Cu, and Zn, which, whenever applied a few times to restricted spaces of land, can bring about significant development of these metals in the soil over the long haul. Biosolids (sewage slop) are essentially regular, amazing items that might be used subsequent to being given by wastewater treatment systems. Land utilization of biosolids materials is a typical procedure in numerous nations that consider the reuse of biosolids given by metropolitan populaces [38–40]. In view of its far and wide acknowledgment and authoritative significance, the expression sewage waste is utilized in an assortment of contexts [41]. In any case, the word biosolids is getting more utilized as a substitute for sewage overflow, since it is considered to better precisely portray the gainful properties inborn in sewage ooze. Most of the 5.6 million dry truck heaps of sewage waste utilized or unloaded in the United States are assessed to be land applied, and biosolids are utilized in provincial regions around the country. Over 30% of sewage ooze is utilized as manure in agriculture in the European social class [42]. The major urban expert in Australia gives around 1,75,000 tons of dry biosolids consistently, and most biosolids applied to agrarian land are presently utilized in arable managing circumstances where they might be united into the dirt. There's likewise a great deal of buzz about the chance of enhancing soil biosolids with other regular assets like sawdust, straw, or nursery squander. There will be suggestions for metal polluting of soils if this inclination proceeds. Biosolids' penchant to debase soils by eliminating substantial metals has started to worry about their use in agrarian activities. Significant metals are probably not going to be foundare Pb, Ni, Cd, Cr, Cu, and Zn, and the metal obsessions are addressed by the nature and the force of the advanced development, similarly as the sort of connection used during the biosolids treatment [43]. Wastewater. The practise of dumping urban and mechanical wastewater and associated effluents on land dates back 400 years and is now commonplace in countless parts of the world [44]. Squander water has been estimated to cover 20 million hectares of fertile land throughout the world. Considerations suggest that horticulture depending on wastewater water systems contributes 50% of the vegetable supply to metropolitan regions in certain Asian and African cities. Ranchers, for the most part, are unconcerned with ecological benefits or risks, and are only interested in increasing their profits and benefits. Despite the fact that metal fixations in wastewater effluents are typically modest, a long-distance water system with such might eventually result in significant metal accumulation in the ground. Industrial Wastes and Metal Mining and Milling Cycles Mining and handling of metal assets, in mix with organizations, has given numerous nations the legacy of inescapable metal contamination transport in soil. Tailings are essentially released into customary dejections, including neighboring wetlands, bringing about expanded obsessions. Contamination of soil has come about because of far and wide Pb and zinc Zn metal mining and filtration, representing a danger to human and organic wellbeing. A significant number of the recuperation techniques utilized here

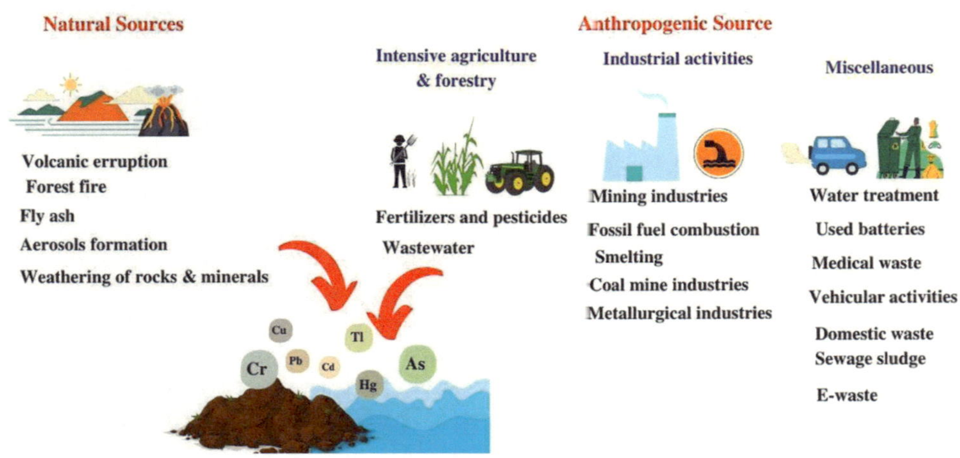

Figure 1.
Source of heavy metal contamination.

are tedious and costly, and they will most likely be unable to reestablish soil utility. Bioavailability has been featured as a soil heavy metal natural danger to individuals. Retention of plant material filled in (normal hierarchy) or quick ingestion (oral bioavailability) of contaminated soil are instances of assimilation courses. Various materials are created by an assortment of organizations, including material, tanning, petrochemicals from unplanned oil spills or the utilization of oil-based items, pesticides, and medication workplaces, and they assume a critical part in the arrangement as shown in (**Figure 1**). Despite the way that some appear on shore, few have agrarian or officer administration benefits. Moreover, a few are conceivably perilous in light of the fact that to the presence of huge metals (Cr, Pb, and Zn) or poisonous normal mixes, and are simply applied to arrive on uncommon events, if by any means. Others are insufficient in plant supplements or do not have the capacity to shape soil [32].

4. Heavy metal uptake and bioaccumulation in food crops

For groundwater and plants, soil serves as both a source and a sink for the presence of Heavy metals [45]. These toxicants get accumulated in the land soil and has become a serious concern as a result of fast advances the agricultural and industrial sectors [46]. Crops are essential part of diet of individuals and are vital source of important nutritional components like minerals and vitamins [47, 48]. Crops grown on contaminated lands with Heavy Metals, these metals accumulate in the plants edible parts, which are then ingested by humans [49] Because heavy metals are resistant from degradation, and have long half-life periods, thus difficult to excrete out. Many metals are hazardous at low doses, Heavy metal poisoning is a major problem in crops [18, 45, 50, 51]. Long-term exposure to heavy metal contaminated crops can result in a variety of health problems, including bone thinning, skin problems, improper endocrine gland function, blood pressure, neoplastic growth, impairment of sexual characteristics, asthma and other respiratory issues, heart diseases, and brain impairments [52, 53]. Heavy metal contamination in crops is a concern worldwide that leads to toxidromes and a variety of illnesses in humans, flora and fauna, when polluted soils and food crops are consumed.

4.1 Metal uptake and transportation pathways

Ingestion and amassing of substantial metals in plant tissues rely on temperature, dampness, natural matter, pH, and supplement accessibility [54]. Heavy metal amassing moreover depends upon plant species, while the viability of plants in engaging metals is directed by either plant take-up or soil-to-plant move factors of the metals [22]. Brought lead step up in soils, for instance, may diminish soil convenience, while very low lead obsession may stifle some fundamental plant capacities like photosynthesis, mitosis, and water absorption, bringing about destructive indications like dull green leaves, shrinking of more prepared leaves, ruined foliage, and hearty hued short roots, among others [55]. Huge metals are possibly unsafe, causing chlorosis, helpless plant development, and low yield, and they might be joined by decreased enhancement take-up, issues in plant absorption, and a diminished ability to fix subatomic nitrogen in leguminous plants [56]. Because of exercises like mineral burrowing, metal transportation, decontaminating and refining, and expulsion of tailings and waste waters around mines, mining and filtering occupations are significant reasons for weighty metal polluting in the environment [57, 58]. Disinfecting of water and soil, phytotoxicity, soil crumbling, and likely dangers to human wellbeing are a portion of the negative normal effects of unnecessary hefty metals tossed about mine and purifying locales [48, 59, 60]. Critical metal pollution of cultivating soils and yields in mining zones has been viewed as an uncommon normal risk [61–63]. Heavy metal take-up by roots from debased soils and surface water, just as immediate exchange of toxins from the climate on plant surfaces, can bring about critical metal defilement of plants [23]. Lead and Cd are suspected malignancy causing synthetics and have been connected to the etiology of an assortment of sicknesses, including cardiovascular, renal, blood, apprehensive, and bone illnesses [64]. Notwithstanding the way that Zn and Cu are fundamental segments, their exorbitant focus in food and feed plants is of incredible concern attributable to their harmfulness to people and living things [65]. Development of yields for human or trained being utilization may possibly prompt the take-up and amassing of these metals in consumable plant leaves, representing a danger to human and living thing wellbeing (**Figure 2**) [66, 67]. Unnecessary dietary gathering of heavy metals like Cd and Pb in the human body may bring about genuine clinical issues [68]. For the greater part, dietary induction is the dominating method of receptiveness, regardless of the way that in profoundly contaminated regions, internal breath can assume a significant part [69]. The significant channel of human receptiveness to generous metals is the soil-to-manage trade of heavy metals. The developing human populace has started an interest for more food [23]. Pesticides, manures, fertilizers, composts, and wastewater have all been utilized all the more frequently in the water framework accordingly [70]. Food crops developed on metal-drained soil can ingest and gather metals in critical amounts to influence food quality and wellbeing [71]. Most nations have given genuine thought to the control of hefty metals in food crops because of soil pollution in country regions [72].

Plants retain fundamental and pointless segments from the soil dependent on fixing inclination and molecule explicit take-up, or by means of scattering [73]. Root assumes a critical part in the take-up of metal particles. Due to the presence of cellulose, gelatin, and glycoproteins, which go about as express molecule exchangers, (TEs) adsorb on the root surface in a cationic setuFp with a negative cell divider [74]. The cations (Zn2+, Mn2+, Cd2+, Fe2+, Pb2+, Ni2+) are open at the root surface and effectively gathered into the root apoplast [75]. In the wake of being accumulated in

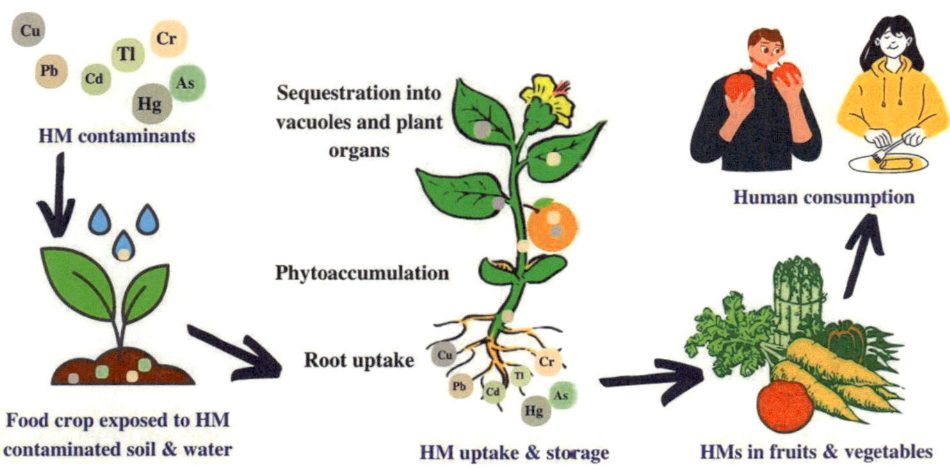

Figure 2.
Heavy metal uptake and transportation to food chain.

the root apoplast, the cations are either held in the root cells or moved radially to the root stele and packed into the xylem and phloem tissues in one of two different ways: apoplastic/reserved transportation or sym-plastic/powerful transportation [76]. The scattering of metal particles in the root cell through the earth plan causes disengaged transportation, while the unique transportation of metal particles happens through the plasma layer, which is hindered by different carriers or transporters [77]. The xylem sap is coordinated upwards by the incident stream, where TEs are moved to the aeronautical tissues. In the event that no redistribution happens, TEs will assemble in photosynthetically powerful leaves. The phloem, another vascular tissue, revamps and supplements the results of photosynthesis across the entire plant body, between the sources and sinks. The scattering of metal particles in the root cell through the earth course of action causes separated transportation, while the powerful transportation of metal particles happens through the plasma layer, which is hindered by different carriers or transporters [77]. The xylem sap is coordinated upwards by the occurrence stream, where TEs are moved to the aeronautical tissues. On the off chance that no redistribution happens, TEs will assemble in photosynthetically powerful leaves. The phloem, another vascular tissue, reworks and supplements the results of photosynthesis across the entire plant body, between the sources and sinks. TEs can be moved from senescing leaves to sinks through the phloem (e.g., creating vegetative parts and creating regular items). Before the xylem sap comes to the mesophyll cells, TEs can likewise relocate to the phloem [78]. Phloem transport comprises of (I) apoplastic stacking into both accomplice cells and sifter parts, just as (ii) unloading at the objective sink tissues [79]. A few metal-limiting mixes, including as nicotianamine and phytochelatins, were demonstrated to be reasonable for shipping TEs in the phloem [80]. Each progression requires a staggering cooperation of chelating designs and metal transporters that influence metal accumulation speed [80]. Metal chelators are connected to a few phases of micronutrient take-up, inside vehicle, and sequestration in the cytosol and subcellular compartments [80]. Metal take-up and remobilization from intracellular compartments into the cytosol is worked with by the ZIP, NRAMP, yellow stripe (YS), and copper transporter (COPT) families, while heavy metal-delivery ATPases (Heavy metalAs), the cation (CDF) family, the cation exchanger (CAX) family, and the multi-drug and destructive compound ejection (MATE) family, just as the plant cadmium resistance [80].

4.2 Metal stress tolerance mechanisms

Heavy metal toxicity causes a wide range of physiological and biochemical changes, and plants must evolve and/or adopt a variety of methods to deal with the detrimental effects of heavy metal toxicity. Plants react through several mechanisms to external stimuli including toxicity to heavy metals. These include (i) external stress stimulus sensing, (ii) signal transduction and signal transmission into the cell and (iii) appropriate actions are taken to offset the negative effects of stress stimuli by modulating the cell's physiological, biochemical, and molecular status. (Singh et.al 2016). Generous metals can instigate DNA strand breakage, nuclear crosslinking, adjustments in innate materials, oxidative pressing factor and harm from ROS and free extremists, just as helpful and hidden layer disintegrating, all of which increment heavymetal phyto-openness and cutoff reap plant growth. Every one of those biochemical, physiological, and genetic changes in plants are indivisibly associated with human prosperity and the advanced lifestyle. Heavy metals likewise produce uncommon physiological changes and opposing impacts at numerous periods of improvement, especially germination and seedlings. Heavy metals antagonistically affect the synthetics and protein profiles engaged with germination (e.g., destructive phosphatases, proteases, and - amylases). Heavy metals, for instance, diminished starch content, restricted enhancement levels, hampered chloroplast PSII, and provoked the declaration of warmth daze proteins and proline [81, 82]. The impacts of substantial metals have been focused on rice [83, 84] as per seed advancement of food yields, and Cd is likely the most considered poison [85]. Regardless of this, restricted examination has zeroed in on multi-metal toxicity in food crops [82, 84]. Co was demonstrated to be the most inconvenient to cauliflower (B. oleracea) as far as hostile effects on biomass and physiological exercises (e.g., foliar Fe, chlorophylls a, b, protein, and catalase action) [86, 87]. Those metals moreover obstructed the development of major parts (e.g., P, S, Mn, and Zn) from the roots to the shoots, with Cr showing the least phytotoxicity [87]. Metal transporters/chelators, for example, phyto-siderophores, are utilized to ship heavy metals and metalloids into the cells of food crops [88–93]. heavy metals and metalloids produce oxidative pressing factor in plants by changing cysteine over to diminished glutathione (GSH) and oxidized glutathione (GSSG) (the extent of GSH/GSSG = oxidative pressing factor or ROS age) [90] and shaping phytochelatins [90, 94]. Saltiness stress can likewise influence the measure of heavy metal pollution in food crops, just as their physiological and biochemical properties [95]. Metal contamination diminished the biomass and chlorophyll substance of explicit vegetables (most strikingly water spinach, trailed by amaranth, leaf mustard, Chinese sprouting cabbage, green capsicum, and tomato); on the other hand, the level of peroxidase, known to be an adversary of oxidative protein, at first extended at low assemblies of the metallic pollutants [87]. With an expansion in heavy metal focuses, tomato, the food least contaminated by metals, turned out to be progressively powerless against pungent pressure [87]. Key cycles in the ability to convey heavy metals incorporate phytochelation and immobilization by lignocellulose and different portions of plants, just as the limit of metals in the vacuole [87].

5. Toxicity in food crops

Yields collect essential TE groups in both appealing and unpalatable portions. When compared to other crops, verdant greens collect large groups of TEs from contaminated soil [96]. Crops absorb Heavy metals from contaminated land and by barometrically measuring particle matter from a variety of origins. Heavy metals are absorbed from the roots and then transported to the various cells of the crops [96]. They are transported to many parts of the plant by multiple channels, resulting in degradation of the developmental characteristics of crops due to change in the metabolic, physiological and biochemical activities of crops [97, 98]. Presence of these harmful components in the harvests can present poisonous indications that has serious outcomes (**Figure 3**).

5.1 Adverse impacts on plant health

The immediate impacts of TEs can cause catalyst inhibition of cytoplasm and underlying oxidative pressure can cause cell damage [99]. Plant is affected in a roundabout way by substituting important nutrients. TEs are present in variety of crops and its associated parts or even the entire plant [80]. Plant roots may easily absorb TEs

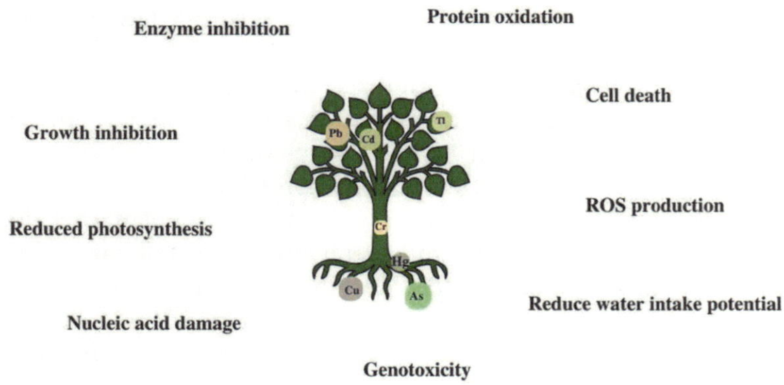

Figure 3.
Heavy metal toxicity in food crops.

and transport them to the edible portion of the plant [100] High convergence of TEs present in the soil has a number of unfavorable effects on crops and, subsequently, human health [80, 101]. Heavy metals like Cd and Pb in crops are not essential, and when they are present in large amounts they accumulate in the crops, it disturbs ingestion and transportation system, disrupting opposition potential, electron move framework, photosynthesis, breath, digestion, enzymatic action, development, and propagation [102]. The presence of the heavy metal also indicated various effects on the crops. Effect of Pb. Cr, Hg and Cd were studied on maize tomato, garlic, rice, and reduced shoot length, reduced nutrient uptake, decrease germination percentage and lower plant protein content was reported by [103].

6. Transportation via the food chain and occurrence in human food

Anthropogenic and natural sources are major supply for heavy metals in the surroundings. Heavy metals contaminate land and leach out to waterbodies, show their effect by persisting in the food cycle, have antagonistic effects on biota since they are very responsive and toxic at low concentrations [104, 105]. Toxic metals gets accumulated by plants from roots. Heavy metal travels through the water stream by apoplast to the internal cells (endodermis). This endodermis function as a protective layer for Heavy metal movement. The casparian strip prevents water stream and the Heavy metal enters the symplastic phase. Heavy metal transportation (in low concentrations) through root to APP have accounted for of immobilization by contrarily charged gelatins inside the root cell divider [106, 107]. Insoluble Heavy metal salts hasten in intercellular spaces of root cells [107, 108]. Essentially, Heavy metal amassing in plasma layers of root cells [107, 109] or in rhizodermal vacuoles and cortical cells of roots. Significant segment of Heavy metal is discharged from internal cells (endodermis) during the detoxification process by the plant. Foundations of hyper accumulator species break up metals in soil [110], increment metal take-up and movement, and make hyper accumulator species to endure higher Heavy metal particles focuses. Other detoxifying components include specific metal take-up, discharge, ligand binding, all contribute to Heavy metal resistance. The transfer of Heavy metal from lower part of plant to upper (root-shoot) is aided by xylem, which is most likely supported by occurrence [111]. On mesquite plants [106], demonstrated X-beam planning and observed high Heavy metal statements in xylem and phloem cells. It was found that Heavy metal traveled to the leaf from the vascular stream via apoplastic pathway [75]. In xylem, Heavy metal can frame edifices with amino/natural acids [112]. In any case, inorganic Heavy metal can likewise be moved. Movement (i.e., Heavy metal present in aeronautical parts/driving roots) can be executed for understanding level of Heavy metal movement [110, 112–114]. Notwithstanding, the admission of Heavy metal tainted plants has been a significant openness to people and creature [113–115]. Consumable/wild plants developed/filled nearby phosphate businesses can be Heavy metal bio-pointers of harmful metals [113]. Occupants and laborers of these businesses/regions might be presented to Heavy metal tainting.

The degree of toxicity of TEs in people relies upon every day admission through utilization of different sorts of vegetables [41, 97, 116, 117]. To understand the human health vulnerabilities from TEs when they consume crops, existing data records, and boundaries are used. Target hazard quotient (THQ), daily dietary intake of metals (DDI), hazard index (HI), daily intake of metals (DIM), and health risk index are among the items on the list (HRI). The THQ list is used to assess the health risks by toxic elements by giving a proper information on metal content from land (soil) and plants [118].

7. Hazardous effects of crops containing heavy metals on health of humans

Heavy metal contaminating eateries has a progression bearing unfavorable consequences for the health of humans because of their pervasive and refractory nature. Unnecessary metals have the capacity to bypass preventive mechanisms like compartmentalization, homeostasis, cellular failure, oxidative breakdown, and ship, resulting in toxic and fatal consequences. The significant manifestations of TEs harmfulness in people are scholarly handicap in youngsters, CNS problems, dementia and sadness in grown-ups, a sleeping disorder, kidney and liver sicknesses, passionate shakiness, and vision unsettling influences [119, 120], expanded the dreariness and death rate. Metal toxicants cause oxidative stress in individuals, which is described by: a) expanded creation of Reactive Oxygen Species/Reactive Nitrogen Species (ROS/RNS) b) exhaustion/debasement of internal enemy of oxidant and free-revolutionary scroungers c) restraint/decrease in digestion and catalysts identified with decontamination process. In spite of the fact that toxicity arising from unexpected or word related openness to generous quantities of metals normally influences organ frameworks, seriousness of the harmfulness relies upon the kind and type of the TEs, openness course and duration and, likewise by person's defenselessness [120, 121].

Harmful substances (metals) ingested in the course of debased crops can result in a variety of long-term illnesses as shown in (**Figure 4**). TEs consumption in polluted eatables might degrade several important nutrients of the body. This depletion is also linked to decreased malnutritional abilities, immunological defenses, hampered psychosocial resources, intrauterine development impediment, and increased threat of upper GI cancer [122]. Extreme sensitivity to Cd may amplify pneumonic effects. Subchronic inward Cd breath might also have a negative influence on the kidneys. Moreover Lead (Pb) toxic has been linked to the Brain/CNS and the GIT in both adolescents and adults [123], the impact may be severe or long-term. Stomach discomfort, desiring trouble, joint pain, pipedreams migraine, hypertension, fatigue, renal brokenness, restlessness, and vertigo are all symptoms of intense Pb openness [124]. Pb over time can cause allergic reactions, chemical imbalances, birth defects,

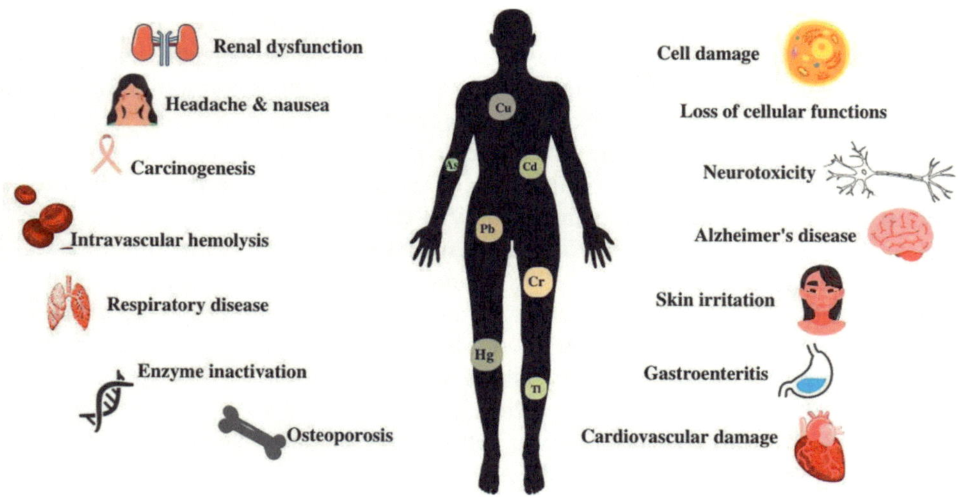

Figure 4.
Heavy metal hazardous effect on human health.

CNS damage, renal damage, solid shortcoming, loss of mobility, psychosis, malnutrition, in severe cases death [125].

Increased centralization of Zn causes disability development and proliferation. Examples include Cr (Cr+3 and Cr+6 effects) which are hazardous to both flora and fauna followed by threats to humans as well [126]. Minimum concentrations of As can lead to irregular heartbeats, low levels of blood plasma proteins, discomfort, nausea and heaving, pricking sensations in upper and lower limbs, accompanied with vein damage. Chronic exposures of As may result in heart disease, high blood pressure, internal cancerous infestations, mental issues, infection of vascular system, aspiratory illness, acne, and Diabetes. Ongoing arsenicosis may cause permanent alterations in major body-organ systems, thus increasing high mortality rate [127].

8. Strategies for food safety regulation and mitigation of the health hazards

Many studies have been conducted on the uptake, collection, and elimination of TEs in model plants at the research facility scale by a few scientists. Despite this, few investigations are conducted in the field in a straightforward manner. As a result, there is a need to look into the gaps in terms of the practicality of ways to reduce TEs in daily food sources and in high-need food items, similarly drinking water which is an essential component in the prevention and elimination of Heavy metals. There have been many advances in molecular biology and biotechnology applications of a few plant and animal species, the altered/controlled ingestion of harmful components, as well as categorizing these components into non-consumable parts, but it is still not fully accomplished [128]. As a result, more unified vegetables-based initiatives are predicted to boost the production of toxic component-free palatable plant parts. Quality exchange/change in articulation of particular receptor/chemical/component carrier may also be used to effectively regulate the retention of hazardous components in vegetable yields. The adjustment of exudates of rootsgives a powerful measure to stifle harmful components in soil. This process canprovide forestalling of the passage of TEs into the jungle. Endeavors can promptly begin in those harvests on which denselyspread population depends for food. One main consideration is the absence of subsidizing [129], with the goal that the consumer sector is fundamental to produce successful financing for TEs research. TEs tainting by and by exists in regular assets, yet, whenever kept up beneath as far as possible, long haul biological system imperativeness can be kept up without influencing the natural way of life.

Ranchers should be educated about the dangers of using excessive amounts of compost and other synthetics, as well as the natural hazards of flooding crops in various water bodies (wastewater, sewage, and mechanical water) so as to balance the requirement for vegetables growth with low TE levels. Ability to manage environmental sources, like soil and water, helps maintaining sustainability for crops cultivation. The information gathered should include wastewater quality indexing data to support for aquatic system water testing at rural areas to establish public health. Capable organizations should complete general wellbeing schooling inside the utilization region to teach people in general the possible impacts of unpredictable waste usage risks and the wellbeing perils related to the use of vegetables developed using untreated/contaminated. It should be capable of measuring TEs contamination & supplement stacking of aquatic system and soils to secure wellbeing, both [130] of ranchers and buyers. Considering the expected harmfulness, diligent nature and combined conduct just as the utilization organic materials and vegetables, it is essential to test the breakdown of food materials to guarantee the number of impurities concerning modern day issues. Standard overview and observing projects of the grouping of TEs in food items have been done for quite a long time in the vast majority of the created nations and similar practices ought to be carried out in agricultural nations.

9. Conclusion

Natural impurities, food handling and safety, and health of humans are interconnected. The presence of Heavy metals in the climate has increased altogether in late many years. Wellsprings of Heavy metal in crops can change in the creating and created world. Heavy metal exchange from ground soil to trim frameworks is unpredictable and requires complex components. Synergistic harmfulness of metals in food crops is also a major metal poisoning and contamination issue. The human wellbeing hazards have been broadly explored on a worldwide scale, however a couple of those works have utilized appropriate epidemiological techniques. To forestall wellbeing chances, existing remediation alternatives center around lessening the convergence of hefty metals in soil and the natural way of life. Fast and precise planning of soil contamination is expected to forestall the exchange of metallic toxins into the natural pecking order and to figure reasonable remediation techniques. Eco-attainable mechanical advancements, for example, nano-devices and the consciousness of ranchers could support neighborhood economies and vocations with certain monetary assurances.

Funding statement

None.

Conflict of interest

The authors declare no conflict of interest.

Author details

Vanisree C.R.[1], Mahipal Singh Sankhla[2*], Prashant Singh[3], Ekta B. Jadhav[4], Rohit Kumar Verma[5], Kumud Kant Awasthi[6], Garima Awasthi[6] and Varad Nagar[2]

1 Department of Forensic Science, Sage University, Indore, Madhya Pradesh, India

2 Department of Forensic Science, Vivekananda Global University, Jaipur, India

3 School of Forensic Science, National Forensic Science University, Gujarat, India

4 Department of Forensic Chemistry and Toxicology, Government Institute of Forensic Science Aurangabad, Maharashtra, India

5 Dr. APJ Abdul Kalam Institute of Forensic Science and Criminology, Bundelkhand University, Jhansi, India

6 Department of Life Sciences, Vivekananda Global University, Jaipur, India

*Address all correspondence to: mahipal4n6@gmail.com

References

[1] Römheld V. Diagnosis of deficiency and toxicity of nutrients. In: Marschner's Mineral Nutrition of Higher Plants. USA: Elsevier, Academic Press; 2012. pp. 299-312

[2] Liu H, Probst A, Liao B. Metal contamination of soils and crops affected by the Chenzhou lead/zinc mine spill (Hunan, China). Science of the Total Environment. 2005;**339**(1-3):153-166

[3] Gadd GM. Metals, minerals and microbes: Geomicrobiology and bioremediation. Microbiology. 2010; **156**(3):609-643

[4] El Hamiani O et al. Metal concentrations in plants from mining areas in South Morocco: Health risks assessment of consumption of edible and aromatic plants. CLEAN–Soil, Air, Water. 2015;**43**(3):399-407

[5] Kohzadi S et al. Concentration, source, and potential human health risk of heavy metals in the commonly consumed medicinal plants. Biological Trace Element Research. 2019;**187**(1):41-50

[6] Parihar K et al. Water quality status of Yamuna River and its toxic effects on humans. Environmental Analysis and Ecological Studies. 2019;**6**(1):597-601

[7] Sankhla MS, Kumar R. Contaminant of Heavy Metals in Groundwater & Its Toxic Effects on Human Health & Environment. International Journal of Environmental Sciences & Natural Resources. 2019;**18**(5):555996

[8] Masindi V, Muedi KL. Environmental contamination by heavy metals. Heavy Metals. 2018;**10**:115-132

[9] Tagesson T et al. Land-atmosphere exchange of methane from soil thawing to soil freezing in a high-Arctic wet tundra ecosystem. Global Change Biology. 2012;**18**(6):1928-1940

[10] Martin YE, Johnson EA. Biogeosciences survey: Studying interactions of the biosphere with the lithosphere, hydrosphere and atmosphere. Progress in Physical Geography. 2012;**36**(6):833-852

[11] Gautam P et al. Thermomechanical analysis of different types of sandstone at elevated temperature. Rock Mechanics and Rock Engineering. 2016;**49**(5):1985-1993

[12] Walker CH, Sibly R, Peakall DB. Principles of Ecotoxicology. CRC Press; 2005

[13] Tchounwou PB et al. Heavy metal toxicity and the environment. Molecular, Clinical and Environmental Toxicology. 2012;**101**:133-164

[14] Villanueva R, Bustamante PJA. Composition in essential and non-essential elements of early stages of cephalopods and dietary effects on the elemental profiles of Octopus vulgaris paralarvae. Aquaculture. 2006;**261**(1):225-240

[15] Tyler G et al. Heavy-metal ecology of terrestrial plants, microorganisms and invertebrates. Water, Air, and Soil Pollution. 1989;**47**(3):189-215

[16] Mapanda F et al. The effect of long-term irrigation using wastewater on heavy metal contents of soils under vegetables in Harare, Zimbabwe. Agriculture, Ecosystems & Environment. 2005;**107**(2-3):151-165

[17] Martín JAR, Arias ML, Corbí JMG. Heavy metals contents in agricultural topsoils in the Ebro basin (Spain). Application of the multivariate geoestatistical methods to study spatial variations. Environmental Pollution. 2006;**144**(3):1001-1012

[18] Arora M et al. Heavy metal accumulation in vegetables irrigated with water from different sources. Food Chemistry. 2008;**111**(4):811-815

[19] Ahmed G et al. Influence of multi-industrial activities on trace metal contamination: An approach towards surface water body in the vicinity of Dhaka Export Processing Zone (DEPZ). Environmental Monitoring and Assessment. 2012;**184**(7):4181-4190

[20] Zakir H et al. Heavy metal contamination in surface soils of some industrial areas of Gazipur. Journal of Chemical, Biological and Physical Sciences. 2015;**5**(2):2191-2206

[21] Ahmad JU, Goni MA. Heavy metal contamination in water, soil, and vegetables of the industrial areas in Dhaka, Bangladesh. Environmental Monitoring and Assessment. 2010; **166**(1):347-357

[22] Khan S et al. Health risks of heavy metals in contaminated soils and food crops irrigated with wastewater in Beijing, China. Environmental Pollution. 2008;**152**(3):686-692

[23] Zhang M-K et al. Use of single extraction methods to predict bioavailability of heavy metals in polluted soils to rice. Communications in Soil Science and Plant Analysis. 2010;**41**(7):820-831

[24] Kirpichtchikova TA et al. Speciation and solubility of heavy metals in contaminated soil using X-ray microfluorescence, EXAFS spectroscopy, chemical extraction, and thermodynamic modeling. Geochimica et Cosmochimica Acta. 2006;**70**(9):2163-2190

[25] Adriano DC. Trace Elements in Terrestrial Environments: Biogeochemistry, Bioavailability, and Risks of Metals. Vol. 860. New York: Springer; 2001

[26] Maslin P, Maier RM. Rhamnolipid-enhanced mineralization of phenanthrene in organic-metal co-contaminated soils. Bioremediation Journal. 2000;**4**(4):295-308

[27] McLaughlin MJ et al. Soil testing for heavy metals. Communications in Soil Science and Plant Analysis. 2000;**31**(11-14):1661-1700

[28] Ling W et al. Use of bentonite to control the release of copper from contaminated soils. Soil Research. 2007;**45**(8):618-623

[29] Zhao Q, Kaluarachchi JJ. Risk assessment at hazardous waste-contaminated sites with variability of population characteristics. Environment International. 2002;**28**(1-2):41-53

[30] D'amore J et al. Methods for speciation of metals in soils: A review. Journal of Environmental Quality. 2005;**34**(5):1707-1745

[31] Kaasalainen M, Yli-Halla M. Use of sequential extraction to assess metal partitioning in soils. Environmental Pollution. 2003;**126**(2):225-233

[32] Basta N, Ryan J, Chaney R. Trace element chemistry in residual-treated soil: Key concepts and metal bioavailability. 2005;**34**:49-63

[33] Verma RK, Sankhla MS, Kumar R. Mercury contamination in water & its impact on public health. International Journal of Forensic Science. 2018;**1**(2)

[34] Lasat M. Phytoextraction of metals from contaminated soil: A review of plant/soil/metal interaction and assessment of pertinent agronomic issues. Journal of Hazardous Substance Research. 1999;**2**(1):5

[35] Jones L, et al. The Fate of Heavy Metals in the Chemistry of Soil Processes. 1981

[36] Awasthi G et al. Arsenic in mushrooms, fish, and animal products. In: Arsenic in Drinking Water and Food. Singapore: Springer; 2020. pp. 307-323

[37] Sumner ME. Analysis, Beneficial use of effluents, wastes, and biosolids. Communications in Soil Science and Plant Analysis. 2000;**31**(11-14):1701-1715

[38] Sankhla MS, Kumar R, Prasad L. Distribution and Contamination Assessment of Potentially Harmful Element Chromium in Water. India: International Medico-Legal Reporter Journal. 2019;**2**:3

[39] Yadav H, Kumar R, Sankhla MS. Residues of pesticides and heavy metals in crops resulting in toxic effects on living organism. Journal of Seybold Report ISSN NO. 2020;**1533**:9211

[40] Sankhla MS et al. Heavy metals contamination in water and their hazardous effect on human health-a review. International Journal of Current Microbiology and Applied Sciences. 2016;**5**(10):759-766

[41] Sankhla MS et al. Arsenic in water contamination & toxic effect on human health: Current scenario of India. Journal of Forensic Sciences & Criminal Investigation (JFSCI). 2018;**10**(2):001-005

[42] Silveira MLA, Alleoni LRF, Guilherme LRG. Biosolids and heavy metals in soils. Scientia Agricola. 2003;**60**:793-806

[43] Mattigod S, Page A. Assessment of metal pollution in soils. In: Thornton I, editor. Chapter 12 Applied Environmental Geochemistry. USA: Academic Press, Elsevier; 1983

[44] Reed SC, Crites RW, Middlebrooks EJ. Natural Systems for Waste Management and Treatment. New York City: McGraw-Hill, Inc.; 1995

[45] Khan S et al. Soil and vegetables enrichment with heavy metals from geological sources in Gilgit, northern Pakistan. Ecotoxicology and Environmental Safety. 2010; **73**(7):1820-1827

[46] Hossni Y. Safety of sewage water in the irrigation of plants. Egypt Horticulture. 1997;**24**:261-270

[47] Grusak MA, DellaPenna D. Improving the nutrient composition of plants to enhance human nutrition and health. Annual Review of Plant Biology. 1999;**50**(1):133-161

[48] Welch RM, Graham RD. A new paradigm for world agriculture: Meeting human needs: Productive, sustainable, nutritious. Field Crops Research. 1999;**60**(1-2):1-10

[49] Ali F et al. An examination of the relationships between physical environment, perceived value, image and behavioural Intentions: A SEM approach towards Malaysian resort hotels. Journal of Hotel and Tourism Management. 2013;**27**(2):9-26

[50] Bahemuka T, Mubofu EB. Heavy metals in edible green vegetables grown along the sites of the Sinza and Msimbazi rivers in Dar es Salaam, Tanzania. Food Chemistry. 1999;**66**(1):63-66

[51] Auch AF et al. Digital DNA-DNA hybridization for microbial species delineation by means of genome-to-genome sequence comparison. Standards in Genomic Sciences. 2010;**2**(1):117-134

[52] Khan K et al. Heavy metals in agricultural soils and crops and their health risks in Swat District, northern Pakistan. Food and Chemical Toxicology. 2013;**58**:449-458

[53] Samiey B, Ashoori F. Adsorptive removal of methylene blue by agar: Effects of NaCl and ethanol. Chemistry Central Journal. 2012;**6**(1):1-13

[54] Guerin TF et al. An application of permeable reactive barrier technology to petroleum hydrocarbon contaminated groundwater. Water Research. 2002;**36**(1):15-24

[55] Bhattacharyya P et al. Fractionation and bioavailability of Pb in municipal solid waste compost and Pb uptake by rice straw and grain under submerged condition in amended soil. Geosciences Journal. 2008;**12**(1):41-45

[56] Guala SD, Vega FA, Covelo EF. The dynamics of heavy metals in plant–soil interactions. Ecological Modelling. 2010;**221**(8):1148-1152

[57] Dudka S, Adriano DC. Environmental impacts of metal ore mining and processing: A review. Journal of Environmental Quality. 1997;**26**(3):590-602

[58] Navarro M et al. Abandoned mine sites as a source of contamination by heavy metals: A case study in a semi-arid zone. Journal of Geochemical Exploration. 2008;**96**(2-3):183-193

[59] Adriano DC et al. Role of assisted natural remediation in environmental cleanup. Geoderma. 2004;**122**(2-4):121-142

[60] Pruvot C et al. Heavy metals in soil, crops and grass as a source of human exposure in the former mining areas (6 pp). Journal of Soils and Sediments. 2006;**6**(4):215-220

[61] Kumar A et al. Nickel in terrestrial biota: Comprehensive review on contamination, toxicity, tolerance and its remediation approaches. Chemosphere. 2021;**275**:129996

[62] Liu W-H et al. Impacts of sewage irrigation on heavy metal distribution and contamination in Beijing, China. Environment International. 2005;**31**(6):805-812

[63] Kachenko AG, Singh B. Heavy metals contamination in vegetables grown in urban and metal smelter contaminated sites in Australia. Water, Air, and Soil Pollution. 2006;**169**(1):101-123

[64] Järup L. Hazards of heavy metal contamination. British Medical Bulletin. 2003;**68**(1):167-182

[65] Kabata-Pendias A, Mukherjee AB. Trace Elements from Soil to Human. Germany: Springer Science & Business Media; 2007

[66] Gupta UC, Gupta SC. Trace element toxicity relationships to crop production and livestock and human health: Implications for management. Communications in Soil Science and Plant Analysis. 1998;**29**(11-14):1491-1522

[67] Lim HS, Lee JS, Chon HT, Sager M. Heavy metal contamination and health risk assessment in the vicinity of the abandoned Songcheon Au–Ag mine in Korea. Journal of Geochemical Exploration. 2008 Feb 1;**96**(2-3):223-230

[68] Dieckow J et al. Land use, tillage, texture and organic matter stock and composition in tropical and subtropical

Brazilian soils. European Journal of Soil Science. 2009;**60**(2):240-249

[69] Tripathi R, Raghunath R, Krishnamoorthy T. Dietary intake of heavy metals in Bombay city, India. Science of the Total Environment. 1997;**208**(3):149-159

[70] Emurotu J, Onianwa PJ. Bioaccumulation of heavy metals in soil and selected food crops cultivated in Kogi State, north central Nigeria. Environmental Systems Research. 2017;**6**(1):1-9

[71] Muchuweti M et al. Heavy metal content of vegetables irrigated with mixtures of wastewater and sewage sludge in Zimbabwe: Implications for human health. Agriculture, Ecosystems & Environment. 2006;**112**(1):41-48

[72] Herawati N et al. Cadmium, copper, and zinc levels in rice and soil of Japan, Indonesia, and China by soil type. Bulletin of Environmental Contamination and Toxicology. 2000;**64**(1):33-39

[73] Peralta-Videa JR et al. The biochemistry of environmental heavy metal uptake by plants: Implications for the food chain. The International Journal of Biochemistry & Cell Biology. 2009;**41**(8-9):1665-1677

[74] Arif N et al. Influence of high and low levels of plant-beneficial heavy metal ions on plant growth and development. Frontiers in Environmental Science. 2016;**4**:69

[75] Krzesłowska M et al. Lead deposited in the cell wall of Funaria hygrometrica protonemata is not stable–a remobilization can occur. Environmental Pollution. 2010;**158**(1):325-338

[76] Hossain MA et al. Molecular mechanism of heavy metal toxicity and tolerance in plants: Central role of glutathione in detoxification of reactive oxygen species and methylglyoxal and in heavy metal chelation. Journal of Botany. 2012;**2012**

[77] Barberon M, Geldner N. Radial transport of nutrients: The plant root as a polarized epithelium. Plant Physiology. 2014;**166**(2):528-537

[78] Page V, Feller U. Heavy metals in crop plants: Transport and redistribution processes on the whole plant level. Agronomy. 2015;**5**(3):447-463

[79] DalCorso G et al. Nutrient metal elements in plants. Metallomics. 2014;**6**(10):1770-1788

[80] Gupta N et al. Trace elements in soil-vegetables interface: translocation, bioaccumulation, toxicity and amelioration-a review. Science of the Total Environment. 2019;**651**:2927-2942

[81] Rai PK. Impacts of particulate matter pollution on plants: Implications for environmental biomonitoring. Ecotoxicology and Environmental Safety. 2016;**129**:120-136

[82] Seneviratne M et al. Heavy metal-induced oxidative stress on seed germination and seedling development: A critical review. Environmental Geochemistry and Health. 2019; **41**(4):1813-1831

[83] He J et al. Protective roles of nitric oxide on seed germination and seedling growth of rice (Oryza sativa L.) under cadmium stress. Ecotoxicology and Environmental Safety. 2014;**108**:114-119

[84] Rizwan M et al. A critical review on effects, tolerance mechanisms and management of cadmium in vegetables. Chemosphere. 2017;**182**:90-105

[85] Prodanovic O et al. Effect of cadmium stress on antioxidative enzymes during the germination of Serbian spruce [Picea omorika (Pan.) Purkyně]. African Journal of Biotechnology. 2012;**11**(52):11377-11385

[86] Chatterjee J, Chatterjee C. Phytotoxicity of cobalt, chromium and copper in cauliflower. Environmental Pollution. 2000;**109**(1):69-74

[87] Rai PK et al. Heavy metals in food crops: Health risks, fate, mechanisms, and management. Environment International. 2019;**125**:365-385

[88] Guerinot ML. The ZIP family of metal transporters. Biochimica et Biophysica Acta (BBA)-Biomembranes. 2000;**1465**(1-2):190-198

[89] Shenker M, Fan TM, Crowley D. Phytosiderophores influence on cadmium mobilization and uptake by wheat and barley plants. Journal of Environmental Quality. 2001;**30**(6):2091-2098

[90] Babula P et al. Uncommon heavy metals, metalloids and their plant toxicity: A review. Organic Farming, Pest Control and Remediation of Soil Pollutants. 2009:**1**;275-317

[91] Rai PK, Tripathi B. Heavy metals in industrial wastewater, soil and vegetables in Lohta village, India. Toxicological and Environ Chemistry. 2008;**90**(2):247-257

[92] Rai P et al. dËBfo (O) Tayl or & Francis. Aquatic Ecosystem Health & Management. 2010;**13**(4):352-361

[93] Rai PK et al. A critical review of ferrate (VI)-based remediation of soil and groundwater. Environmental research. 2018;**160**:420-448

[94] Supalkova V et al. Electroanalysis of plant thiols. Sensors. 2007;**7**(6):932-959

[95] Li Q et al. Toxic effects of heavy metals and their accumulation in vegetables grown in a saline soil. Ecotoxicology and Environmental Safety. 2010;**73**(1):84-88

[96] Vymazal J. Removal of nutrients in various types of constructed wetlands. Science of the Total Environment. 2007;**380**(1-3):48-65

[97] Agrawal S et al. Bioaccumulation of heavy metals in vegetables: A threat to human health. Terrestrial and Aquatic Environmental Toxicology. 2007;**1**(2):13-23

[98] Singh S et al. Heavy metal tolerance in plants: Role of transcriptomics, proteomics, metabolomics, and ionomics. Frontiers in Plant Science. 2016;**6**:1143

[99] Jadia CD, Fulekar M. Phytoremediation of heavy metals: Recent techniques. African Journal of Biotechnology. 2009;**8**(6)

[100] Jolly YN, Islam A, Akbar S. Transfer of metals from soil to vegetables and possible health risk assessment. Springer Plus. 2013;**2**(1):1-8

[101] Hu J et al. Bioaccessibility, dietary exposure and human risk assessment of heavy metals from market vegetables in Hong Kong revealed with an in vitro gastrointestinal model. Chemosphere. 2013;**91**(4):455-461

[102] Qinsong X, Guoxin S. The toxic effects of single Cd and interaction of Cd with Zn on some physiological index of [Oenanthe javanica (Blume) DC]. Nanjing shi da xue bao. Zi ran ke xue ban= Nanjing Shida Xuebao. 2000;**23**(4):97-100

[103] Gautam PK et al. Heavy metals in the environment: Fate, transport, toxicity and remediation technologies. Nova Science Publishers. 2016;**60**:101-130

[104] Golubkina N et al. Effects of crop system and genotype on yield, quality, antioxidants and chemical composition of organically grown leek. Advances in Horticultural Science. 2019;**33**(2):263-270

[105] Kumar A et al. Lead toxicity: Health hazards, influence on food chain, and sustainable remediation approaches. International Journal of Environmental Research and Public Health. 2020;**17**(7):2179

[106] Arias JA et al. Effects of Glomus deserticola inoculation on Prosopis: Enhancing chromium and lead uptake and translocation as confirmed by X-ray mapping, ICP-OES and TEM techniques. Environmental and Experimental Botany. 2010;**68**(2):139-148

[107] Zhang C et al. Transfer of lead (Pb) in the soil-plant-mealybug-ladybird beetle food chain, a comparison between two host plants. Ecotoxicology and environmental safety. 2017;**143**:289-295

[108] Meyers DE et al. Uptake and localisation of lead in the root system of Brassica juncea. Environmental Pollution. 2008;**153**(2):323-332

[109] Jiang W, Liu D. Pb-induced cellular defense system in the root meristematic cells of Allium sativum L. BMC Plant Biology. 2010;**10**(1):1-8

[110] López-Grimau V, Gutierrez MJC. Decolourisation of simulated reactive dyebath effluents by electrochemical oxidation assisted by UV light. Chemosphere. 2006;**62**(1):106-112

[111] Liao Y et al. Effect of transpiration on Pb uptake by lettuce and on water soluble low molecular weight organic acids in rhizosphere. Chemosphere. 2006;**65**(2):343-351

[112] Maestri E et al. Metal tolerance and hyperaccumulation: Costs and trade-offs between traits and environment. Environmental and Experimental Botany. 2010;**68**(1):1-13

[113] Saba D et al. Bioaccessibility of lead in Dittrichia viscosa plants and risk assessment of human exposure around a fertilizer industry in Lebanon. Journal of environmental management. 2019;**250**:109537

[114] Barbillon A et al. Health risks assessment in three urban farms of Paris Region for different scenarios of urban agricultural users: A case of soil trace metals contamination. Agricultural Sciences. 2019;**10**(3):352-370

[115] Liu X et al. Human health risk assessment of heavy metals in soil–vegetable system: A multi-medium analysis. Science of the Total Environment. 2013;**463**:530-540

[116] Sonone SS et al. Water contamination by heavy metals and their toxic effect on aquaculture and human health through food Chain. Letters in NanoBioScience. 2020;**10**(2):2148-2166

[117] Singh Sankhla M, Kumar R, Prasad L. Seasonal variations of lead and chromium concentrations in the water samples from Yamuna River in Delhi, India. Iranian Journal of Toxicology. 2021;**15**(2):109-114

[118] Harmanescu M et al. Heavy metals health risk assessment for population via consumption of vegetables grown in old mining area; a case study: Banat County, Romania. Chemistry Central Journal. 2011;**5**(1):1-10

[119] Flora S, Mittal M, Mehta A. Heavy metal induced oxidative stress & its possible reversal by chelation therapy.

Indian Journal of Medical Research. 2008;**128**(4):501

[120] Jan A, Ali A, Haq Q. Glutathione as an antioxidant in inorganic mercury induced nephrotoxicity. Journal of Postgraduate Medicine. 2011;**57**(1):72

[121] Duruibe JO, Ogwuegbu M, Egwurugwu J. Heavy metal pollution and human biotoxic effects. International Journal of Physical Sciences. 2007;**2**(5):112-118

[122] Iyengar GV, Nair P. Global outlook on nutrition and the environment: Meeting the challenges of the next millennium. Science of the Total Environment. 2000;**249**(1-3):331-346

[123] Markowitz M. Lead poisoning. Pediatrics in Review. 2000;**21**(10):327-335

[124] Jaishankar M et al. Toxicity, mechanism and health effects of some heavy metals. Interdisciplinary Toxicology. 2014;7(2):60

[125] Martin S, Griswold W. Human health effects of heavy metals. Environmental Science and Technology Briefs for Citizens. 2009;**15**:1-6

[126] Monalisa M, Kumar PH. Effect of ionic and chelate assisted hexavalent chromium on mung bean seedlings (Vigna radiata L. wilczek. var k-851) during seedling growth. Journal of Stress Physiology & Biochemistry. 2013;**9**(2)

[127] Mazumder DG. Chronic arsenic toxicity & human health. Indian Journal of Medical Research. 2008;**128**(4):436-447

[128] Sankhla MS, Kumar R, Prasad L. Variation of chromium concentration in Yamuna River (Delhi) water due to change in temperature and humidity. Journal of Seybold Report ISSN NO. 2020;**1533**:9211

[129] Mosa KA et al. Potential biotechnological strategies for the cleanup of heavy metals and metalloids. Frontiers in Plant Science. 2016;7:303

[130] Verma N et al. Green filter development: An innovative technique for removal of heavy metals from water. ARC Journal of Forensic Science. 2020;**5**(1):7-12

Chapter 5

Soil-Skeleton and Soil-Water Heavy Metal Contamination by Finite Element Modelling with Freundlich Isotherm Adsorption Parameters

Nguyen Van Hoang

Abstract

World research results indicate that untreated leachate contains high contents of heavy metals (HM) that are likely to pollute the soil and groundwater (GW) environment and contribute to the increase of HMs in soil and GW. The Freundlich isotherm adsorption parameters are essential to soil input parameters for modelling of HMs' transport to access the soil skeleton and soil pore water contamination by HMs. Finite element (FE) modelling of advection-dispersion transport of HMs by GW movement along with Freundlich isotherm adsorption parameters which continuously change with space in the model domain and with time is sophisticated to accurately evaluate the HMs' concentrations in soil skeleton and pore water. The chapter describes the background of the existing isotherm adsorption theory, the adaptation of the Freundlich isotherm adsorption in the soil skeleton and soil pore water contamination by HMs, method of determination of the Freundlich isotherm adsorption parameters, the FE procedure of modelling of advection-dispersion transport of HMs by GW movement in general and along with Freundlich isotherm adsorption parameters in particular. A case study modelling has been demonstrated.

Keywords: heavy metal (HM), Freundlich, isotherm, adsorption, geochemical background (GB), finite element (FE)

1. Introduction

In the world, in many study areas, most of the heavy metal (HM) content in the samples did not exceed the local standards, which can be mentioned as the study of [1] conducted a determination of soil samples of HM components in the North-Western area of Thessaloniki, North Greece near the insanitary landfill. Soil samples were taken at a depth of 2.5–17.5 m. Heavy metals Cd, Cr, Cu, Ni, Pb and Zn were analysed, however, although this area has a high degree of industrialisation, the soil is not contaminated by HMs. Research by Agamuthu and Fauziah [2] conducted sampling of typical soil at different locations in two landfills of Panchang Bedena and Kelana Jaya (Malaysia) to analyse HM content. Soil sampling depth is 2 m–35 m. The analysis results of samples at the Panchang Bedena landfill showed that all the analysed HMs have lower concentrations than the Dutch standard [3]. Opaluwa et al. [4] studied HMs in the soil at a depth of 0–15 cm and leaves on the campus of the Agriculture Faculty of National Polytechnic University, Nasarawa State, Nigeria and the areas near landfills and got the soil concentration of metals As, Cd, Co, Cu, Fe, Ni, Pb and Zn lower than the levels allowed by the World Health Organisation. Similarly, the translocation of HMs from the polluted soil to the aboveground parts of plants and lichens leads to a 1.5- to 5-fold increase in the content of HMs in all species, which fortunately neither exceed the toxicity threshold and nor prevent their growth in the experimental plots as by the work of Lyanguzova et al. [5]. According to Piyada and Suksaman [6], the landfill on the edge of Nai Muang Phichai district in Uttaradit Province, Thailand is one of the most polluted landfills in the world, leachate has a high content of HMs. Samples of soil and groundwater from the landfill and in the vicinity of the landfill are collected and analysed for HM content. Fortunately, the concentrations of Cd, Pb, Cu, Zn and Fe are all low, within the limits of soil quality standards. Research by Siti et al. [7], in Selangor state, Malaysia, there are 20 landfills, including the Ampar Tenang landfill closed in January 2010. However, the landfill is not covered with protective soil according to operational design standards, and before that the garbage is dumped directly onto the ground without any insulating material. Surface soil is relatively polluted by As, Pb, Fe, Cu and Al. As and Pb concentrations are greater than the allowable levels which are 5.90 mg/kg and 31.0 mg/kg, respectively. In addition, only Cu tends to decrease concentration with depth. Kamarudin et al. [8] studied the distribution of HMs in underground aquifers in the solid waste treatment area in Taiping, Perak, Malaysia. Soil samples were taken in 6 boreholes at a depth of 6 m–30 m, and a sampling distance with depth was 1 m. HMs Pb, Mn, Cr, Fe, Zn and Cd were analysed by ICP-MS. The concentrations of Pb, Mn, Fe and Zn are quite high, exceeding the allowable concentrations in the drinking water standard.

It can be seen that most of the HM contents in the soil environment of the landfill do not exceed the allowable levels. In cases where the content of HMs exceeds the standards, there will be an insanitary landfill that does not operate properly with waste burial techniques.

From the above review of several studies, it can be seen that untreated leachate, containing high levels of HMs, is a clear pollution source. Best. Leachate from landfills is capable of polluting the soil and groundwater environment if there are no measures for collection and treatment and to prevent the release of leachate to the surrounding environment. Most studies show that soil at the depth of 2 m and greater is not contaminated with HMs at a level greater than the allowable levels for agricultural land. However, the transport of HMs in the upper soil layer is extremely slow, especially thanks to the effect of adsorption.

2. Modelling of heavy metal transport in soil by groundwater movement

2.1 Theory of heavy metal advection-dispersion transport with soil adsorption

The general two-dimensional partial differential equation of the contaminant transport by advection-dispersion is as follows [9]:

$$D_x \frac{\partial^2 C}{\partial x^2} + D_y \frac{\partial^2 C}{\partial y^2} - v_x \frac{\partial C}{\partial x} - v_y \frac{\partial C}{\partial y} + Q = \frac{1-n}{n} \rho_s \frac{\partial q_e}{\partial t} + \frac{\partial C}{\partial t} \tag{1}$$

where C is the contaminant concentration (M/L^3, e.g., mg/L); t is the time (d); D_x and D_y are the hydrodynamic dispersion coefficient in x and y direction, respectively (m^2/d); x and y are the distances (m), v_x and v_y are the seepage velocity in x and y direction, respectively (m/d); Q is the distributed source of contaminant (mg/d); ρ_s is the solid particle density (note that $\rho_d = \rho_s(1\text{-}n)$ in which ρ_d is the unit weight of the dry soil); n is the soil porosity and q_e is the contaminant mass adsorbed per adsorbent mass (mg/kg).

Solid particles are capable of adsorption of dissolved ions of HMs in the soil pore water. The two most common models used to represent the adsorption isotherm are Freundlich and Langmuir isotherms [10]. The Freundlich isotherm is the most common isotherm model, used to describe physical adsorption in a solid-liquids system [11]. Besides, the Langmuir isotherm includes the maximum adsorption capacity of the considered soil, which requires a further special study for the study site.

Freundlich's adsorption isotherm is used in this study and is described as follows [12, 13] (refer to Patiha et al.):

$$q_e = K_F C^{1/\eta} \tag{2}$$

where C is the concentration in solution at equilibrium (mg/L); K_F and $1/\eta$ are fitting constants [13] and K_F is termed as the Freundlich coefficient (adsorption coefficient) (the unit for the Freundlich constant is $mg^{\frac{1-\eta}{\eta}}\ l^{\frac{1}{\eta}}/kg$) and $1/\eta$ is the adsorption intensity (dimensionless). The value of K_F is obtained from the intercept and $1/\eta$ from the slope of the logarithmic plot of $\log q_e$ versus $\log C$ of the equation:

$$log\, q_e = log\, K_F + \frac{1}{\eta} log\, C \tag{3}$$

$$\frac{dq_e}{dC} = K_F \frac{1}{\eta} C^{\frac{1-\eta}{\eta}} = K_d \tag{4}$$

where K_d is the distribution coefficient.

K_F is the Freundlich constant and $1/\eta$ depends on the linearity of the isotherm and varies between 0 and 1. Only when $1/\eta = 1$, the isotherm is linear and $K_F = K_d$.

From (4) the source term $\frac{\rho_s}{n} \frac{\partial q_e}{\partial t}$ in (1) is:

$$\frac{1-n}{n} \rho_s \frac{\partial q_e}{\partial t} = \frac{1-n}{n} \rho_s K_F \frac{1}{\eta} C^{\frac{1-\eta}{\eta}} \frac{\partial C}{\partial t} \tag{5}$$

Therefore, the Eq. (1) may be written in the following form:

$$D_x \frac{\partial^2 C}{\partial x^2} + D_y \frac{\partial^2 C}{\partial y^2} + D_z \frac{\partial^2 C}{\partial x^2} - v_x \frac{\partial C}{\partial x} - v_y \frac{\partial C}{\partial y} - v_z \frac{\partial C}{\partial z} = \left(1 + \frac{1-n}{n} \rho_s K_F \frac{1}{\eta} C^{\frac{1-\eta}{\eta}}\right) \frac{\partial C}{\partial t} \tag{6}$$

The so-called coefficient of retardation (retardation factor) R is also used:

$$R = 1 + \frac{1-n}{n} \rho K_F \frac{1}{\eta} C^{\frac{1}{\eta}-1} = 1 + \frac{1-n}{n} \rho K_d \tag{7}$$

$$D_x \frac{\partial^2 C}{\partial x^2} + D_y \frac{\partial^2 C}{\partial y^2} - v_x \frac{\partial C}{\partial x} - v_y \frac{\partial C}{\partial y} + Q = R \frac{\partial C}{\partial t} \quad (8)$$

The retardation factor R is always greater or equal to 1. It is equal to 1 when $1/\eta$ = 1.

The partial differential equation of the contaminant transport by advection-dispersion equation is subject to initial and boundary conditions for a particular problem in reality over a certain domain. The initial condition defines the known contaminant concentration over the whole domain at the initial time $t = t_0$:

$$C = C_o(x, y, z) \quad (9)$$

The boundary condition (BC) would be one of the following kinds:

- The first kind BC (the Dirichlet BC) defines a known concentration on the boundary:

$$C = C_c \text{ on} \Gamma_c \quad (10)$$

- The second kind BC (the Neumann BC) defines a known gradient of contaminant concentration across the boundary:

$$\frac{\partial C}{\partial n} = J_C \text{ on} \Gamma_{JC} \quad (11)$$

- The third kind BC (the Cauchy BC) defines a known rate of contaminant flux through the boundary:

$$v_n C - D \frac{\partial C}{\partial n} = q_c \text{ one} \Gamma_{qvc} \quad (12)$$

where v_n is the velocity normal to the boundary and C is the contaminant concentration at the boundary.

Eq. (8) has an analytical solution only for simple domain configurations, unchanged boundary conditions and constant spatial and temporal values of parameters, i.e., hydrodynamic dispersion coefficient, seepage velocity and retardation factor. Among the transport parameters, the retardation factor is the most sensitive and variable value in time and space as it is a non-linear function of the HM concentrations. This issue always needs to be kept in mind in numerical simulation of solute transport in groundwater in a soil medium with adsorption ability. Numerical methods, e.g., the finite element method (FEM), are capable of solving the equation for any domain configuration, spatial and temporal changing boundary conditions and parameters' values.

Due to the adsorption isotherm, to more accurately estimate retardation and contaminant transport other than the use of a single value is required in accordance with the relationship in Eq. (7). However, defining transport in terms of a retardation coefficient based on nonlinear adsorption could be complicated. Therefore, Coles [14] examined how the Freundlich model can be used to predict retardation by presenting a simpler way of accounting for nonlinear adsorption and by employing a more appropriate parameter than the Freundlich constant. The linear distribution coefficient K_d, was used by the author to calculate the retardation factor R. Based on the results, the author concluded that the actual ratio of adsorbate to adsorbent may be smaller by a factor of about 10 at higher contaminant concentrations, it could be safer and more accurate to make use of the unified sorption variable K_F to calculate R. Since K_F changes constantly with C and using a constantly changing K_F would be complicated, one solution is to select a small number of discrete values of K_F that can be used to approximate and slightly underestimate the actual values of K_F and each of these values can be used to calculate R over the range of contaminant concentrations that they are applicable.

2.2 FEM of the heavy metal advection-dispersion transport with adsorption by the soil

Let the domain Ω bed be divided into a number of elements E with a total number of nodes M. Let us temporarily not consider the right-hand side term $R\partial C/\partial t$ of Eq. (8), take the weighting of Eq. (8) and let it be zero [15]:

$$\int_\Omega \left(D_x \frac{\partial^2 C}{\partial x^2} + D_y \frac{\partial^2 C}{\partial y^2} D - v_x \frac{\partial C}{\partial x} - v_y \frac{\partial C}{\partial y} + Q \right) W_\ell dxdy = 0 \qquad \ell = 1, 2, \ldots, M \quad (13)$$

where W_ℓ is the weighting function.

Using the Green lemma:

$$\int_\Omega \left(D_x \frac{\partial^2 C}{\partial x^2} + D_y \frac{\partial^2 C}{\partial y^2} \right) W_\ell dxdy = -\int_\Omega \left(D_x \frac{\partial W_\ell}{\partial x} \frac{\partial C}{\partial x} + D_y \frac{\partial W_\ell}{\partial y} \frac{\partial C}{\partial y} \right) dxdy + \int_\Gamma \left(D_x \frac{\partial C}{\partial x} n_x + D_y \frac{\partial C}{\partial y} n_y \right) W_\ell d\Gamma \quad (14)$$

The integral $\int_\Gamma$ is present only for the elements having sides in boundary Γ_{qc} or Γ_{qvc} which are generally denoted as Γ_q ta có:

$$-\int_\Omega \left(D_x \frac{\partial W_\ell}{\partial x} \frac{\partial C}{\partial x} + D_y \frac{\partial W_\ell}{\partial y} \frac{\partial C}{\partial y} \right) dxdy - \int_\Omega \left(v_x \frac{\partial C}{\partial x} + v_y \frac{\partial C}{\partial y} \right) W_\ell dxdy = +\int_\Gamma \left(D_x \frac{\partial C}{\partial x} n_x + D_y \frac{\partial C}{\partial y} n_y \right) W_\ell d\Gamma = 0 \quad (15)$$

With the approximation function of the contaminant concentration is as follows [15]:

$$C \approx \hat{C} = \sum_{m=1}^{M} C_m N_m \quad (16)$$

where C_m is the approximation of the contaminant concentration at node m and N_m is the shape functions.

Equation (15) becomes:

$$-\int_\Omega \left(D_x \frac{\partial W_\ell}{\partial x} \frac{\partial N_m}{\partial x} C_m + D_y \frac{\partial W_\ell}{\partial y} \frac{\partial N_m}{\partial y} C_m \right) dxdy - \int_\Omega \left(v_x \frac{\partial N_m}{\partial x} W_\ell C_m + v_y \frac{\partial N_m}{\partial y} W_\ell C_m \right) dxdy = \int_{\Gamma_q} (\bar{q}_x W_\ell n_x + \bar{q}_x W n_y) d\Gamma = 0 \quad (17)$$

$$K = -\int_\Omega \left(D_x \frac{\partial W_\ell}{\partial x} \frac{\partial N_m}{\partial x} + D_y \frac{\partial W_\ell}{\partial y} \frac{\partial N_m}{\partial y} \right) dxdy - \int_\Omega \left(v_x \frac{\partial N_m}{\partial x} W_\ell + v_y \frac{\partial N_m}{\partial y} W_\ell \right) dxdy \quad (18)$$

$$F = -\int_{\Gamma_q} (\bar{q}_c W_\ell n_x + \bar{q}_c W_\ell n_y) d\Gamma \tag{19}$$

$$\mathrm{KC} = \mathrm{F} \tag{20}$$

The shape functions Nm and weighting functions W_ℓ can be linear or higher order functions. For unsteady problems, i.e., $R\partial C/\partial t \neq 0$ we have:

$$E\left(R\frac{dC}{dt}\right) + KC = F \tag{21}$$

The square matrix E is:

$$E = \int_\Omega R W_\ell N_m dxdy \tag{22}$$

Eq. (21) has the following general form in regard to temporal derivative:

$$\left(\theta K + \frac{E}{\Delta t}\right) C^{t+\Delta t} = \left(-(1-\theta)K + \frac{E}{\Delta t}\right) C^t + \frac{1}{2}\left(F^t + F^{t+\Delta t}\right) \tag{23}$$

The typical schemes are:

i. Forward difference (θ = 0):

$$R\frac{[E]}{\Delta t}\{C^{t+\Delta t}\} + \left([K] - R\frac{[E]}{\Delta t}\right)\{C^t\} = \{F^t\} \tag{24}$$

ii. Backward difference (θ = 1):

$$\left([K] + \frac{[E]}{\Delta t}\right)\{C^{t+\Delta t}\} - \frac{[E]}{\Delta t}\{C^t\} = \{F^{t+\Delta t}\} \tag{25}$$

iii. Central difference (the Crank–Nicolson scheme) (θ = 0.5):

$$\left(\frac{1}{2}[K] + \frac{[E]}{\Delta t}\right)\{C^{t+\Delta t}\} + \left(\frac{1}{2}[K] - \frac{[E]}{\Delta t}\right)\{C^t\} = \frac{1}{2}\left(\{F^t\} + \{F^{t+\Delta t}\}\right) \tag{26}$$

Let us consider two-dimensional in xy direction problems. The domain is divided into a mesh of triangular or quadrangular finite elements. For the mesh of rectangular elements (**Figure 1**) which have the side of h_x and h_y.

In the above equations, the matrix K at the element level is a square matrix Ne $\times$ Ne in which Ne is the number of the vertices of the elements (square matrix 3 $\times$ 3 or 4 $\times$ 4 for triangular or quadrangular elements, respectively). As an illustration, Galerkin FEM with linear shape functions and for a rectangular element with nodes i, j, k and l numbered counter-clockwise (**Figure 1**) which has sides of hxe and hye the matrices K, E and F are determined as follows. Since each term of the matrix is very long, each column containing Ne rows is to be written (columns 1, 2, 3, 4 are denoting nodes i, j, k and l, respectively, and rows 1, 2, 3, 4 are denoting nodes i, j, k and l, respectively):

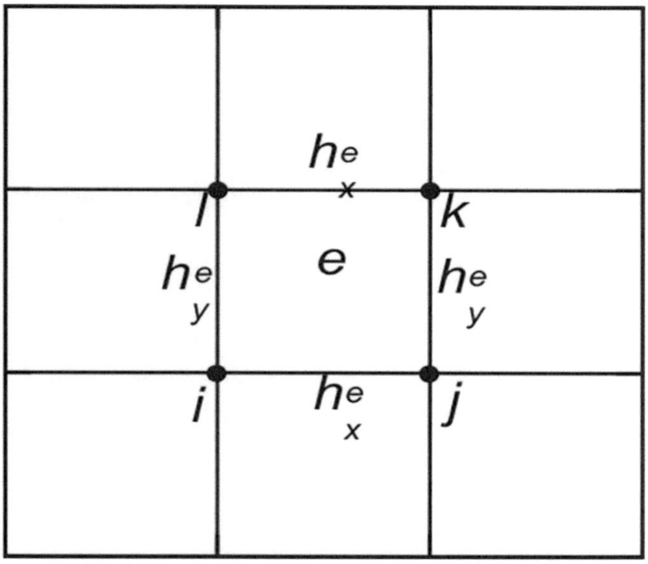

Figure 1.
A mesh of rectangular elements.

The four-row terms of column i are:

$$K^e_{(i,j,k,l)i} = \begin{bmatrix} \int_0^{h_y^e}\int_0^{h_x^e}\left[W_i v_x \frac{\partial N_i}{\partial x} + W_i v_y \frac{\partial N_i}{\partial y} + D_x \frac{\partial W_i}{\partial x}\frac{\partial N_i}{\partial x} + D_y \frac{\partial W_i}{\partial y}\frac{\partial N_i}{\partial y}\right] dxdy \\ \int_0^{h_y^e}\int_0^{h_x^e}\left[W_j v_x \frac{\partial N_i}{\partial x} + W_j v_y \frac{\partial N_i}{\partial y} + D_x \frac{\partial W_j}{\partial x}\frac{\partial N_i}{\partial x} + D_y \frac{\partial W_j}{\partial y}\frac{\partial N_i}{\partial y}\right] dxdy \\ \int_0^{h_y^e}\int_0^{h_x^e}\left[W_k v_x \frac{\partial N_i}{\partial x} + W_k v_y \frac{\partial N_i}{\partial y} + D_x \frac{\partial W_k}{\partial x}\frac{\partial N_i}{\partial x} + D_y \frac{\partial W_k}{\partial y}\frac{\partial N_i}{\partial y}\right] dxdy \\ \int_0^{h_y^e}\int_0^{h_x^e}\left[W_l v_x \frac{\partial N_i}{\partial x} + W_l v_y \frac{\partial N_i}{\partial y} + D_x \frac{\partial W_l}{\partial x}\frac{\partial N_i}{\partial x} + D_y \frac{\partial W_l}{\partial y}\frac{\partial N_i}{\partial y}\right] dxdy \end{bmatrix} \tag{27}$$

The four-row terms of column j are:

$$K^e_{(i,j,k,l)j} = \begin{matrix} \int_0^{h_y^e}\int_0^{h_x^e}\left[W_i v_x \frac{\partial N_j}{\partial x} + W_i v_y \frac{\partial N_j}{\partial y} + D_x \frac{\partial W_i}{\partial x}\frac{\partial N_j}{\partial x} + D_y \frac{\partial W_i}{\partial y}\frac{\partial N_j}{\partial y}\right] dxdy \\ \int_0^{h_y^e}\int_0^{h_x^e}\left[W_j v_x \frac{\partial N_j}{\partial x} + W_j v_y \frac{\partial N_j}{\partial y} + D_x \frac{\partial W_j}{\partial x}\frac{\partial N_j}{\partial x} + D_y \frac{\partial W_j}{\partial y}\frac{\partial N_j}{\partial y}\right] dxdy \\ \int_0^{h_y^e}\int_0^{h_x^e}\left[W_k v_x \frac{\partial N_j}{\partial x} + W_k v_y \frac{\partial N_j}{\partial y} + D_x \frac{\partial W_k}{\partial x}\frac{\partial N_j}{\partial x} + D_y \frac{\partial W_k}{\partial y}\frac{\partial N_j}{\partial y}\right] dxdy \\ \int_0^{h_y^e}\int_0^{h_x^e}\left[W_l v_x \frac{\partial N_j}{\partial x} + W_l v_y \frac{\partial N_j}{\partial y} + D_x \frac{\partial W_l}{\partial x}\frac{\partial N_j}{\partial x} + D_y \frac{\partial W_l}{\partial y}\frac{\partial N_j}{\partial y}\right] dxdy \end{matrix} \tag{28}$$

The four-row terms of column k are:

$$K^e_{(i,j,k,l)k} = \begin{matrix} \int_0^{h_y^e}\int_0^{h_x^e}\left[W_i v_x \frac{\partial N_k}{\partial x} + W_i v_y \frac{\partial N_k}{\partial y} + D_x \frac{\partial W_i}{\partial x}\frac{\partial N_k}{\partial x} + D_y \frac{\partial W_i}{\partial y}\frac{\partial N_k}{\partial y}\right] dxdy \\ \int_0^{h_y^e}\int_0^{h_x^e}\left[W_j v_x \frac{\partial N_k}{\partial x} + W_j v_y \frac{\partial N_k}{\partial y} + D_x \frac{\partial W_j}{\partial x}\frac{\partial N_k}{\partial x} + D_y \frac{\partial W_j}{\partial y}\frac{\partial N_k}{\partial y}\right] dxdy \\ \int_0^{h_y^e}\int_0^{h_x^e}\left[W_k v_x \frac{\partial N_k}{\partial x} + W_k v_y \frac{\partial N_k}{\partial y} + D_x \frac{\partial W_k}{\partial x}\frac{\partial N_k}{\partial x} + D_y \frac{\partial W_k}{\partial y}\frac{\partial N_k}{\partial y}\right] dxdy \\ \int_0^{h_y^e}\int_0^{h_x^e}\left[W_l v_x \frac{\partial N_k}{\partial x} + W_l v_y \frac{\partial N_k}{\partial y} + D_x \frac{\partial W_l}{\partial x}\frac{\partial N_k}{\partial x} + D_y \frac{\partial W_l}{\partial y}\frac{\partial N_k}{\partial y}\right] dxdy \end{matrix} \tag{29}$$

The four-row terms of column l are:

$$K^e_{(i,j,k,l)l} = \left.\begin{matrix} \int_0^{h_y^e}\int_0^{h_x^e}\left[W_i v_x \frac{\partial N_l}{\partial x} + W_i v_y \frac{\partial N_l}{\partial y} + D_x \frac{\partial W_i}{\partial x}\frac{\partial N_l}{\partial x} + D_y \frac{\partial W_i}{\partial y}\frac{\partial N_l}{\partial y}\right] dxdy \\ \int_0^{h_y^e}\int_0^{h_x^e}\left[W_j v_x \frac{\partial N_l}{\partial x} + W_j v_y \frac{\partial N_l}{\partial y} + D_x \frac{\partial W_j}{\partial x}\frac{\partial N_l}{\partial x} + D_y \frac{\partial W_j}{\partial y}\frac{\partial N_l}{\partial y}\right] dxdy \\ \int_0^{h_y^e}\int_0^{h_x^e}\left[W_k v_x \frac{\partial N_l}{\partial x} + W_k v_y \frac{\partial N_l}{\partial y} + D_x \frac{\partial W_k}{\partial x}\frac{\partial N_l}{\partial x} + D_y \frac{\partial W_k}{\partial y}\frac{\partial N_l}{\partial y}\right] dxdy \\ \int_0^{h_y^e}\int_0^{h_x^e}\left[W_l v_x \frac{\partial N_l}{\partial x} + W_l v_y \frac{\partial N_l}{\partial y} + D_x \frac{\partial W_l}{\partial x}\frac{\partial N_l}{\partial x} + D_y \frac{\partial W_l}{\partial y}\frac{\partial N_l}{\partial y}\right] dxdy \end{matrix}\right] \tag{30}$$

The Galerkin FEM with linear shape functions results in:
The four-row terms of column I, the Eq. (27) become:

$$K^e_{(i,j,k,l)i} = \begin{bmatrix} \left[D_x^e\left\{\frac{(h_y^e-y)}{h_x^e h_y^e}\right\}^2 + D_y^e\left\{\frac{(h_x^e-x)}{h_x^e h_y^e}\right\}^2\right] + \left[\frac{(h_x^e-x)(h_y^e-y)}{h_x^e h_y^e}\left(-v_x^e\frac{(h_y^e-y)}{h_x^e h_y^e} - v_y^e\frac{(h_x^e-x)}{h_x^e h_y^e}\right)\right] \\ \left[-D_x^e\frac{(h_y^e-y)}{h_x^e h_y^e}\frac{(h_y^e-y)}{h_x^e h_y^e} - D_y^e\frac{x}{h_x^e h_y^e}\frac{(h_x^e-x)}{h_x^e h_y^e}\right] + \left[\frac{x(h_y^e-y)}{h_x^e h_y^e}\left(-v_x^e\frac{(h_y^e-y)}{h_x^e h_y^e} - v_y^e\frac{(h_x^e-x)}{h_x^e h_y^e}\right)\right] \\ \left[-D_x^e\frac{y}{h_x^e h_y^e}\frac{(h_y^e-y)}{h_x^e h_y^e} - D_y^e\frac{x}{h_x^e h_y^e}\frac{(h_x^e-x)}{h_x^e h_y^e}\right] + \left[\frac{xy}{h_x^e h_y^e}\left(-v_x^e\frac{(h_y^e-y)}{h_x^e h_y^e} - v_y^e\frac{(h_x^e-x)}{h_x^e h_y^e}\right)\right] \\ \left[D_x^e\frac{y}{h_x^e h_y^e}\frac{(h_y^e-y)}{h_x^e h_y^e} - D_y^e\frac{(h_x^e-x)}{h_x^e h_y^e}\frac{(h_x^e-x)}{h_x^e h_y^e}\right] + \left[\frac{(h_x^e-x)y}{h_x^e h_y^e}\left(-v_x^e\frac{(h_y^e-y)}{h_x^e h_y^e} - v_y^e\frac{(h_x^e-x)}{h_x^e h_y^e}\right)\right] \end{bmatrix} \tag{31}$$

The four-row terms of column j, the Eq. (28) become:

$$K^e_{(i,j,k,l)j} = \begin{array}{l} \frac{1}{3h^e_x h^e_y}\left[-D^e_x \left(h^e_x\right)^2 + D^e_y \left(h^e_y\right)^2\right] + \frac{1}{12}\left(-2v^e_x h^e_y + v^e_y h^e_x\right) \\ \frac{1}{3h^e_x h^e_y}\left[D^e_x \left(h^e_x\right)^2 + D^e_y \left(h^e_y\right)^2\right] + \frac{1}{12}\left(-2v^e_x h^e_y + 2v^e_y h^e_x\right) \\ \frac{1}{3h^e_x h^e_y}\left[\frac{1}{2}D^e_x \left(h^e_x\right)^2 - D^e_y \left(h^e_y\right)^2\right] + \frac{1}{12}\left(-v^e_x h^e_y + 2v^e_y h^e_x\right) \\ \frac{1}{3h^e_x h^e_y}\left[-\frac{1}{2}D^e_x \left(h^e_x\right)^2 - \frac{1}{2}D^e_y \left(h^e_y\right)^2\right] + \frac{1}{12}\left(-v^e_x h^e_y + v^e_y h^e_x\right) \end{array} \tag{32}$$

The four-row terms of column k, the Eq. (29) become:

$$K^e_{(i,j,k,l)k} = \begin{array}{l} \frac{1}{3h^e_x h^e_y}\left[-\frac{1}{2}D^e_x \left(h^e_x\right)^2 - \frac{1}{2}D^e_y \left(h^e_y\right)^2\right] + \frac{1}{12}\left(-v^e_x h^e_y - v^e_y h^e_x\right) \\ \frac{1}{3h^e_x h^e_y}\left[\frac{1}{2}D^e_x \left(h^e_x\right)^2 - D^e_y \left(h^e_y\right)^2\right] + \frac{1}{12}\left(-v^e_x h^e_y - 2v^e_y h^e_x\right) \\ \frac{1}{3h^e_x h^e_y}\left[D^e_x \left(h^e_x\right)^2 + D^e_y \left(h^e_y\right)^2\right] + \frac{1}{12}\left(-2v^e_x h^e_y - 2v^e_y h^e_x\right) \\ \frac{1}{3h^e_x h^e_y}\left[-D^e_x \left(h^e_x\right)^2 + \frac{1}{2}D^e_y \left(h^e_y\right)^2\right] + \frac{1}{12}\left(-2v^e_x h^e_y - v^e_y h^e_x\right) \end{array} \tag{33}$$

The four-row terms of column l, the Eq. (30) become:

$$K^e_{(i,j,k,l)l} = \left.\begin{array}{l} \frac{1}{3h^e_x h^e_y}\left[\frac{1}{2}D^e_x \left(h^e_x\right)^2 - D^e_y \left(h^e_y\right)^2\right] + \frac{1}{12}\left(v^e_x h^e_y - 2v^e_y h^e_x\right) \\ \frac{1}{3h^e_x h^e_y}\left[-\frac{1}{2}D^e_x \left(h^e_x\right)^2 - \frac{1}{2}D^e_y \left(h^e_y\right)^2\right] + \frac{1}{12}\left(v^e_x h^e_y - v^e_y h^e_x\right) \\ \frac{1}{3h^e_x h^e_y}\left[-D^e_x \left(h^e_x\right)^2 + \frac{1}{2}D^e_y \left(h^e_y\right)^2\right] + \frac{1}{12}\left(2v^e_x h^e_y - v^e_y h^e_x\right) \\ \frac{1}{3h^e_x h^e_y}\left[D^e_x \left(h^e_x\right)^2 + D^e_y \left(h^e_y\right)^2\right] + \frac{1}{12}\left(2v^e_x h^e_y - 2v^e_y h^e_x\right) \end{array}\right] \tag{34}$$

The contribution of the loading vector $\mathbf{f}$ is determined by taking the integral over the element. For example, for node i at a side along the boundary:

$$\begin{aligned} F^e_i &= \int_0^{h^e_x}\int_0^{h^e_y} \frac{Q^e}{h^e_x h^e_y}\left(h^e_x - x\right)\left(h^e_y - y\right)dxdy - \underline{\int_0^{h^e_x} \frac{q^e}{h^e_x}\left(h^e_x - x\right)dx} - \underline{\int_0^{h^e_y} \frac{q^e}{h^e_y}\left(h^e_y - y\right)dy} \\ &= \frac{1}{4}Q^e h^e_x h^e_y - \underline{\frac{1}{2}q^e h^e_x} - \underline{\frac{1}{2}q^e h^e_y} \end{aligned} \tag{35}$$

The underlined terms are existing only if the side i, j and l are lying in boundary Γ_q, and therefore they are not existing for the inside elements:

The matrix E is:

$$\begin{array}{lcccc} \text{node}: & i & j & k & l \end{array}$$

$$E^1 = R_e \begin{bmatrix} \int_\Omega W_i N_i dxdy & \int_\Omega W_i N_j dxdy & \int_\Omega W_i N_k dxdy & \int_\Omega W_i N_\ell dxdy \\ \int_\Omega W_j N_i dxdy & \int_\Omega W_j N_j dxdy & \int_\Omega W_j N_k dxdy & \int_\Omega W_j N_\ell dxdy \\ \int_\Omega W_k N_i dxdy & \int_\Omega W_k N_j dxdy & \int_\Omega W_j N_k dxdy & \int_\Omega W_j N_\ell dxdy \\ \int_\Omega W_\ell N_i dxdy & \int_\Omega W_\ell N_j dxdy & \int_\Omega W_\ell N_k dxdy & \int_\Omega W_\ell N_\ell dxdy \end{bmatrix} \begin{bmatrix} C_i \\ C_j \\ C_k \\ C_\ell \end{bmatrix} \tag{36}$$

Putting the weighting function W and shape function N into Eq. (36) results in:

$$\begin{array}{lcc} \text{node}: & i & j \end{array}$$

$$E^1 = R_e \left[\begin{array}{ll} \int_\Omega \frac{(h_x^e - x)(h_y^e - y)}{h_x^e h_y^e} \frac{(h_x^e - x)(h_y^e - y)}{h_x^e h_y^e} dxdy & \int_\Omega \frac{(h_x^e - x)(h_y^e - y)}{h_x^e h_y^e} \frac{x(h_y^e - y)}{h_x^e h_y^e} dxdy \\ \int_\Omega \frac{x(h_y^e - y)}{h_x^e h_y^e} \frac{(h_x^e - x)(h_y^e - y)}{h_x^e h_y^e} dxdy & \int_\Omega \frac{x(h_y^e - y)}{h_x^e h_y^e} \frac{x(h_y^e - y)}{h_x^e h_y^e} dxdy \\ \int_\Omega \frac{xy}{h_x^e h_y^e} \frac{(h_x^e - x)(h_y^e - y)}{h_x^e h_y^e} dxdy & \int_\Omega \frac{xy}{h_x^e h_y^e} \frac{x(h_y^e - y)}{h_x^e h_y^e} dxdy \\ \int_\Omega \frac{(h_x^e - x)y}{h_x^e h_y^e} \frac{(h_x^e - x)(h_y^e - y)}{h_x^e h_y^e} dxdy & \int_\Omega \frac{(h_x^e - x)y}{h_x^e h_y^e} \frac{x(h_y^e - y)}{h_x^e h_y^e} dxdy \end{array} \right. \tag{37}$$

$$\begin{array}{lcc} \text{node}: & k & l \end{array}$$

$$\left. \begin{array}{ll} \int_\Omega \frac{(h_x^e - x)(h_y^e - y)}{h_x^e h_y^e} \frac{xy}{h_x^e h_y^e} dxdy & \int_\Omega \frac{(h_x^e - x)(h_y^e - y)}{h_x^e h_y^e} \frac{(h_x^e - x)y}{h_x^e h_y^e} dxdy \\ \int_\Omega \frac{x(h_y^e - y)}{h_x^e h_y^e} \frac{xy}{h_x^e h_y^e} dxdy & \int_\Omega \frac{x(h_y^e - y)}{h_x^e h_y^e} \frac{(h_x^e - x)y}{h_x^e h_y^e} dxdy \\ \int_\Omega \frac{xy}{h_x^e h_y^e} \frac{xy}{h_x^e h_y^e} dxdy & \int_\Omega \frac{xy}{h_x^e h_y^e} \frac{(h_x^e - x)y}{h_x^e h_y^e} dxdy \\ \int_\Omega \frac{(h_x^e - x)y}{h_x^e h_y^e} \frac{xy}{h_x^e h_y^e} dxdy & \int_\Omega \frac{(h_x^e - x)y}{h_x^e h_y^e} \frac{(h_x^e - x)y}{h_x^e h_y^e} dxdy \end{array} \right] \begin{bmatrix} C_i \\ C_j \\ C_k \\ C_\ell \end{bmatrix} \tag{38}$$

By assembling all the element matrices K, E and F a global system of equations can be obtained the solutions of which are the approximated contaminant concentrations at all nodes.

For linear elements, the element sizes and time steps need to be selected based on the following criteria [16]:

$$Pe = \frac{v_x h_x^e}{D_x} \leq 2; Pe = \frac{v_y h_y^e}{D_y} \leq 2; Cr = \frac{v_x \Delta t}{h_x^e} \leq 1; Cr = \frac{v_y \Delta t}{h_y^e} \leq 1 \tag{39}$$

K_F	$1/\eta$	R^2	K_F	$1/\eta$	R^2	K_F	$1/\eta$	R^2	K_F	$1/\eta$	R^2
With q_e in mg/g											
	Cr			Cu			Pb			Zn	
0.264	0.260	0.999	0.131	0.450	0.991	0.300	0.290	0.996	0.050	0.569	0.999
	As			Cd			Hg				
0.404	0.340	0.989	0.080	0.420	0.964	0.510	0.300	0.985			

Table 1.
Freundlich coefficient KF and adsorption intensity $1/\eta$ for the soil of the study of He et al. [17].

2.3 Soil heavy metal adsorption parameters

The adsorption capacities of HMs change with physical parameters such as pH, temperature etc. The adsorption of heavy metals As, Cu, Zn, Pb, Cr, Cd and Hg on the soil at different pH was experimentally investigated by He et al. [17]. The adsorption capacities of Cr decreased with increasing pH, which may be caused by the unique physical properties of Cr. The adsorption capacities of the remaining HMs are increased with increasing pH. One of the reasons is that the increase in pH effectively reduces the concentration of H^+ in the solution. In solution with pH greater than 7, all the ions H^+ are in par with ion OH^-. Therefore, HM ions with a positive charge can more effectively be absorbed by the soil colloids. It means the adsorption capacities of HM ions increased with increasing pH value.

One of the aspects of the influence of pH on the adsorption of HMs by soil particles is that pH has an influence on the solubility of HMs in solution [18] and controls various adsorption reactions on the surface of solid particles, and the increase in pH, which promoted an increase in the adsorption point of the soil colloid since soil colloids generally have a negative charge [19]. The chapter will deal with the HM adsorption capacity at pH around 7.

To investigate the effect of temperature on the adsorption of As, Cu, Zn, Pb, Cr, Cd and Hg, He et al. [17] used soil samples at a different temperature from 30–50°C. The data obtained by the authors show the increase of adsorption capacities of HMs in the soil material with increasing temperature.

The experiment data for Cr, Cu, Pb and Zn by He et al. [17] at the temperature of 25°C have been extracted from the authors' publication's figures. The least squared error method was used by the authors of this study for determining the Freundlich coefficient K_F and adsorption intensity $1/\eta$, the results of which are also presented in **Table 1** which is showing a very high correlation of the experiment data point and the Freundlich isotherm coefficients K_F and $1/\eta$ of the fitting curves (**Figure 2**).

The Freundlich isotherm coefficients KF and 1/η of silty soils were also studied by some other authors. The study of Noppadoland [20] investigates the adsorption of the most common HMs (Cu, Ni, and Zn) by various soils. Fifteen soil samples were collected from various areas of North-Eastern Thailand.

They were excavated from different depths, ranging from 20 cm to 50 cm below the soil surface. The average soil pH is about 6.5. The areas near watercourses, communities and industries were selected as sites from which the soil samples were taken. The authors have received the following average values of the Freundlich isotherm coefficients K_F and $1/\eta$ of the soils: K_F = 0.348 (mg/g) and $1/\eta$ = 0.235 (σ = 0.059) for Zn, K_F = 0.462 and $1/\eta$ = 0.320 for Cu (σ = 0.054), K_F = 0.279 and $1/\eta$ = 0.437 (σ = 0.059) for Ni (with the q_e in mg/g and C unit in mg/L).

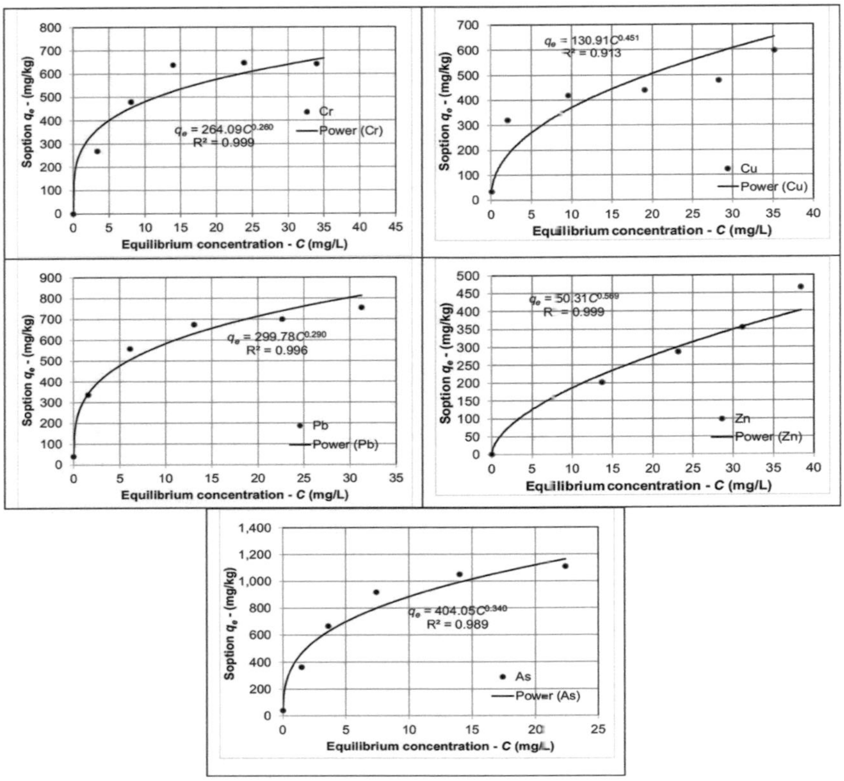

Figure 2.
Adsorption capacities of heavy metals adsorbed on soil material at equilibrium concentration.

Soils in some regions of North-Western Spain have been the subject of agricultural management practices involving the use of fertilisers and various types of organic waste containing HMs. Although such practices have facilitated crop growth, they have also raised the natural contents in HMs of the soils. Therefore, Emma et al. [21] researched the ability of the soils with high concentrations of Cr and Ni to adsorb and retain Cd, Cu, Ni, Pb and Zn. The soil pH is about 6.5 and the experiments' temperature is 25°C. They have obtained the following Freundlich coefficients: K_F = 1.560 (mg/g) and $1/\eta$ = 0.327 for Cu, K_F = 0.363 and $1/\eta$ = 0.441 for Ni, K_F = 1.363 and $1/\eta$ = 0.351 for Pb, K_F = 0.463 and $1/\eta$ = 0.426 for Zn, K_F = 0.540 and $1/\eta$ = 0.293 for Cd (with the q_e in mg/g and C unit in mg/L).

Claudia et al. [22] carried out a specific adsorption evaluation through the amounts of adsorbed Cu, Pb, Cr, Ni and Zn after desorption experiments in ten different soils. The HM adsorption isotherm Freundlich parameters at temperature 25°C and for the neutral pH soils are as follows: K_F = 2.540 (mg/g) and $1/\eta$ = 0.91 for Cu, K_F = 0.702 and $1/\eta$ = 0.510 for Ni, K_F = 0.998 and $1/\eta$ = 0.440 for Pb, K_F = 1.016 and $1/\eta$ = 0.440 for Zn (with the q_e in mg/g and C unit in mg/L).

2.4 Dispersion parameters

The coefficient of longitudinal hydrodynamic dispersion D_L in the water flow direction which is the coefficient of hydrodynamic dispersion coefficient D in Eq. (1) consists of two components: the coefficient of mechanical dispersion

D' and the coefficient of molecular diffusion in a porous medium D^*_d, i.e., $D_L = D' + D^*_d$ [9].

The coefficient of mechanical dispersion D' depends upon the microstructure of the soil, the seepage velocity and the molecular dispersion in water as follows by Jacob and Arnold [9]:

$$D'_{ij} = a_L v f(Pe, \xi); Pe = \frac{Lv}{D_d} \tag{40}$$

in which: v is the seepage velocity (m/d); Pe is the Peclet number; L is the characteristic length of the pores (m); D_d is the molecular dispersion in water; ξ is the ratio between the pores' size and the characteristic length through the pores; $f(P_e,\xi) = P_e/(P_e+2 + 4\xi^2)$ is a function which is expressing the transport of the HMs or solutes via molecular dispersion between the neighbouring flow streams at the macro scale, and in most cases $f(\mathrm{Pe},\xi) \cong 1$; a_L is the longitudinal dispersivity.

For a one-directional groundwater flow, the coefficient of mechanical dispersion D' is the multiplication of longitudinal dispersivity (a_L) and seepage velocity [9]. The longitudinal dispersivity is of the order of the average soil particle [9], e.g., particle size d_{50}. The coefficient of molecular diffusion in a porous medium D^*_d is as follows [23]:

$$D_d^* = \frac{D_d}{nF_R} \tag{41}$$

in which: F_R is the formation factor which is specified by the geophysicists as the ratio between soil resistivity and water resistivity.

The formation factor F_R ranges from 0.1 (for clay) to 0.7 (for sand) [23], and always less than 1.

The coefficient of molecular diffusion in water D_d is:

$$D_d = \frac{RT}{N}\frac{1}{6\pi r\mu} \tag{42}$$

In which: R is the gas constant; K is temperature unit Kelvin; N is the Avogadro number; T is the temperature (K); μ is the water viscosity; r is the average radius of the HM or solute.

The coefficients of molecular diffusion coefficients of inorganic cations and anions in water D_d may be found in the book by Henry [24].

Jacob and Arnold [9] have divided dispersion and diffusion into five zones (Figure 7 in [9]) in accordance to the Pectlet number, for which the roles of the molecular diffusion and the hydrodynamic dispersion are described. Zone I is corresponding to very slow water movement with the Pectlet number less than 0.4 so that the molecular diffusion predominates and the mechanical dispersion (D') is negligible, i.e., $D_L \approx D^*_d$. In our case, the Pectlet number is equal to 0.0011, for which the hydrodynamic dispersion D is approximately equal to the molecular diffusion in saturated porous medium D^*_d.

3. The FE modelling application to Kieu Ky landfill, Hanoi, Vietnam

Kieu Ky waste landfill is located in Gia Lam district in the South-East of the Center of Hanoi in the Bac Bo plain, the second largest plain in Vietnam. The waste landfill facility has an area of 13 ha consisting of composting compartments, a leachate reservoir and five landfill cells (**Figure 3**). The landfill cells have bottoms at the depth of 4.5 m and the thickness of dumped waste of 5 m–15 m (**Figure 4**). The facility handled 175 tons of solid waste in a day. It is operated from 2002 to 2020. The area is covered by Holocene formation, under which a rich and with good quality Pleistocene aquifer is underlying.

Kieu Ky landfill area has a natural ground surface of elevation around 4.5 m above mean sea level. The local shallow geological and hydrogeological conditions are as follows (**Figure 4**): (1) Surface cultivated soil of about 0.8 m in thickness, which consists of grey-yellow silt with some small construction solid waste pieces, and (2)

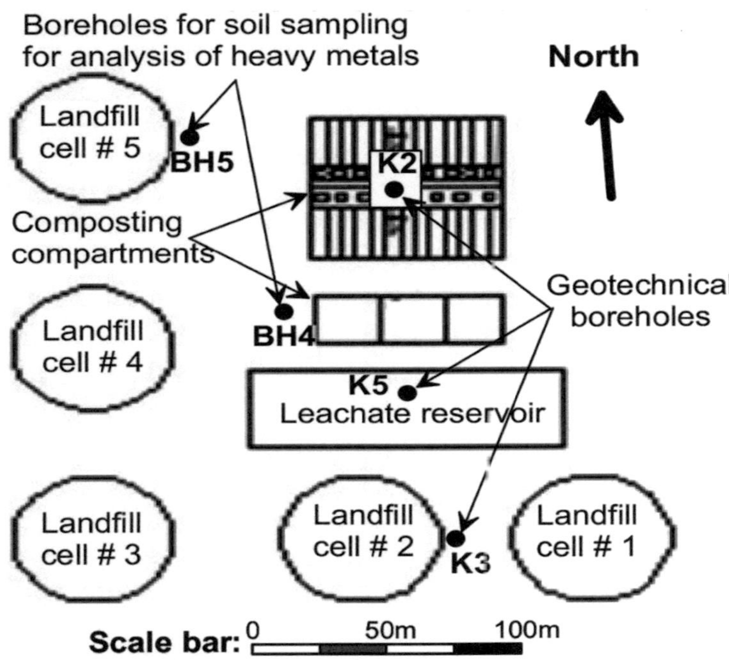

Figure 3.
The layout of landfills, geotechnical boreholes and boreholes for soil quality sampling.

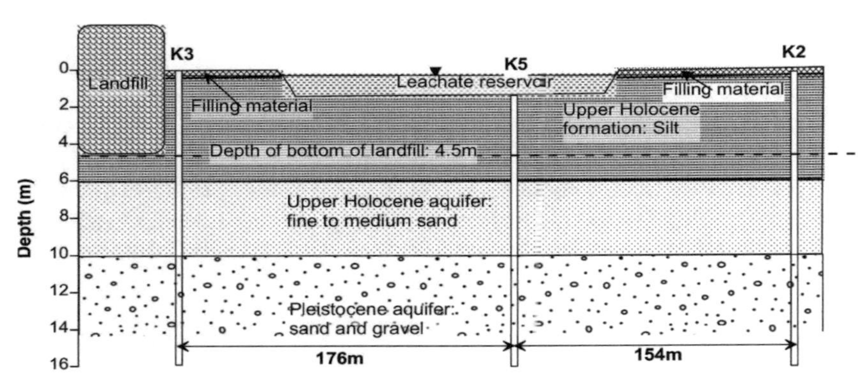

Figure 4.
The soil profile of Kieu Ky landfill site.

Layer of Upper Holocene silt of grey-yellow, grey-green and grey-brown colours, the thickness is around 6 m. The silt's porosity (n) and hydraulic conductivity (K) have been determined and are 0.455 m/d and 0.0045 m/d, respectively.

Two model domains (MD) have been selected: one is the natural soil profile next to the landfill (from the ground surface to the depth of 6 m, i.e. to the groundwater aquifer surface, with the length of 6 m) (MD1) and the second one is the soil profile beneath the bottom of the landfill (from the depth 4.5 m to the surface of the groundwater aquifer with the length of 1.5 m) (MD2). The three characteristic values (minimum, average and maximum) of the Freundlich isotherm adsorption parameters are considered in the two selected model domains.

The hydraulic conditions of the two model domains are determined based on **Figure 5** and on that the water level of the Upper Holocene aquifer is 2 m below the ground surface, the level of leachate and the water level of the leachate pond are the same. Domain 1 is a former rice field and almost is constantly wet. This creates a saturated soil profile. Direct leakage of leachate from the landfills to the land slot to supply HMs to penetrate the soil profile. Domain 2 is underneath the bottoms of the

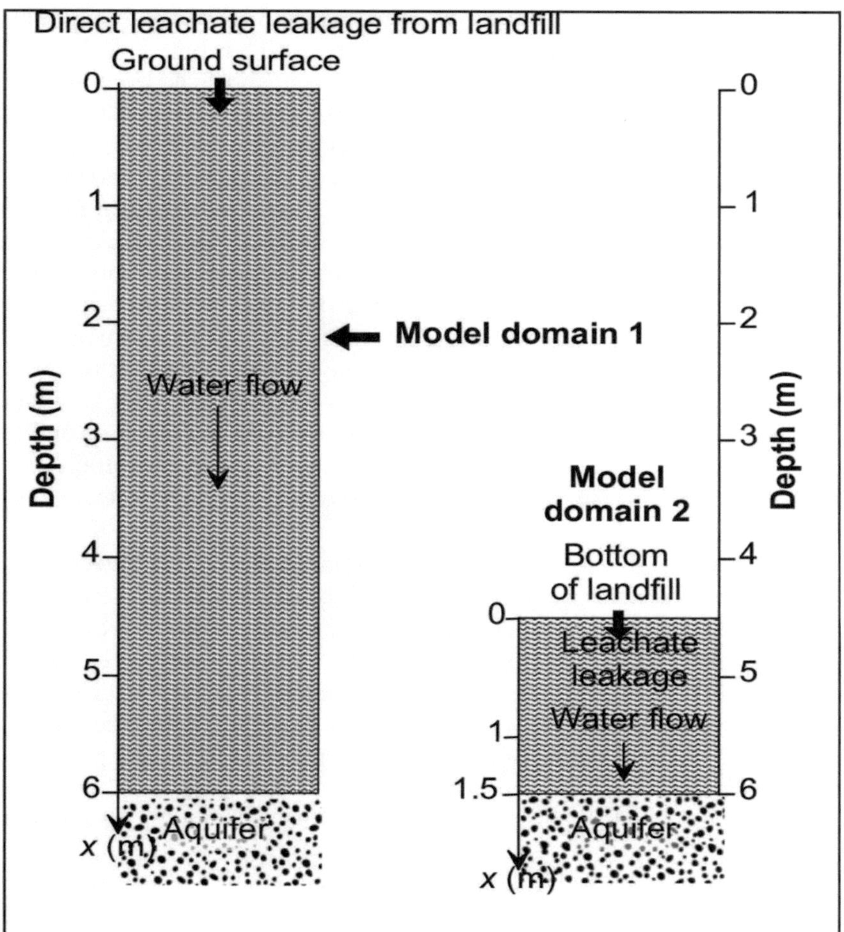

Figure 5.
Typical model domains in the study site.

landfills, which are lower than the groundwater level of the beneath aquifer. Similar to domain 1, this also creates a saturated soil profile. Besides, it is most likely that some landfill leachate may leak into the domain. The seepage velocity of which is determined by Darcy law through the hydraulic gradient and soil hydraulic conductivity. The soil hydraulic conductivity in the vertical direction was determined by the laboratory permeameter. Subsurface soil samples have been collected for laboratory experiments for the determination of saturated hydraulic conductivity.

3.1 Parameter calibration

The problem of aquifer parameter calibration has been studied extensively. In modelling of groundwater flow and transport, besides the specification of the aquifer geometry and its boundary conditions, the determination of aquifer's geohydraulic parameters, e.g., hydraulic conductivity, porosity, dispersivity, source or sink and prescribed boundary fluxes is necessary. The inverse problem of parameter calibration can be defined as the optimal determination of the parameters based on the observation data of the dependent variables, such as hydraulic heads or solute concentrations, collected in space and time. The inverse methods have been classified by Neuman [25] into two groups: indirect and direct. The indirect approach is based on the output error criterion, where the accuracy of the parameters is improved by an iterative process until the model response is close enough to the observed one. The direct approach is based on the use of super-determinate equations derived from rearranging the discretisation equations in such a way that the parameters are considered as unknown variables and their optimal values are such that minimise the residuals of the equations in a certain sense. The modern inverse techniques are often imbedded with the numerical models, usually finite difference and finite element models. All the soil HMs relevant transport parameters may be calibrated simultaneously. However, this would result in a high uncertainty of the obtained calibrated parameters as the overall modelling results may have a good optimisation error while the calibrated parameters are not reliable as they are beyond the physical limits. Therefore, some parameters are better determined by experimental tests and the remaining parameters are calibrated by inverse analyses. This procedure is particularly suitable for the soil adsorption parameters and dispersion parameters of low permeable soils.

Generally, the objective function (E(k)) to be minimised in the inverse analysis can be expressed as the sum of weighted squares of the differences between the model responses and the observation ones and the sum weighted squares of the difference between the estimated model parameter and prior parameter. The indirect method using this kind of objective function is called regularised Output Least Squares (OLS). If the second term of sum weighted squares of the difference between the estimated model parameter and the prior parameter has vanished, e.g., the regulation coefficient is equal to zero, the method is called generalised OLS. In the latter, if the optimal weighting coefficients all are equal to the unit, the method becomes original OLS.

The numerical methods in the solution of OLS problems are unconstrained nonlinear optimisation, which includes search method, gradient method and second-order method (Quasi-Newton methods). Within the chapter, one-dimensional dispersion testing for the determination of soil dispersivity by Quasi-Newton methods is described for demonstration.

3.1.1 One-D dispersion testing for determination of soil dispersivity by Quasi-Newton methods

A tracer column test is illustrated in **Figure 6** in which a constant tracer concentration is maintained in the left boundary (a relative concentration of 1 is usually used) and a constant zero-concentration in the other boundary.

The special and temporal concentrations are monitored, for which the observed (C^{obs}) and model (C^{mod}) concentration at time t are presented in **Figure 7**. One-D dispersion determination of the soil dispersivity by Quasi-Newton methods are described as follows.

The most common criterion in the evaluation of the difference between the model estimated and observed variables is the least squared root given as:

$$Min\ E(p) = \sum_{l=1}^{L} \left[C_l^{mod}(p) - C_l^{obs}\right]^2 \tag{43}$$

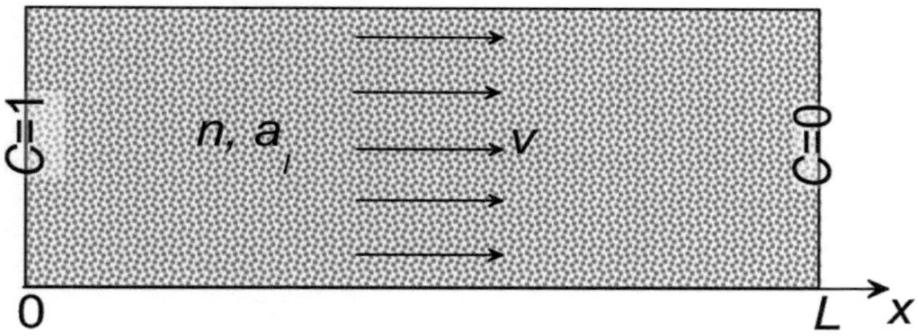

Figure 6.
A tracer column test scheme.

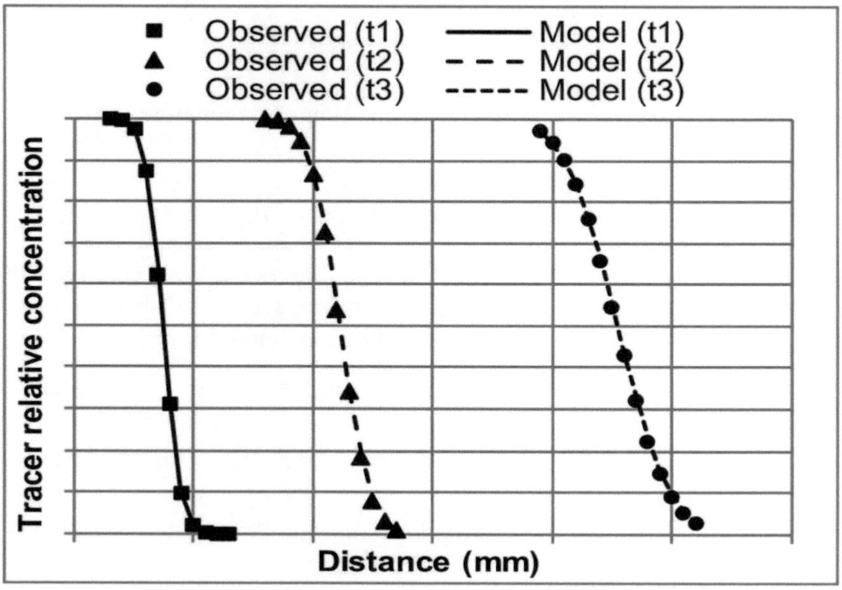

Figure 7.
Plots of observed vs. model tracer concentration.

where $E(\mathbf{p})$ - objective function; L- number of observed variables; C_l^{mod}- model estimated values of the concentration; C_l^{obs}- observed measured values of the concentration; $\mathbf{p}$- parameters of the physical medium (hydrodynamic dispersion coefficient as a function of pore velocity, porosity and dispersivity).

Let us consider the following multi-dimensional optimisation problem:

$$Min\ E(p), p \in p_{ct} \tag{44}$$

where $\mathbf{p}_{\text{ct}}$ - a set of possible values of parameter variables $\mathbf{p}$. This set of parameter values may be selected based on the existing data of the parameters of similar physical materials, of materials at the same locations, statistical data etc.

If the objective function $E(\mathbf{p})$ has a second-order derivative then the necessary and sufficient conditions for $\hat{p}$ to be the stationary point, i.e., $E(\mathbf{p})$ has a local extreme value [26]:

Gradient $g = \nabla E(\mathbf{p}) = 0$ at $\mathbf{p}$, i.e.:

$$\left|\frac{\partial E}{\partial p_m}\right|_{\hat{p}} = 0; (m = 1, 2, \ldots, M) \tag{45}$$

where M is the number of parameter variables.

Hessian matrix $G = \nabla^2 E(\mathbf{p})$:

$$G = \begin{vmatrix} \frac{\partial^2 E}{\partial p_1^2} & \frac{\partial^2 E}{\partial p_1 \partial p_2} & \cdots & \frac{\partial^2 E}{\partial p_1 \partial p_M} \\ \frac{\partial^2 E}{\partial p_2 \partial p_1} & \frac{\partial^2 E}{\partial p_2^2} & \cdots & \frac{\partial^2 E}{\partial p_2 \partial p_M} \\ \cdots\cdots & \cdots\cdots & \cdots\cdots & \cdots\cdots \\ \frac{\partial^2 E}{\partial p_M \partial p_1} & \frac{\partial^2 E}{\partial p_M \partial p_2} & \cdots & \frac{\partial^2 E}{\partial p_M^2} \end{vmatrix} \tag{46}$$

is a positive definite matrix at $\hat{p}$.

The optimisation algorithms in the determination of parameters consist of the following steps:

Selection of the initial values of parameters $\mathbf{p}_0$.

Determination of the search sequence: $\mathbf{p}_0, \mathbf{p}_1, \mathbf{p}_2, \ldots, \mathbf{p}_n \ldots$ in such a way that $E(\mathbf{p}_{n+1}) < E(\mathbf{p}_n)$ for all n.

Checking the convergence criterion. If the convergence is observed, then the local extremes have been reached and the parameter values are considered to be estimated.

Commonly, the search sequence has the following general form:

$$p_{n+1} = p_n + \lambda_n d_n \tag{47}$$

where $\mathbf{d}_n$- vector of displacement directions; λ_n- step size (that must be most optimally selected).

Three main groups of optimisation algorithms may be classified for solving optimisation problems: (1) Search method, when only the values of the objective function are considered, (2) Gradient method, when the gradients of the objective function are utilised and (3) Second-order method, if the second derivatives of the objective functions are utilised. The Quasi-Newton method belongs to the third group.

Suppose there is a set of initial values of parameters $\mathbf{p}_0$, it is required to find out the search sequence $\mathbf{p}_0, \mathbf{p}_1, \mathbf{p}_2, \ldots, \mathbf{p}_n$ so that $E(\mathbf{p}_{n+1}) < E(\mathbf{p}_n)$ for all n. Gradient vector $\mathbf{g}(\mathbf{p}_{n+1})$ at vicinity of $\mathbf{p}_n$ may be determined as follows:

$$g(p_{n+1}) \approx g_n + G_n \Delta p; \Delta p = p - p_n; g_n = g(p_n); G_n = G(p_n) \tag{48}$$

The necessary condition of extreme existence is $\mathbf{g}(\mathbf{p}_{n+1}) \approx 0$. This can be done if $\mathbf{p}_{n+1} = \mathbf{p}_n + \Delta\, \mathbf{p}_n$, where $\Delta\mathbf{p}_n$ are the solution of the following equation:

$$g_n + G_n \Delta p = \mathbf{0} \tag{49}$$

$$\Delta p_n = \Delta p = -g_n G_n^{-1} \Rightarrow p_{n+1} = p_n - g_n G_n^{-1} \tag{50}$$

This process has to be repeated until the convergence criterion is reached. Thus, the displacement direction $\mathbf{d}_n$ is equal to $-\mathbf{G}_n^{-1}\mathbf{g}_n$ and the optimal search step λ_n is always equal to 1. This method is called the Newton method.

In Quasi-Newton methods, the matrix $\mathbf{G}_n^{-1}$ is replaced by symmetric positive $\mathbf{H}_n$, which is updated from iteration to iteration. The following steps are included in the n iteration:

Initiate search direction:

$$d_n = -H_n g_n \tag{51}$$

Definition of the next search point:

$$p_{n+1} = p_n + \lambda_n d_n \tag{52}$$

This may be done by any line search method such as blanket method, golden section search, Fibonacci section search, quadratic interpolation method.

Replacement of matrix $\mathbf{H}_n$ by $\mathbf{H}_{n+1}$.

Initial Hessian matrix $\mathbf{H}_1$ can be any symmetric positive definite and the simplest one is a unit matrix $\mathbf{I}$. Matrices $\mathbf{H}_{n+1}$ have been proposed by different authors. Davidson, Fletcher and Powell have proposed the following [26]:

$$H_{n+1} = H_n + \frac{\Delta p_n \Delta p_n^T}{\Delta p_n^T \Delta g_n} - \frac{H_n \Delta g_n \Delta g_n^T H_n}{\Delta p_n^T H_n \Delta g_n} \tag{53}$$

where: $\Delta\mathbf{p}_n = \mathbf{p}_{n+1} - \mathbf{p}_n$, $\Delta\mathbf{g}_n = \mathbf{g}_{n+1} - \mathbf{g}_n$, and superscript T indicates transposed matrix.

The parameters estimation finishes when either of the following criteria is observed:

$$|\mathbf{p}_n - \mathbf{p}_{n-1}| < \varepsilon_1 \& |E(\mathbf{p}_n)| < \varepsilon_2 \tag{54}$$

$$|\mathbf{p}_n - \mathbf{p}_{n-1}| < \varepsilon_1 \& |E(\mathbf{p}_n) - E(\mathbf{p}_{n-1})| < \varepsilon_3 \tag{55}$$

where ε_1, ε_2 and ε_3 are given small arbitrary positive values.

The block scheme of the parameter estimation process is given in **Figure 8**.

3.1.2 Determination of Freundlich's adsorption isotherm parameters

The soil samples were taken from borehole BH5 in April 2016, which is 15 years from the operation of landfill cell No. 5 (**Figure 3**). The hydrodynamic dispersion coefficient was determined based on the above-described values of the coefficient of

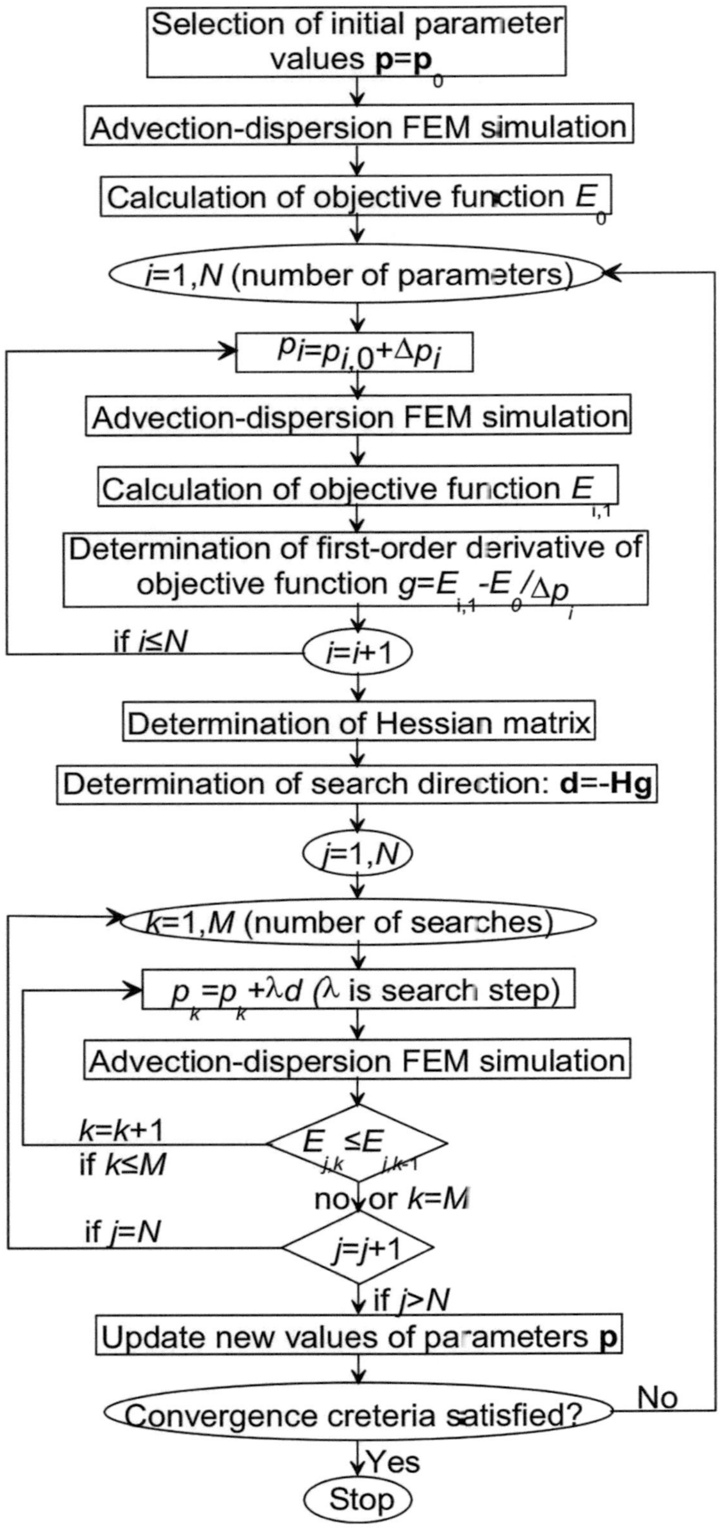

Figure 8.
Block-scheme of inverse analysis by Quasi-Newton method.

molecular diffusion, the soil porosity and the formation factor in paragraph 2.4 which were used as the input parameters. The element size and time step need to be not greater than 0.63 m and 422 days, respectively. Element size of 0.01 m and a time step of 1 day were used in this modelling for having sufficient data points along with a short distance of the concentration breakthrough curve.

Freundlich's adsorption isotherm parameters are calibrated with the obtained HMs' contents in soil taken from BH5. The trial and error method of calibration is used. **Table 2** summarised the calibration results, i.e., the values of Freundlich's adsorption isotherm parameters and mean error between the analysed and model HMs' contents in the soil. **Figure 9** illustrates the calibrated model HMs' concentrations versus the analysed HMs' concentrations. In general, the calibration models have a good fitting with a relative error of less than 7%, except the zinc.

3.2 The FE model results

As the analysis results presented in **Figure 5**, four heavy metals Cr, Cu, Pb and Zn expose high concentrations on the surface 1–2 m of the soil layer. Modelling the transport of those four HMs was carried out. The breakthrough curves of concentrations of the four HMs in soil and pore water in MD1 are presented in **Figure 10**, where the allowable limits [27, 28] are also indicated. Thanks to the HMs' adsorbability of the soil, only the upper layer of the soil horizon would be contaminated with HMs at levels higher than the allowable limits for agricultural land. For a period of 30 years, the soil would be contaminated in the upper 1 m, 2 m and 3 m by Cr, Zn and Pb,

Metal	K_F	$1/\eta$	Mean error (mg/g)	Relative mean error (%)
Cr	0.264	0.260	0.0051	2.76
Cu	0.131	0.450	0.0033	5.03
Pb	0.073	0.850	0.0069	6.81
Zn	0.144	0.279	0.0312	19.41

Table 2.
The calibrated Freundlich's adsorption isotherm parameters.

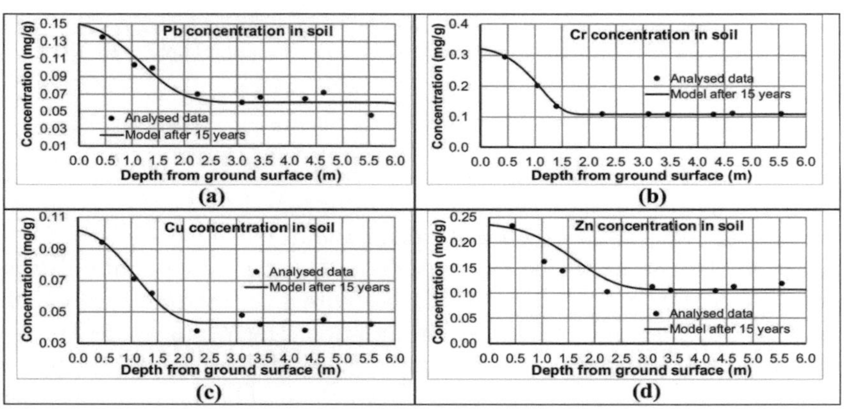

Figure 9.
Analysed and modelled results with the calibrated Freundlich's adsorption isotherm parameters. (a) Lead in soil. (b) Chromium in soil water. (c) Copper in soil. (d) Zinc in soil water.

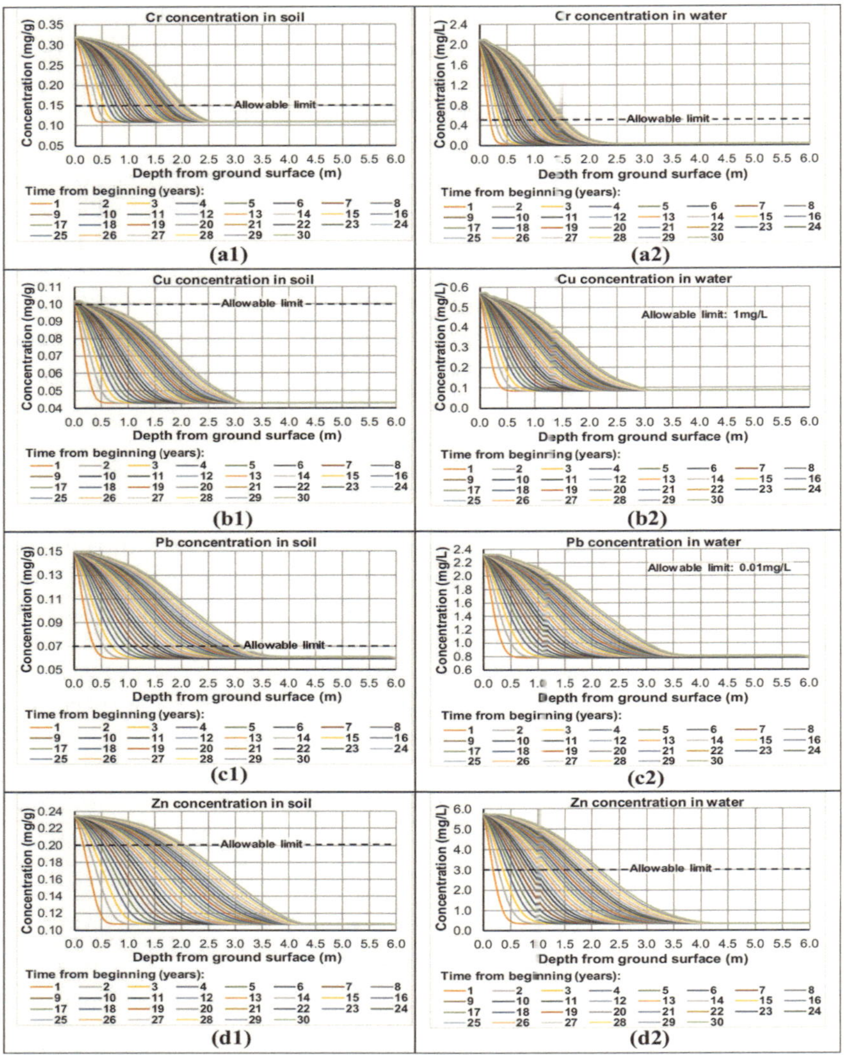

Figure 10.
Heavy metal concentrations prediction by FEM for 30 years - MD1. (a1) Chromium in soil. (a2) Chromium in pore water. (b1) Copper in soil. (b2) Copper in pore water. (c1) Lead in soil. (c2) Lead in pore water. (d1) Zinc in soil. (d2) Zinc in pore water.

respectively (**Figure 10**: a1, c1 and d1). The concentrations of Cr, Pb and Zn in the soil pore water are higher than allowable limits in the upper 1.5 m, 6 m (i.e., the whole soil layer) and 2.2 m, respectively (**Figure 10**: a2, c2 and d2).

Since the soil layer is under the landfill cells and leachate pond, only HMs in the soil pore water in MD2 are described here. MD2 with a very short length (1.5 m) presents a more problematic contamination situation. The MD2's results are described here. Since the 27th year from the beginning of the landfill operation, the pore water with a concentration of Cr greater than the allowable limit begins to discharge into the upper Holocene aquifer (**Figure 11a**). The situation is more severe regarding Pb: the pore water with Pb concentration greater than the allowable limit begins to discharge into the upper Holocene aquifer from the 9th year (**Figure 11b**). The Arsenic

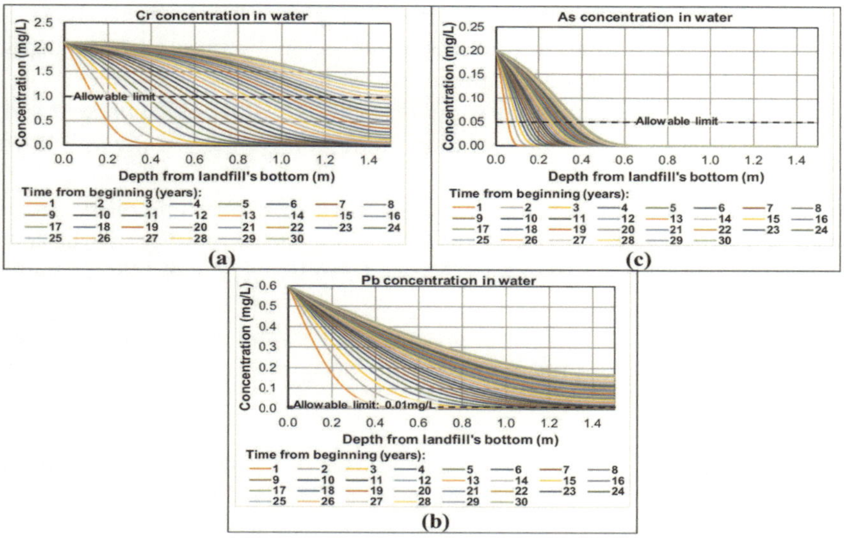

Figure 11.
Heavy metal concentrations in pore water prediction by FEM for 30 years: MD2. (a) Chromium in pore water - MD2. (b) Lead in pore water - MD2. (c) Arsenic in pore water - MD2.

concentration greater than the allowable limit in pore exists only in the upper 0.4 m after 30 years of the landfill operation (**Figure 11c**).

4. Conclusions

FE modelling of advection-dispersion transport of HMs by GW movement along with Freundlich isotherm adsorption parameters which continuously change with space in the model domain and with time is sophisticated, but is capable of accurately evaluating the HMs' concentrations in soil skeleton and pore water. The chapter describes the background of the existing isotherm adsorption theory. The chapter has provided a detailed mathematical formulations of the FEM in solving the advection-dispersion contaminant transport in soil water. It also demonstrates that the Freundlich isotherm adsorption parameters are essential to soil input parameters for modelling of HMs' transport to access the soil skeleton and soil pore water contamination by HMs. In designing the experiments for the determination of the Freundlich isotherm adsorption parameters, the range of the HMs' concentrations in water is suggested to be corresponding to the actual HMs' concentrations under study. Besides, the background of the existing isotherm adsorption theory, the adaptation of the Freundlich isotherm adsorption in the soil skeleton and soil pore water contamination by HMs has been introduced.

The methodology has been applied to a case study of Kieu Ky waste landfill in Hanoi, Vietnam. The transport of HMs in soil water is determined not only by hydrodynamic dispersion but also largely by the adsorption of the metals by the soil. With the use of the collected interpreted isotherm adsorption parameters, the magnitudes of soil and soil water contamination by HMs from the waste leachate are very much different from each other due to both the HMs' concentrations in leachate and the soil isotherm adsorption parameters. Unlike the pollutant transport in aquifers with coarse grain size particles like sand and gravel without clay materials with nearly-zero adsorption, the transport of pollutants in silty soils essentially requires adsorption parameters to have reliable modelling results.

The application modelling results show that HMs Cr, Pb and Zn present soil, soil pore water and groundwater contamination vulnerability, specifically as follows.

- Soil contamination with Cr, Pb and Zn by the direct spreading of the metals with dust and leachate from the waste landfills. For a period of 30 years, the soil would be contaminated in the upper 1 m, 2 m and 3 m by Cr, Zn and Pb, respectively

- The concentrations of Cr, Pb and Zn in the pore water in the silt layer are higher than allowable limits in the upper 1.5 m, 6 m (i.e., the whole soil layer) and 2.2 m, respectively.

- Since the 9th and 27th year from the beginning of the landfill operation, the pore water with Pb and Cr concentrations greater than the allowable limits begins to discharge into the upper Holocene aquifer, respectively.

- From a quarter of a century from the landfill operation start, Cr and Zn in the soil water would reach the Upper Holocene aquifer to pollute the aquifer. The waste leachate would cause the Upper Holocene aquifer polluted with Cr.

Conflict of interest

The author declares no conflict of interest.

Author details

Nguyen Van Hoang
Institute of Geological Sciences under Vietnam Academy of Science and Technology (VAST), Hanoi, Vietnam

*Address all correspondence to: N_V_Hoang_VDC@yahoo.com

References

[1] Kasassi A, Rakimbei P, Karagiannidis A, Zabanitou A, Tsiouvaras K, Nastis A, et al. Soil contamination by heavy metals: Measurement from a closed unlined landfill. Bioresource Technology. 2008; **99**:8578-8584

[2] Agamuthu P, Fauziah SH. Challenges and issues in moving towards sustainable landfilling in a transitory country - Malaysia. Waste Management & Research. 2011;**29**(1):13-19. DOI: 10.1177/0734242X10383080

[3] Dutch Ministry of Housing, Spatial Planning and Environment. Dutch Standards: Circular on Target Values and Intervention Values for Remediation. Netherlands: Netherlands Government Gazette; 2000

[4] Opaluwa OD, Aremu MO, Ogbo LO, Abiola KA, Odiba IE, Abubakar MM, et al. Heavy metal concentrations in soils, plant leaves and crops are grown around dumpsites in Lafia metropolis, Nasarawa State, Nigeria. Advances in Applied Science Research. 2012;**3**(2):780-784

[5] Lyanguzova IV, Bondarenko MS, Belyaeva AI, Kataeva MN, Barkan VS, Lyanguzov AY. Migration of Heavy Metals from Polluted Soil to Plants and Lichens under Conditions of Field Experiment on the Kola Peninsula. Russian Journal of Ecology. 2020;**51**:528-540. DOI: 10.1134/S1067413620060053

[6] Piyada W, Suksaman S. Assessment of Heavy Metal Distribution in Soil and Groundwater Surrounding Municipal Solid Waste Dumpsite in Nai Muang Subdistrict Administrative Organization, Amphur Phichai Uttaradit. NU Science Journal. 2013;**10**(1):18-29

[7] Siti N, Sumiani Y, Chua YP. Soil chemistry and pollution study of a closed landfill site at Ampar Tenang, Selangor Malaysia. Waste Management & Research. 2013;**31**(6):599-612. DOI: 10.1177/0734242X13482031

[8] Kamarudin S, Mohd T, Ismail A, Mohamed H. Heavy metals profiles in a groundwater system at a solid waste disposal site, Taiping, Perak. Bulletin of the Geological Society of Malaysia. 2012; **58**:9-14

[9] Jacob B, Arnold V. Modelling Groundwater Flow and Pollution. Dordrecht, Holland: D. Reidel Publishing Company; 1987. p. 414

[10] Ijagbemi CO, Baek MH, Kim DS. Montmorillonite surface properties and sorption characteristics for heavy metal removal from aqueous solutions. Journal of Hazardous Materials. 2009;**66**(1): 538-546. DOI: 10.1016/j.jhazmat.2008.11.085

[11] Suna EA, Fazal MA, Onay TT, Craig WH. Determination of solid waste sorption capacity for selected heavy metals in landfill. Journal of Hazardous Materials. 2005;**121**(1–3):223-232. DOI: 10.1016/j.jhazmat.2005.02.011

[12] Freundlich HMF. Over the Adsorption in Solution. The Journal of Physical Chemistry. 1906;**57**:385-471

[13] Patiha M, Firdaus S, Wahyuningsih K, Nugrahaningtyas D. Derivation and constants determination of the Freundlich and (fractal) Langmuir adsorption isotherms from kinetics. Materials Science and Engineering. 2018; **333**:Hidayat Y, 012010

[14] Coles CA. Estimating retardation from the Freundlich isotherm for modelling contaminant transport. Engineering Geology. 2007;**87**:35-40

[15] Zienkiewicz OC, Morgan K. Finite Elements and Approximation. John Wiley and Sons; 1983. p. 328

[16] Huyakorn PS, Pinder GF. Computational Methods in Subsurface Flow. New York, NY: Academic Press; 1983. p. 473

[17] He G, Zhang Z, Wu X, Cui M, Zhang J, Huang X. Adsorption of Heavy Metals on Soil Collected from Lixisol of Typical Karst Areas in the Presence of $CaCO_3$ and Soil Clay and Their Competition Behavior. Sustainability. 2020;**12**:7315. DOI: 10.3390/su12187315

[18] Gramss G, Voigt KD, Bublitz F, Bergmann H. Increased solubility of (heavy) metals in the soil during microbial transformations of sucrose and casein amendments. Journal of Basic Microbiology. 2003;**43**: 483-498

[19] Liu GN, Wang J, Xue W, Zhao JB, Wang J. Effect of the size of variable charge soil particles on cadmium accumulation and adsorption. Journal of Soil Sediments. 2017;**17**:2810-2821

[20] Noppadol S, Pongsakorn P. Adsorption behavior of heavy metals on various soils. Polish Journal of Environmental Studies. 2014;**23**(3): 853-865

[21] Emma FC, Flora AV, Luisa A. Heavy metal sorption and desorption capacity of soils containing endogenous contaminants. Journal of Hazardous Materials. 2007;**143**:419-430

[22] Claudia CC, Manuel CC, Manuel AE, David FC, Flora AV. Specific adsorption of heavy metals in soils: Individual and competitive experiments. Agronomy. 2020;**10**:1113. DOI: 10.3390/agronomy10081113

[23] Ghislain de Marsily. Quantitative Hydrogeology-Groundwater Hydrology for Engineers. Orlando, FL: Academic Press; 1986. p. 440

[24] Henry VK, editor. CRC Handbook of Thermophysical and Thermochemical Data. Boca Raton: CRC Press Inc; 1994

[25] Neuman SP. Calibration of distributed parameter groundwater flow model viewed as a multiple-objective decision process under uncertainty. Water Resources Research. 1973;**9**(4): 1006-1021

[26] Ledermann W. In: Robert F, editor. Handbook of Applicable Mathematics. Volume III: Numerical Methods. Churchhouse: John Wiley & Sons; 1983. p. 584

[27] MoNRE. QCVN 40:2011/BTNMT National technical regulation on industrial wastewater. Hanoi, Vietnam: MONRE Publisher; 2011 (in Vietnamese)

[28] MoNRE. QCVN 09:2008/BTNMT National technical regulation on underground water quality. Hanoi, Vietnam. MONRE Publisher; 2008 (in Vietnamese)

Chapter 6

Microwave Vitrification of Hazardous Sludge by Şırnak Asphaltite Slime, Shale and Sewage Sludge – Landfill Management by Vitrified Char Sand

Yildırım İsmail Tosun

Abstract

Mazıdağı Etibakır Cu ore leaching waste stockpiles, land soil and groundwater in the field should be controlled for seepages to avoid the acidic flow of solute containing the heavy metals of Pb, Cu and Zn. The heavy metal-associated liquor from Electrowinning Plant and Sulphuric acid Dissolution units threats the neighborhood in the town. Cu and Co are recovered by electrolysis and acidic solutions of Cu leaching are spent. According to this concern of waste management, a method commonly used, "vitrification of sludge," among others such as special cementing or bituminous pasting may be used in the waste disposal and even ground stabilization. However, mixing that with glassy powder and further vitrification furnace heating yield a vitrified form of waste sludge, covered by glazed matter avoiding contact with water in the landfill. In this study, sodium silicate is used as a binder in the vitrification mixture with the sludge at 14% water under microwave radiation. The dissolved contents of Pb Zn and Fe in the yielded vitrified briquette are determined. Additionally, the strength of vitrified briquettes is investigated in terms of vitrification parameters of microwave radiation.

Keywords: microwave radiation, hazardous salt slurries, metal encapsulation, sorption, hazard assessment, vitrification treatment, vitrification sorbent, hybrid sorbent, waste sludge, hazard salt, radioactive salt slurries, sludge treatment, heavy metal, salt composts, shale

1. Introduction

The disposal of hazardous sludge is much significant in landfill waste management, covering and dumping. The reactive chemistry of sludge threats ecology even in a landfill [1–4]. There are many hazardous wastes such as the muddy by-product from the heat-treated steel manufacturing with CN baths [1–4], textile painting [1–4] and tanning sludge metal peroxide salts [5, 6], radioactive fuel waste sludge [7, 8], heavy peat of pulp washing industry [1–4].

The runoff mine phosphate ores are calcined for the production of reactive phosphate compound before conversion of phosphoric acid in the production of superphosphate fertilizer in Mazıdağı Etibakır Plant [9, 10]. However, the Cu ores from Küre and Şirvan are transferred to Mazıdağı Etibakır Cu ore stockpiles field and dissociated from Electrowinning Plant and Sulphuric acid Dissolution units. Cu is recovered by electrolysis and spent acid solutions of dissolution and electrowinning are decanted and sludge effluent is collected in the two different tailing ponds [1–4]. Under the atmospheric conditions of hard wintertime, during the time of heavy raining months in Mazıdağı, Mardin province of Southeast Anatolian Region, phosphate plant production facilities located need clean irrigation water and a hundred meter away, the freshwater reservoir of Mazıdağı town is located. The economical value of this reservoir reduces the cost of living, agricultural irrigation and animal farming in the town with a low population of about 7400 [1–4]. Effective wastewater management of high capacity of Cu dissolution plant will not threaten the scarce freshwater potential of the town and provide the much clean ecology. The spent sludge of sulfuric acid in Cu leaching is advantageous for the Co recovery process. There is a resulting heavy sludge waste of electrolysis rich with Fe, Pb, Cd and Zn. This black metal sludge is used for extraction of metals such as Au and Co. However, the recycling dissolution results in a high solute level of Pb, Hg Zn, Cd and Fe during recovering Co. This vitrification method provided a new idea for the hazardous sludge disposal in recycling plants with char/coal slime use by sludge waste.

In this study, the sludge samples of tailing ponds 1 and 2 were economically heated by microwave oven as briquettes of 50 mm size homogenously mixed with sodium silicate in the 16% porous structure and even Şırnak asphaltite slime mixed reduces metal contamination in the wet sludge of the tailing ponds. The development of compaction stress reduces the porosity of briquettes and provides a much higher strength for vitrified block formation and much possible inert-vitrified briquette yield for landfilling. Particularly in this study, the Şırnak asphaltite slimes and oak wood char subjected to the fine screening under 0.2 mm and carbon ability over Pb and Fe contamination were investigated as weight rate.

1.1 Hazardous acidic leaching waste sludges

Neutralization of acidic waste effluents is washed, and settled precipitated metal sulfates and lime iron hydrates form sludge in the oxide micron-sized hydrates in a muddy state. However, the filtrated matter containing the sulfate part of the reaction [7–9] may cause redox effect oxidation. Then

$$H_2S + 4H_2O \rightarrow {SO_4}^{2-} + 10H^+ + 8e^-. \tag{1}$$

$$PbS + 2Fe^{3+} + 3{SO_4}^{2-} \sim +3/2O_2 + H_2O \rightarrow PbSO_4 + 2Fe^{2+} + 2H^{4+} + 3SO4^{2-}. \tag{2}$$

$$ZnS + 2Fe^{3+} + 3{SO_4}^{2-} \sim +3/2O_2 + H_2O \rightarrow ZnSO_4 + 2Fe^{2+} + 2H^{4+} + 3SO4^{2-}. \tag{3}$$

The heavy metal contents such as Pb and Hg are dissolved in the use of lower pH acidic solutions of H2SO4 or HNO3 in the electrolysis mud recycling leaching end as regards Pb heavy metal contamination is followed by equation, where HNO_3^{-2} nitrate concentration in the effluent

$$\frac{dc_{Pb}}{dt} = k_i c^{tin}.dc.f_i\left(HNO_3^{-2}\right)^{tin} \quad (4)$$

$$2CaO(Fe_2O_3/Al_2O_3.MgO\ Si_8O_{20}(OH)_4.nH_2O + 2NaS_2O_4^{2-} + 2H_2O \leftrightarrow 2NaCaO(FeO\ Fe_2O_3.\ Al_2O_3MgO.Si_8O_{20}\ (OH)_4.nH_2O + SO_3^{2-} + 4H^+ \quad (5)$$

$$Clay - Fe^{3+} + 4NO3/SO_4^{2-} \leftrightarrow Clay - Ferrous + 2S_2O_4^{2-} + H_2O \rightarrow 2SO_3^{2-} + S_2O_3^{2-} + 2H^+ \quad (6)$$

$$Fe^{2+} + Pb/Zn/MnO_4^{2-} + HNO_3 + 4\ H_2O \rightarrow (Fe_X2Cr_{1-x})(OH)_3 + 5OH- \quad (7)$$

The dissolution kinetics of soil mud particle for Pb, MnO_2 heavy metal is followed by Eq. (7)

where CPb, CaO, MnO_2 dissolution mg/l, k the rate of digesting of lead, i is the reaction style, t is time.

The digesting amount of heavy metal in aliquate of solute of tailing pond as regarding sludge contamination is managed by equation, where n is the kinetic order type as given below

$$\frac{dc_{ACaO}}{dt} = k_i c^{tPb} dc \quad (8)$$

İndustrial hazardous waste effluents threaten the agricultural land and freshwater reservoirs in a high-risk concern with relationship between the collection of wastewater through the sewage network of urbanization, the hazardous sludge treatment, transmission to decantation, disposal treatment and discharge style. Hazardous sludge and effluent management with projection on neutralization and decantation never avoid the harmful end of the sewage output, and toxic substances still exist. Regarding hydroelectricity dams, animal farming freshwater lakes are located near highly populated cities, where water management taxes, loans and low-interest loans use discounts as well as other financial support mechanisms. Hazardous sludge and effluent discharge management are so much extra critical in the way of financial consideration for public health and ecology [10–15].

1.2 Hazardous sludge

The industrial effluents with hazardous sludges are subjected to clean filtering in a continuous flow system. The hazardous effluent is decanted and followed to a best sorption process and the resulted sludge of sorption and filtered neutralization sludges show that a high amount of lime and hazardous metal salts are suitable for the vitrification of hazardous sludges for disposal. At landfill areas, the vitrified products should protect their form without cracking and digesting the hazardous content down to the irrigation or freshwater limits defined by the ecology legislation, with mg/l metal Fe.

The Fe analyses were performed with the sludge original samples and vitrified samples of different weight rate vitrifying binder to determine the duration period as leaching kinetics of Fe and Zn metals on the dissolved briquettes. The results show that Fe oxides, hydroxides, sulfates, Zn oxides, hydroxides and carbonates sulfates were dependent on ion exchange ability with lime Ca, the Zn retention occurs by crystallization as hydrozincite, and $Zn_5(OH)^-(CO_3)^{2-}$ and ferric hydroxide crystallizes on neutralization sludge lime coated as ferric hydrate /zinc oxide hydrocarbonate.

Material	Alkali leaching solution	Solubility constant, logK
Quartz	$NH_4/Na/KOH+UO_2= KO_2UO_2 + 2H+$	–5.5
Chlorite	$Na/KOH+UO_2= KO{\cdot}UO_2 + 2H+$ $Na/KOH+UO_2= KO{\cdot}UO_2 + 2H+$	–4.7
Muscovite	$Ca/Fe(OH)_2+UO_2= KO_2{\cdot}UO_2 + 2H+$	–5.75
	$NH4/Na/KOH+UO_2= KO_2UO_2 + 2H+$ $NH4/Fe(OH)_2+UO_2= KO_2{\cdot}UO_2 + 2H+$	–0.55
Albite	$NH4/Fe(OH)_2+UO_2= KO_2{\cdot}UO_2 + 2H+$	–1.5
Ferrihydrite	$NH4/Fe(OH)_2+UO_2+HCO_3= KO_2{\cdot}UO_2 + 2H+$ $NH4/Fe(OH)_2+UO_2+HCO_3= KO_2{\cdot}UO_2 + 2H+$	–0.32
Limonite		–0.62
Ferrihydrite		–0.62
Limonite		–0.42

Table 1.
Values of samples at thermal dissolution [16] over 100°C.

However, Pb sulfate and hydroxide cover are also observed in the sludges at less rates depending on the electropotential of pH values over porosity of sludge formed from clusters of ferric iron oxide and lime solids (**Table 1**) [16].

Sulfite oxidation kinetic rate is developed as given below Eq. (9),

$$r = k\sqrt{S/}(SO3)^{2-}\left\{(O)^{2-} - \left[(O)^{2-}/(SO3)^{2-}K_{sol}\right]^2\right\}, \tag{9}$$

where K_{sol} is dissolution equilibrium constant, and SO_3^{2-} and O^{2-} solute concentrations of sulfite and oxygen are dissolved.

The mass diffusion of cracked bonds of hydrocarbon aromatic apolar reactive sites raises the kinetic rate of dissolution through porous coal texture, while asphaltite massive texture avoids the dissolution.

Specification and sorption for risk assessment, [17–19] modeling and application for hazardous waste management should be considered over legitimate rules regarding:

radioactive decay

complexation reactions; hydrolysis, dissociations, association polymerization, oxidation redox reactions may cover hazardous components in those reactions

precipitation

co-precipitation, inclusions, surface precipitation

physical and chemical sorption on surface

formative solids

ion exchange

extraction

colloid formation

biosorption

1.3 Vitrification silicate

The microwave vitrification studies of Cu, Zn, Pb and Fe were conducted on tailing pond sludges of pools 1 and 2 of Mazıdağı Etibakır Cu recovery plant to determine the efficiency of briquetting and strength of blocks. The resulted blocks are in the form of vitrified conventional heating and microwave heating during retention time. The melting capacity of Na silicate for these sludges commonly presents the dissolution ability in the landfill area.

The window glass production technology includes the crushing-grinding, screening, washing and sand flotation unit following a very fine 200 micron and Fe % grade of sand decreased to below 0.5%, while vitrification does not need clean sand, and even dirty recycled glass waste is evaluated [20–23]. Thus, to provide vitrified matter, the cement retort kiln or firing grate furnace produces sintered waste material, homogeneous, vitrified and suitable landfill slag material. The suitable slag by-product without landfilling can be used as aggregate in asphalt road pavement and masonry stone production with low costs. Asphalt pasting of hazardous sludges is also becoming another covering method to avoid heavy metal contamination [24–28]. Şırnak asphaltite slime is already below 100-micron size and so easily mixed to cover sludge in microwave radiation. The recycled bottle waste glass and the broken window should be easily evaluated as aggregate. The shale waste of Şırnak asphaltite coal quarries reaching over 7 million tons may be evaluated as vitrification binder following grinding [29]. This vitrifying method costs less. In this study, negative effects on the vitrification quality and capacity are determined. Şırnak asphaltite slime properties are also important for vitrified briquette breakage and porosity, and surface area change.

Instead of the use of conventional grate firing, microwave vitrification is becoming advantageous in internal surface covering by inner volume heating by radiation of sludge fine solids mixed with Na silicate fine and coal slimes avoiding contamination.

1.4 Microwave heating

Microwave radiation conducts the waves through the material as radio wave frequency in tri-band microwave frequency (UHF: 300 MHz to 3 GHz), super high frequency (SHF: 3 GHz and 30 GHz) and extremely high frequency (EHF 30 GHz to 300 GHz) [11]. The microwaves pass through the whole inner depth of the diamagnetic solid texture [30–34]. The iron oxides such as wustite, magnetite and hematite can be heated in 20–30 sec at 2–3 mm size, while plastic materials isolate the waves [31]. The metallic salts such as Pb and Zn oxide or semi-metallic sulfides behave similarly as ferrous solids with high emissivity ın electromagnetic energy [32], in which solid temperature increases the temperature of whole sample volume, unlike conventional heating [33, 34].

Mineral packed in solid-densed form is easily heated under the radiation of microwave with high-frequency vibrations of inner atomic layers in mineral crystal and thermal energy increase conducts the heat from core to surface of particle grain. The heat-covered surface raises the temperature and creates an effect of melting of surface and sintering particles in the microwave vitrification of oxide solids. The studies showed that iron-bearing ores, roasting of sulfides, refractory gold concentrate oxidation, and activated carbon regeneration can be accomplished by microwave radiation in the shortest time periods between 30 sec. and 3 min [31]. The microwave heating slightly affects the calcination of limestone rock in 30 min.

Microwave act on minerals was determined to be sufficient [33–43]. Microwave interaction parameters on mineral crystals, microwave penetration level, the vibration of mineral grains, grain boundary heating, and heat absorption were managed. The thermal effects vary according to microwave-radiated mineral species [44]. The least penetration of mineral grains of quartz is given in **Table 2** and has 79°C a temperature change.

Mineral type	Maximum temperature, °C	Time, min
Albite	69	7
Chromite	155	7
Chalcopyrite	920	1
Cinabarite	144	8,5
Galenite	956	7
Hematite	1082	7
Magnetite	1258	2,75
Marble	74	4,25
Molibdenite	192	7
Ortochlase	67	7
Pyrite	1019	6.75
Pyhrotite	586	1.75
Quartz	79	7
Sphalerite	88	7
Zircon	52	7

Table 2.
Microwave temperature effect on minerals [45].

Quality of vitrified briquettes—efficiency of vitrification.

High-intensity microwave radiation provides high-thermal inner particle surface melting Na silicate over 300°C such as low-temperature glazing. The microwave act as a sintering bound of particles of hazardous oxide and sulfate salts of dissolved sludge with heavy metals such as Pb, Cu, Zn, Fe. The vitrified glassy product contains 16–12% Na coming from melting Na silicate behaving highly transparent liquid interactive conduction heating. A high duration period will also recrystallize Na silicate binding phase. The strength of briquettes will be reduced by breaking the act of lime and Ca hydrates, and carbonates. The addition of Şırnak asphaltite slime and Şırnak shale as clay stone fines was examined in this study. The effect of microwave radiation on vitrification ability and the qualities of briquettes of this mixture was investigated.

Industrial hazardous sludges and wastewater effluents from the metal coating, Zn-galvanizing effluents and other hydrometallurgical processes generally contained high levels of heavy metals such as Pb, Zn, Cr, Hg, Cd, Fe [46–48]. The vitrified matter encapsulates this hazardous salty solid-precipitated matter in a mixture of bound silicate cover. Hence, the hazard of heavy metal dissolution is avoided. The dissolution of vitrified heavy metal salts shows the quality of vitrification. Current encapsulation technology is also advantageous for radioactive sludge However, the other methods following precipitation, ionic exchange and covering or melting in synthetic resins, plastics require high-cost processing and operational costs even still create waste disposal issues. The vitrification method is usually capable of proving the limits of legislation of below 0.1 and 3 mg/l for Pb, Zn, Cd and Cd metal values [49–51]. The hazardous metal precipitation is not sufficient in neutralization and decantation down to the legal limits because organic and inorganic complex compounds allowed effluent levels above those regarding the solubility of the metal hydroxides [52–57]. Recycling of heavy metals based on vitrification may also be suggested as an alternative approach.

2. Methods

The studied waste deposit tailing ponds cover a surface area of approximately 9 decares, and they are 3 m deep at the northeast disposal area, and lowering 3–5 m deep at the eastern Mazıdağı water reservoir end. It is estimated that a total 1–0.5 million tonnes of hazardous leaching ferrite slime wastes are deposited from adjacent sulfuric acid unit that is grayish-black in color, and now found mostly in the north part of the Mazıdağ Etibakır plants. Groundwater investigation was carried out by 5 boreholes drilled 2-meter depth in the sludge slime deposit and 15 samples were collected at 1 m and 2 m depth from the muds for the analysis of pH, electrical conductivity, leachable Pb, Fe, Zn, Cu, Cd and SO_4. **Table 3** gives the chemical analysis taken from ponds 1 and 2.

The amount of binder is investigated in the strength of briquetted blocks at 10-mm cubic forms as reported on different weight rates of sludge waste. The contamination characteristics in 1 M HCl and H_2SO_4 soluble acid solutions are investigated with standard wastewater tests over the resulted effluents at the end of 1-h boiling. The sludge below <150-micron size is mixed with the sodium silicate sand fraction (–0.6mm + 150 microns) by microwave melting.

The Pb, Fe and Zn metal cations studied were sludge effluents were analyzed UV spectrophotometer from calibrated standards of 1–100 mg/l and leaching was carried out using HNO_3 5N for a leaching period of 1 h with 75 ml of solution in the microwave. The sludge samples received from Mazıdağı are subjected analyses. The metal values of sludge in effluent analyses showed that sludge had 130 mg/l for Zn and 325 mg/l for Fe and 28 mg/l for Pb at pH: 5–6, 15% wet solute content.

Batch flasks using 10 gr solid samples dried as sludge slime were dried in the microwave oven at 10 minutes and then settled in 10-mm-diameter steel molds and compressed under 2-ton loading. The Na silicate mixed sample mixtures by Ermenek lignite/Şırnak asphaltite/Gediz lignite slime, shale and hazardous sludge to homogenization mixer in the experiments.

2.1 Particle size distribution: asphaltite slime/lignite slimes

This compost is especially sorbent used wastewater treatment. The fly ash compost granules are used as hazardous industrial wastewaters and the sorbent types and

Effluent, mg/l	Pond 1	Pond 2	Sludge1	Sludge2	Groundwater well
Hg	8.11	4.71	52.3	74.11	4.71
Pb	10.58	14.53	73.2	128.58	11.53
Fe%	4.33	7.62	5.91	9.3	0.52
K+Na	227.52	338.46	1328.7	1748.52	58.6
Cd	24.72	19.56	144.1	184.72	19.56
Mn	33.3	24.1	274.2	463.3	24.1
Cu	27.2	30.2	715.7	997.2	10.2
As	31.10	22.44	92.8	232.10	2.44
SO_4%	4.57	9.37	5.9	8.67	0.45

Table 3.
The compositions of sludge samples of Mazıdağı Etibakır Leaching Plant Tailing Ponds.

desirable properties of activated and cleaned sorbents in the experiments are given in **Table 4** and chemistry is given in **Table 5** for geothermal waters used in neutralization.

M is mass of aggregate is, the void is affected by compaction of briquetting and binder distribution, and especially, melted asphalt and polymer distribution are controlled by volume % of compaction. The bulk elasticity will also be controlled by the amount of polymer-bound as a volume.where, γg=_density of aggregate, g/cm3; *V(r)* and *dN(r)* are the volume and particle amount of aggregate in the size region of integration of cumulative pile from *r, to r+dr*), respectively. *V*r volumetric equation is,

$$dMr = \gamma g \, \mathrm{V(r)dN(r)} \quad (10)$$

$$V(\mathrm{r}) = k\, r^3 \quad (11)$$

where *k* is the shape factor.
Slime particle size distribution
Particle size distribution is defined by aggregate crushing matter,

$$\mathrm{u(x, d_f, c)} = (\chi/\mathrm{d_f})\,\left(1 + \mathrm{k/d_f}\,(\mathrm{x} - \chi)\right)^{-1/\mathrm{k}} \quad (12)$$

$$Rss(n) = f(n)xW(n)\sum_{m=1}^{n} 1/(1 - r)^m \quad (13)$$

Sample	Şırnak asphaltite slime	Ermenek lignite slime	Gediz lignite slime	Oak wood char	Şırnak coal shale
SiO_2	23.53	19.42	14.14	0.1	17.53
Al_2O_3	10.23	6.53	12.61	0.1	13.61
Fe_2O_3	14.59	8.48	7.34	0.1	9.67
CaO	16.48	11.23	10.13		19.48
MgO	5.20	5.28	4.77	0.1	4.28
K_2O	4.41	2.53	3.22	0.2	2.51
Na_2O	3.35	2.24	1.71	0.1	2.35
Ignition loss	26.19	50.11	36.43	60.9	26.09
C/H/S	39.32	59.21	42.20	99	9.67

Table 4.
The ash chemical analysis values of vitrification mixture filling materials of Şırnak province and lignite slimes.

Coal type	Şırnak asphaltite slime	Ermenek lignite slime	Gediz lignite slime	Oak wood
C	20.3	42	29	24
H	2.0	2.1	2.3	2
S	6.1	2,2	3.6	0.2
Ash	67.7	24	33	0.7
Moisture	1.9	24	29	72

Table 5.
The elemental analysis values of vitrification carbon mixture filling materials, Şırnak asphaltite slime, lignite.

2.2 Briquetting of mixture prior to vitrification

The mixed slime and sludge with Na silicate and Şırnak asphaltite slime microwave heated following pressing in 50-mm mold under 3-tons load. The squeezed matter reduced wet solute at a 10% weight rate. The surface area and porosity are measured by the Rigden flow meter in this work. It is generally used for cement surface area measurement. This slime and sludge mixing nearly had a low surface area, as measured by the single-point BET N_2 adsorption method, of 11 m^2/g. Sludge used was at a particle size < 0.1 mm for vitrification and ranging below 0.2 and 1 mm for Şırnak asphaltite slime and Şırnak shale.

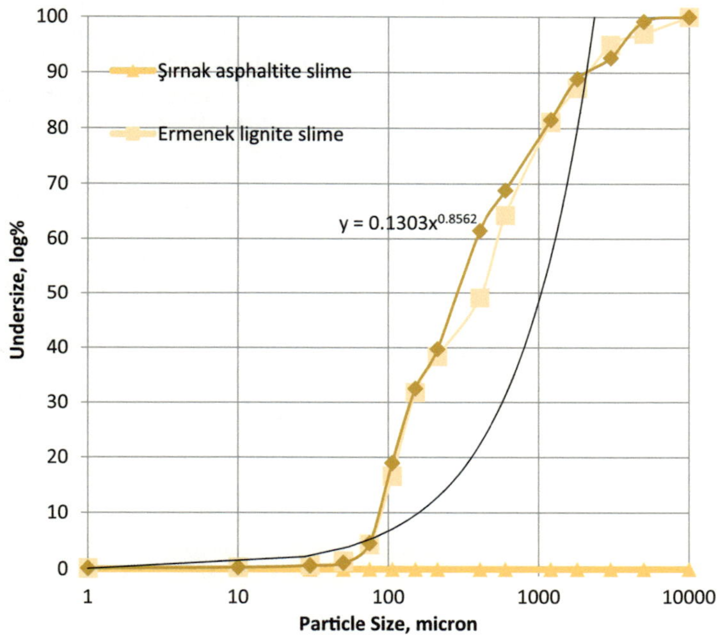

Figure 1.
The ash and Şırnak asphaltite slime distribution regarding void in gradation in ASTM standard.

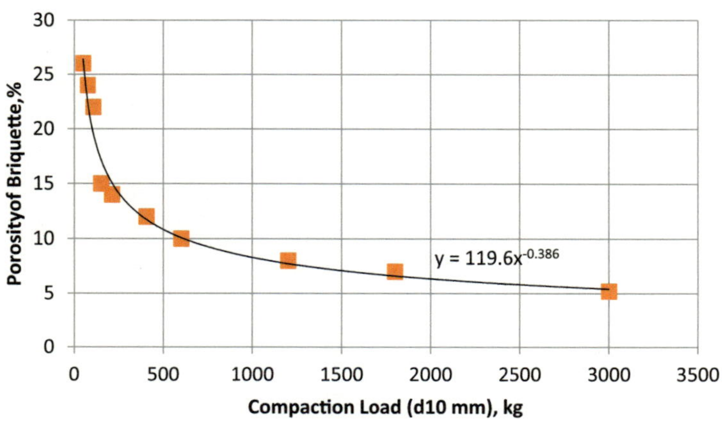

Figure 2.
The compaction limits of sludge regarding void in fineness gradation.

Figure 1 shows the slime sample fineness and size distribution as solid sludge compost increasing the compaction ability and even fine homogenous mixing. **Figure 2** illustrates the compaction ability of sludge compost under a loading press of 20 kN at size 10 mm diameters.

3. Results and discussions

3.1 Microwave-vitrified matter

The collected sludges were analyzed by XRD and the chemical analysis results show that high ferric oxide and ferrous sulfur, over 83% with 9% wet weight rate the rest was carbonate hydrates consisting mainly of sillimanite (ZnCO3H2O), Zn-cupric carbonates such as malachite FexZnyCuy CO3H2O, Zn-ferrous hydrate (Zn, Cu) FeSO4 H2O, gypsium (CaSO4 2H2O) and chalcanthite (CuS045H20). Their chemical oxide and hydrate distribution are given in **Tables 6** and 7.

The mixed slime and sludge with Na silicate and Şırnak asphaltite briquettes comprised that below composition.

The chemical change irons in iron sulfates into the sludge contacted binder to salt phases easily melted by salt fluxing effect with Na silicate. However, sintering iron silicate and lead silicate melts occurred in the sludge vitrification end. The zinc and copper ferrite bounds to Na silicate melts present whisker style needle-like fillings in the briquette texture. The dominant spherical sludge lime hydrates and iron sulfates are not completely wetted by binder Na silicate causing weakness in the briquette

Mixture components	Binder +Sludge1	Şırnak asphaltite slime +oak wood char+ sludge1/2	Ermenek lignite + oak wood char + sludge 1/2	Gediz lignite +oak wood char+ sludge 1/2
Na_2Si_2O6	15			
Fe_2O_3	4	42	29	24
PbO	7	2.1	2.3	2
ZnO	2	2.2	3.6	0.2
CaO	17	24	33	0.7
MgO	2	24	29	72
Zn/Cu ferri hydrate	4	42	29	24
Al_2O_3	11	2.1	2.3	2
SiO_2	22	2.2	3.6	0.2
Na+K	5	24	33	0.7
SO_4	4	24	29	72
H_2O	12			
C				
Total	99			

Table 6.
The chemical analysis values of vitrification carbon mixture-filling materials: Şırnak asphaltite slime and lignite of lime sludge.

Vitrified components	Binder +Sludge2	Şırnak asphaltite slime +oak wood char	Ermenek lignite + oak wood char	Gediz lignite +oak wood char
Na_2Si_2O6	15			
$ZnCO_3/CuCO_3$	6	42	29	24
Fe_2O_3	40	2.1	2.3	2
PbO	7	2.2	3.6	0.2
ZnO/CuO SiO_2	10	24	33	0.7
CaO	10	24	29	72
MgO	2	42	29	24
Zn/Cu ferrite	4	2.1	2.3	2
$FeSO_4$	1	2.2	3.6	0.2
$CaSO_4$	2	24	33	0.7
Na+K	5	24	29	72
SO_4	5		29	
H_2O	12		2.3	
C		23.5	43.6	34
Total	99	99	99	99

Table 7.
The chemical analysis values of vitrification carbon mixture filling materials: Şırnak asphaltite slime and lignite of salty sludge.

structure. The dissolved iron amount relatively changed to below 6% in final vitrified matter dissolution (**Figure 3**).

Gypsum is generally looks like lamella morphology, and frequently presents as plate layer formings adhering close to residual ferric iron oxide. The compacted briquette strength increased by binder addition of 15% till continued to 20% weight rate as illustrated in **Figure 4**.

3.2 Acid digestion of vitrified matter

Lead sulfate salts over lime hydrate surface in the vitrified matrix as solid is dissolved in 1 N HNO_3 and 1 M H_2SO_4 solution on digestion bath of 100-ml flasks at 30-minute boiling period. The heavy metals such as Pb Zn and Fe hydroxides replace CaO until the solution reaches the composition expected for the equilibrium of $FeSO_4$ nH_2O /$PbSO_4$ nH_2O or $ZnCO_3$ .n H_2O/ $CuCO_3$ .n H_2O/$CdCO_3$.n H_2O.

Higher dissolved Cd/Zn in the effluent caused Zn carbonate precipitation in the sludge with iron sulfa-hydroxide. The surface solid of over lime was high oxidation potential, while solute concentrate of Fe was high for inhibiting precipitation as the second phase of the sorption process. This process involves mixed heavy metal precipitation followed by slow formation of microcrystalline dirty solids solute.

These tests also showed that microwave melting or sintering sludge mud in briquetted form provided much more heat effect on the sand matter with less amount of weight rate of 12% and highly reactive and dissolving Pb and iron amount reduced to 1 and 3.2 mg/l in the porosity of briquetted sludge at 16% in the vitrified matter dissolution.

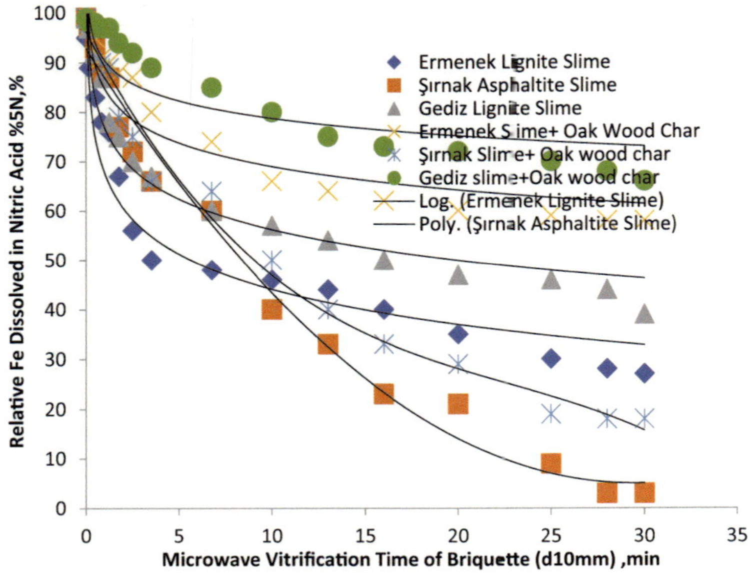

Figure 3.
The ash and Şırnak asphaltite slime RRS distribution regarding gradation factor in 0.45 sieve.

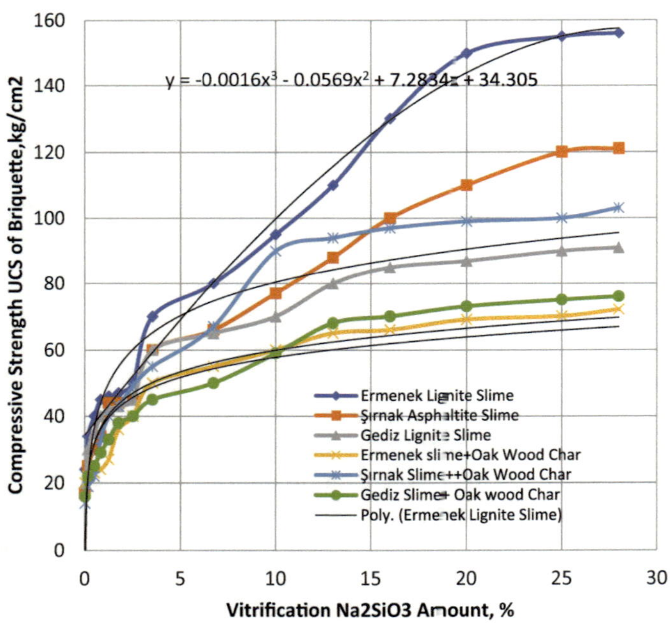

Figure 4.
The ash and Şırnak asphaltite slime RRS distribution regarding gradation factor in 0.45 sieve.

Toxic intermediates may be sorped by char and shale clay may be generated precipitation heavy metals as organic complexes from barrier-integrity vitrification, effective silicate barriers and homogenous mixing with active carbon were found to be quite difficult due to surface wetting manner of carbon.

The microwave digestion with 1M acid hot solutions avoided disposal of hazardous sludge to landfill following decantation techniques in paste thickeners used flocculants, polyelectrolytes, chelants, inorganic acids or surfactants related to sludge particle size and type. The compaction porosity decreased by solvent use in compaction to 5 % under loading as shown in **Figure 5**.

The microwave vitrification of industrial waste slurries and hazardous sludge creates a safe working environment while absorbant fly ash is used. The neutralization of hazardous slurries with fly ash vitrification treatment needs just Na silicate as a binder at a 10% weight rate for hazardous sludge of the Mazıdağı Plant of more than 220,000 tons/year.

3.3 Microwave dissolution of vitrified sludge

This work was carried out a leaching method of hazardous sludges under microwave-radiated digestion of vitrified product briquette of Şırnak asphaltite and shale mixture for heavy metal sorption and reduction following vitrification. Şırnak fly ash, coal char and the shale may be used in vitrification as absorbant even improves low acidic digestion.

To evaluate the impact of shale on five different size fractions of the treatment with microwave and heated for 3 minutes at 500°C Sirnak shale samples, vitrified briquette dissolution and fe dissolution rate were determined as % the rate of vitrified matter as the efficiency of vitrification success by weight of briquette.

Dissolved Fe concentrates of sludges in the microwave interaction are illustrated in **Figure 6** as the Şırnak asphaltite slime and wood char from below 0.5 mm size fraction was observed.

The compaction of sludge at 15% wetness is becoming advantageous as mentioned below:

- wet compaction provides the prompted compaction sliding of wet liquid by adhesing the solid matter over the wetness of 15% grade,

- compaction loading extrusion molds, where the high-load intrusion to helical drive the forming rope style briquettes,

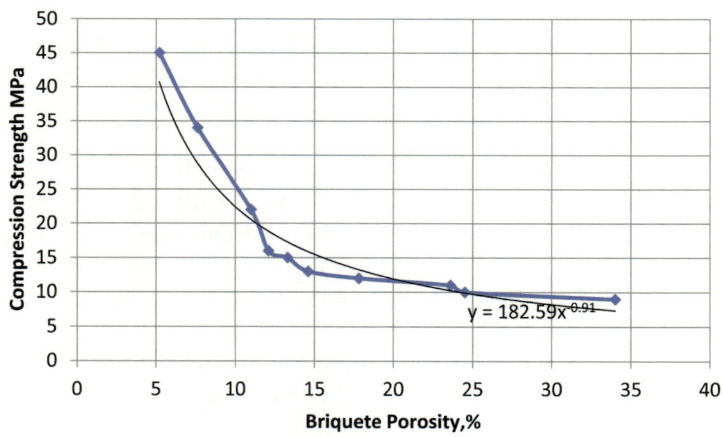

Figure 5.
The optimum binder porosity of compost by increased solvent use as volume weight of the briquette containing 1% kerosene.

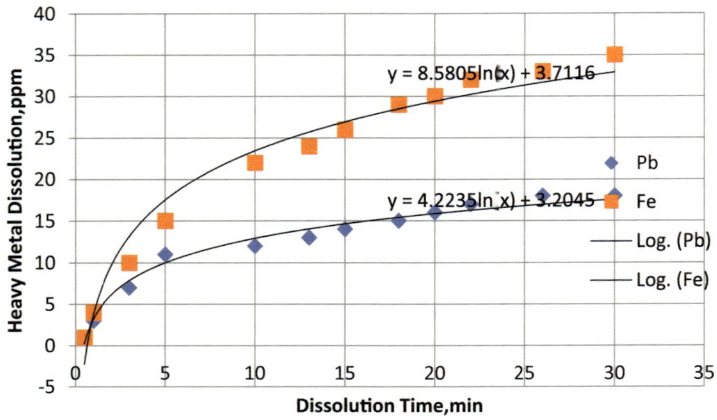

Figure 6.
The dissolution of Fe and Pb content-maximum ppm at 1L volume efflluent, 10g weight of the briquette sample containing 15 solid rate digestions.

- Microwave treatment or polymer extrusion at a high temperature of at least 200°C in sludges to be able drive extrusion,
- Less operating and control cost during extrusion.

Oxide melting ability was so effective in metal silicate formation manner. The lower-temperature activities of metal salts also improve melted metal silicate crystallization as efficient criteria in the vitrification hazard glazing of sludge grains including certain salt content. The oxidation and digestion effect of vitrified sludge briquette in nitric acid solution is seen in **Figure 6**.

As seen in **Figure 6**, the 5 N HNO_3 solutions show the contamination change, high level precipitated suspensions obtained using the precipitation-siphoning technique, depending on the salt concentration added at 10 g-100 mg.

4. Conclusions

The slag type and fluxing matter of vitrified Na silicate waste briquettes may be evaluated as aggregates in concrete or as filler in masonry products or as concrete debris use in foundations. The special vitrified waste briquettes may be evaluated in colored glass bottle production.

The phosphate-contained vitrified matters may be evaluated in the green house sands or as sand in the soil remediation of the local agricultural land. In the landfill deposition for hazardous-type wastes with weak vitrified sludges, avoiding costing of disposal asphalt or bituminous tar type binding is beneficial for inhibiting to solute contact in the waste dumping so that decaying with acidic solutions or digesting of sludge associated with environmental act is not permitted.

Basic Na silicate vitrification by 15% weight rate is sufficient for vitrification of acidic waste sludges with a filling carbon source in a weight rate of 15%. The carbon source of asphaltite slime and oak wood char decrease the digestion of heavy metals into the nitric acid solution from vitrified texture. The longer vitrification time over 30 minutes produced better strength of over 40 MPa compressive strength for 15% Na silicate added hazardous sludge and 15% carbon filler source of Şırnak asphaltite slime.

Following 900 W microwave radiation at 20–22 min type microwave laboratory equipment, the strength of vitrified briquettes as the quality was tested. The strength increased from original values of 22 MPa to 10, 4 and 1% for Şırnak asphaltite slime, Şırnak shale and Sewage sludge. The sufficient microwave duration of 30 min showed a much efficient sintering method for hazardous sludge vitrification as designed.

Zn is decomposed as the original form of $Zn_5(OH)\ 2(CO3)$ on vitrification grain surface to Na ZnO silicate form causing higher strength evolution and dissolution process is negligible, bearing in acid-digesting medium.

Heap leaching applications for gold and copper productions in the area are used for planning the feasibility reports regarding environmental contamination showing some degree of contamination and certain collection pools and seepage area will be highly contaminated by atmospherical dry conditions. In the pH measurements made, the pH value of 5.3–6.3 in washing hazardous waste sludges finally at the last washing pound to 6.3, depending on the sulfate and heavy metal of salt sludge content, was observed.

Soil washing and chelate-decayed solvents tend to destroy the soil profile and should be performed to recover metals from heavily polluted industrial sites and in case, no other methods can be applied. *In situ* tar or asphalt barrier layers near the aquifer are a very promising technique for the soil protection and the aquifer may not be toxified even sludges with toxic heavy metals.

Microwave vitrification using sodium silicate binder at 15% weight rate with 15% carbon filler decreased heavy metal flow in seepages of 5N nitric acid solutions with complete elimination to below 1ppm levels as shown reductions in Pb and Fe at 237% performance. In vitrified dissolution, effluent had the 24 ppm Fe and 5 ppm Pb values, in which Pb reduction rates of sorption at Langmuir model with 5N nitrate washed were 0.01 ppm/min.l Pb and total Fe reduction rate 0.03 ppm/min.l, respectively.

Author details

Yildırım İsmail Tosun
Engineering Faculty, Mining Engineering Department, Şırnak University, Şırnak, Turkey

*Address all correspondence to: yildirimismailtosun@gmail.com

References

[1] Leverenz H. Source reduction: Quantity and toxicity: Part 6a, quantity reduction. In: Kreith F, Tchobanoglous G, editors. Handbook of Solid Waste Management. 2nd ed. New York: McGraw Hill; 2002

[2] Geiser K. Source reduction: Quantity and toxicity: Part 6b, toxicity reduction. In: Kreith F, Tchobanoglous G, editors. Handbook of Solid Waste Management. 2nd ed. New York: McGraw Hill; 2002

[3] Anonymous. Transforming Waste Into Glass: Vitrification Processes to Stabilize Radioactive Waste. Paris: VEOLIA; 2021 Available from: https://www.nuclearsolutions.veolia.com/en/our-expertise/solutions/transforming-waste-glass-vitrification-processes-stabilize-radioactive

[4] Anonymous. EPA 625 V-92-002, Handbook: Vitrification Technologies for Treatment of Hazardous and Radioactive Waste. Washington, DC: United States Environmental Protection Agency, Office of Research and Development; 1992

[5] Brenniman GR. Other special wastes: Part 11e, computer and other electronic solid waste. In: Kreith F, Tchobanoglous G, editors. Handbook of Solid Waste Management. 2nd ed. New York: McGraw Hill; 2002

[6] Diaz LF. Composting of municipal solid wastes. In: Kreith F, Tchobanoglous G, editors. Handbook of Solid Waste Management. 2nd ed. New York: McGraw Hill; 2002

[7] Brunner CR. Waste-to-energy combustion: Part 13a, incineration technologies. In: Kreith F, Tchobanoglous G, editors. Handbook of Solid Waste Management. 2nd ed. New York: McGraw Hill; 2002

[8] Shah KL. Basics of Solid and Hazardous Waste Management Technology. Chicago: Prentice Hall; 1999

[9] Pichtel J. Waste Management Practices: Municipal, Hazardous, and Industrial. 2nd ed. Boca Raton: CRC Press; 2014

[10] U.S. Environmental Protection Agency. Quality Criteria for Water. Washington, DC: Government Printing Office; 1976

[11] Van Kessel JS, Reeves JB III, Meisinger JJ. Nitrogen and carbon mineralization of potential manure components. Journal of Environmental Quality. 2000;**29**:1669-1677

[12] Anonymous. Agricultural and Forest Developement Bureau of Mardin City. Mardin: Agricultural and Forest Developement Bureau of Mardin City; 2021 Available from: https://www.tarimorman.gov.tr/Duyuru/1478/Kirsal-Kalkinma-Destekleri-Kapsaminda-Tarima-Dayali-Ekonomik-Yatirimlarin-Desteklenmesi-Ve-Kirsal-Ekonomik-Altyapi-Yatirimlarinin-Desteklenmesi-2021-2022-Basvuru-Donemi-Uygulama-Rehberi-Yayinlandi

[13] Anonymous. Water Source Report. Mardin: Agricultural and Forest Developement Bureau of Mardin City; 2021 Available from: https://www.tarimorman.gov.tr/

[14] Kanagamani K, Geethamani P, Narmatha M. Hazardous waste management. In: Sarvajayakesavalu S, Charoensudjai P, editors. Environmental Issues and Sustainable Development. London: IntechOpen; 2020. DOI: 10.5772/intechopen.94080

[15] LaGrega MD, Buckingham PL, Evans JC. Land disposal. In: Hazardous Waste

Management. 2nd ed. Illinois: Waveland Press; 2001. pp. 812-864

[16] Singh IB, Chaturvedi K, Mochhale RK, Yegneswaran AH, 2007. Thermal treatment of toxic metal of industrial hazardous waste withfly ash and clay. Journal of Hazardous Materials 141(1); 215-222.

[17] Fatta D, Canna-Michaelidou S, Michael C, Georgious ED, Chritodoulidou M, Achilleos A, Vasquez M, 2007. Organochlorine and organophosphoric insecticides, herbicides and heavy metal residue in industrial waste waters in cyprus. Journal of Hazardous Materials 145(1-2): 169-179.

[18] Anonymous. EC Regulation No 1013/2006 of the european parliament and of the council of 14 june 2006 on shipments of waste. Official Journal of the European Union. 2006:1-98

[19] Anonymous. Handbook: Vitrification Technologies for Treatment of Hazardous and Radioactive Waste. Cincinati, Ohio: EPA, National Service Center for Environmental Publications (NSCEP); 1992

[20] Jouan A, Boen R, Merlin S, Pujadas V. New development for medium and low level waste vitrification. In: Nuthos-5, Beijing, 14–18 April 1997. 1997

[21] Ojovan MI, Lee WE. Self sustaining vitrification for immobilisation of radioactive and toxic waste. Glass Technology. 2003;**44**:218-224

[22] Karlina OK, Varlakova GA, Ojovan MI, Tivanski VM, Klimov VL, Pavlova GY, et al. Ash and soil conditioning using exothermic metallic compositions. Materials Research Society Symposium Proceedings. 2001;**663**:65-70

[23] European Council. European Waste Catalogue, Council Decision 2000/532/EC. Official Journal of the European Communities. 2000;**L226**:3-24

[24] Räbiger K, Keldenich K, Scheffer J. Experience in operation of a pilot plant melting residual substances. Glass Science and Technology. 1995;**68**:84-90

[25] Frugifer P, Godon N, Vernaz E, Larché F. Influence of composition variations on the initial alteration rate of vitrified domestic waste incineration fly-ash. Waste Management. 2002;**22**: 137-142

[26] Piepel G, Redgate T. Mixture techniques for reducing the number of components applied for modeling waste glass sodium release. Journal of American Ceramic Society. 1997;**80**: 3038-3044

[27] Bingham PA, Hand RJ. Vitrification of toxic wastes: A brief review. Advances in Applied Ceramics. 2006;**105**(1):21-31. DOI: 10.1179/174367606X81687

[28] Lee WE, Ojovan MI, Stennett MC, Hyatt NC. Immobilisation of radioactive waste in glasses, glass composite materials and ceramics. Advances in Applied Ceramics. 2006;**105**(1):3-12. DOI: 10.1179/174367606X81669

[29] Bingham PA, Connelly AJ, Hyatt NC, Hand RJ. Corrosion of glass contact refractories for the vitrification of radioactive wastes: A review. International Materials Reviews. 2011; **56**(4):226-242. DOI: 10.1179/1743280410Y.0000000005

[30] Kikuchi R. Vitrification process for treatment of sewage sludge and incineration ash. Journal of the Air & Waste Management Association. 1998; **48**(11):1112-1115. DOI: 10.1080/10473289.1998.10463766

[31] Terrones-Saeta JM, Suárez-Macías J, Castañón AM, Corpas-Iglesias FA. Evaluation of copper leaching for subsequent recovery from the waste dumps of the Linares mining district and their use for construction materials. Metals. 2021;**11**(8):1328. DOI: 10.3390/met11081328

[32] Calaminus B, Stahlberg B. Continuous in-line gasification/vitrification process for thermal waste treatment: Process technology and current status of projects. Waste Management. 1998;**18**(6-8):547-556

[33] Hetland J, Lynum S, Santen S. Thermo-chemical multi-processing of solid waste materials featuring recovery of gas heat, metals and minerals with complete volume reduction and residue stabilization. In: Proceedings of the 8th Conference on Environmental Science and Technology, Lemnos Island, Greece, 8–10 September 2003. New York: Curran Associates Inc.; 2003

[34] Bonizzoni G. Design of a plasma torch for toxic waste treatments. In: Benocci R, Bonizzoni G, Sindoni E, editors. Proceedings of the International School of Plasma Physics workshop on the Thermal Plasmas for Hazardous Waste Treatment. Varenna, Italy; 1995

[35] Paul SF. Review of the thermal plasma research and development for hazardous waste remediation in the United States. In: Benocci R, Bonizzoni G, Sindoni E, editors. Proceedings of the International School of Plasma Physics workshop on the Thermal Plasmas for Hazardous Waste Treatment. Varenna, Italy; 1995

[36] Colombo P, Brustain G, Bernardo E, Scarinci G. Inertization and reuse of waste materials by vitrification and fabrication of glass-based products. Current Opinion in Solid State and Materials Science. 2003;**548**(7):225-239

[37] Cubas ALV, Machado MDM, Machado MDM, Gross F, Magnago RF, Moecke EHS, et al. Inertization of heavy metals present in galvanic sludge by DC thermal plasma. Environment Science & Technology. 2014;**48**(5):2853-2861

[38] Cubas ALV, Machado MDM, Machado MDMM, Dutra ARDA, Moecke EHS, Fiedler HD, et al. Final treatment of spent batteries by thermal plasma. Environment Management. 2015;**159**:202-208

[39] Du CM, Chao S, Gong XJ, Ting W, Xiange W. Plasma methods for metal recovery from metal-containing waste. Waste Management. 2018;**77**(2018):373-387

[40] Du CM, Huang Y, Gong XJ, Wei XG. Plasma purification of halogen volatile organic compounds. IEEE Transaction on Plasma Science. 2018;**46**(4):838-858

[41] Haque K. Microwave energy for mineral treatment Processes-A brief review. International Journal of Mineral Processing. 1999;**57**(1):1-24

[42] Jacob J, Chia LHL, Boey FYC. Review-thermal and non-thermal interaction of microwave radiation with materials. Journal of Materials Science. 1995;**30**(21):5321-5327

[43] Kelly RM, Rowson NA. Microwave the reduction of oxidised ilmenite concentrates. Minerals Engineering. 1995;**8**(11):1427-1438

[44] Kingman SW, Vorster W, Rowson NA. The influence of mineralogy on microwave assisted grinding. Minerals Engineering. 1999;**13**(3):313-327

[45] Tosun YI. Chemical activated coal grinding of Turkish coals by acidic mine water. In: Geo 14th Conference of SGEM 2014. Bulgaria: SGEM; 2014. pp. 853-857. DOI: 10.5593/sgem2014b13

[46] Veasey TJ, Fitzgibbon K. Thermally assisted liberation-A review. Minerals Engineering. 1990;**3**(1/2): 181-185

[47] Walkiewicz JW, Kazonich G, McGill SL. Microwave heating and compounds: Characteristics of selected minerals. Minerals and Metallurgical Processing. 1988;**5**(1):39-42

[48] Walkiewicz JW, Clark AE, McGill SL. Microwave assisted grinding. IEEE Transactions on Industry Applications. 1991;**27**(2):239-243

[49] Allen SJ, Whitten LJ, Murray M, Duggan O. The adsorption of pollutants by peat, lignite and activated chars. Journal of Chemical Technology & Biotechnology. 1997;**68**(1997):442-452

[50] Álvarez-Ayuso E, Nugteren HW. Purification of chromium(VI) finishing wastewaters using calcined and uncalcined Mg-Al-CO3-hydrotalcite. Water Research. 2005;**39**(2005): 2535-2525

[51] Gregg SJ, Sing KSW. Adsorption, Surface Area and Porosity. London: Academic Press; 1982. p. 52

[52] Lopez-Gonzalez JD, Deitz VR. Surface changes in an original and activated bentonite. Journal of Research of the National Bureau of Standards. 1952;**48**(4):325

[53] Srasra E, Bergaya F, van Damme H, Arguib NK. Surface properties of an activated bentonite. Decolorization of rape-seed oil. Applied Clay Science. 1989;**4**:411-421

[54] Reedy CR, Nagendrappa G, Prakash BSJ. Surface acidity study of Mn+-montmorillonite clay catalysts by FT-IR spectroscopy. Catalysis Communications. 2007;**8**(3):241-246

[55] Rodriguez MAV, Barrios MS, Gonzalez JDL, Munoz MAB. Acid activation of a ferrous saponite (griffithite): Physico-chemical characterization and surface area of the products obtained. Clays and Clay Minerals. 1994;**42**:724-730

[56] Tabak A, Afsin B. Firmly adsorbed ammonia and pyridine species at activated kaolinite surfaces. Adsorption Science & Technology. 2001;**18**:673-679

[57] Tanabe K, Hölderich W. Industrial application of solid acid-base cataysts. Applied Catalysis A: General. 1999;**181**: 399-434

Chapter 7

Heavy Metals in Cosmetics

Tamara Attard and Everaldo Attard

Abstract

This review provides a comprehensive insight into the content of five heavy metals found in cosmetics and their effects at the site of application and on several organs via the dermal route of administration. Regulation of these products is very limited with little information on limits of these metals at the disposition of manufacturers. Lead and arsenic are considered to be contaminants in cosmetics whereas cadmium may be present as a coloring agent besides being a contaminant. Nickel is notorious for its association with allergic chronic dermatitis. Though mercury is rarely present in cosmetics, it has been found in significant quantities in skin lightening creams. A multi-variate meta-analysis was conducted to determine the relationships between the five metals and any possible differences between the sixteen categories of formulations used for facial and body skin care and cosmetic purposes. In general, lipsticks, eye shadows, face paints, make-up foundation and skin lightening creams exhibited a high amount of heavy metals superior to the levels of these metals in other facial and body products. The outcome of this analysis urges regulators and manufacturers to consider routine monitoring for the presence of these metals in cosmetics.

Keywords: lead, cadmium, nickel, mercury, arsenic, metallokinetics, metallodynamics

1. Introduction

A cosmetic is any product that is intended to be applied superficially to the human body to keep the treated part in good health. In this process, the cosmetic should not alter the physiological functioning of the body [1]. The use of cosmetics has been practiced since antiquity as apart from cleansing, cosmetics also beautify and alter the appearance hence making the individual more appealing and attractive. There are a plethora of ingredients that were and are used in the formulation of cosmetics. The general intended purposes of cosmetics have not changed throughout centuries and millennia, the formulation of these cosmetics has undergone significant transformations, some of which include the processing and the ingredients used for their formulation. The processing has changed from a domestic/small scale production to cater for a small number of individuals to industrial/large scale production to cater for a wider consumer population with the use of worldwide advertising and social media. On the other hand, the constitution of cosmetics has also changed with time. With industrialization and the use of petrochemically-derived substances, the cosmetic formulation changed from one based on natural products to one which is mainly based on petrochemicals. However, during these past decades, there was a change in the constitution of cosmetics, to include more natural ingredients, due to the great interest and concern by consumers.

As a consequence, the massive production of cosmetic products with a wide range of ingredients, has raised several health and safety concerns. Nowadays, cosmetics are generally regulated [2]. One major concern is that cosmetics overlap in use and functionality with topical medicines. Several regulatory bodies attempted to devise a proper definition for cosmetic products with the intent to segregate cosmetic products from topical medicinal products. Within the European Union (EU), Council Directive 93/35/EEC [3] amending Council Directive 76/768/EEC, a definition for cosmetic products was laid down in article 1 of the directive. In the first part of the definition, the external body parts which may be treated with cosmetics are mentioned. Other body parts are excluded and this eludes to the understanding that cosmetic products should not be applied to these other body parts. The second part is related to the 'activities' which are allowed for a product to be considered as a cosmetic. These distinguish cosmetics from topical medicinal products which are intended for the control or treatment of conditions or else in making a medical diagnosis [4]. However, whereas topical medicinal products are meticulously scrutinized before their placement on the market, cosmetics do not undergo rigorous testing. Nevertheless, for cosmetic products, the manufacturers, distributors and importers are responsible for the safety of cosmetic products being placed on the market [5]. The latter regulation also states clearly what ingredients are prohibited for their presence in cosmetic products. Amongst the prohibited ingredients several heavy metals are also included. Whereas some metals and their salts are completely prohibited (e.g., tin, arsenic, cadmium, nickel and lead), other metals and their salts are either allowed with a specific limit or else only specific salts for such metals are allowed (e.g., cobalt, chromium, gold, mercury and selenium amongst others). Such additions may not be intentional as the addition of some minerals may originate from a natural source. Heavy metals, such as cadmium (Cd), lead (Pb), nickel (Ni), arsenic (As) and mercury (Hg) were also detected in numerous other raw materials which can be used for the production of cosmetics considered as natural products. These include honey [6], argan oil [7], and olive oil [8] as well as citrus essential oils [9].

Because of this, some authorities also impose limits on the presence of certain metals in cosmetics. For example, The Cosmetic Ingredient Review Expert Panel established by Food and Drug Administration (FDA) in the USA issued limits on As (5 ppm), Pb (5 ppm) and other heavy metals (20 ppm) [10]. The World Health Organisation (WHO) set limits for Pb (10 ppm), Cd (0.3 ppm) and Hg (1 ppm). The EU's limits for Pb, Cd and chromium are 0.5, 0.5 and 1.0 ppm, respectively, while the Canadian authorities set limits for Pb (10 ppm), Cd (3 ppm) and Hg (3 ppm) [11]. However, there tends to be inconsistency in the type of metals and the limits for the metals by different authorities. As a consequence, this lack of harmonization leads to confusion amongst authorities as regulators, several manufacturers as producers and the general population as consumers. Despite all this, several researchers investigated the potential presence of heavy metals in a wide range of products. This review aims at compiling a large number of studies related to the presence of heavy metals in cosmetics and their potential harm in human beings.

2. Heavy metals

Heavy metals are elements that are primarily found in the d and p-blocks of the periodic table showing a metallic character and an ability to form salts. Some sources specify that heavy metal should have a high density [12]. Such metals include lead, cadmium, nickel, mercury and arsenic amongst others.

The main concern is that heavy metals are ubiquitous and are present in several matrices both living and non-living. Particularly in living matrices, heavy metals may interfere with beneficial metals some of the latter being replaced by heavy metals that would result in the erratic physiological functioning of bodily systems. This is not an issue related only to mammals and humans, but such erratic behavior, in terms of morbidity and mortality has been observed in other animals, such as insects [13], and also in plants. Some of these heavy metals accumulate in biological systems and one source for such accumulation is the daily and/or repeated use of cosmetic products such as face powders, lipsticks and eye shadows [2]. Some heavy metals are commonly found in cosmetics [14–20].

3. Metallokinetics and metallodynamics within the body

The absorption, distribution, metabolism, excretion and interaction of heavy metals with bodily systems are complex processes that are not yet fully understood. This is even more complex when considering that some cosmetics are applied and rinsed shortly after (such as toothpastes, shampoos and conditioners and cleansers), others are applied and allowed for a few minutes to hours (such as body creams, lotions and facial makeup) and those that are applied and remain in contact with the skin for several hours (such as nail polish and hair dyes). The kinetics and dynamics of metals present in these cosmetic products, vary significantly in their fate and their extent of effects [21].

With the application of cosmetic products, the mode of entry of heavy metals in the body is via dermal or topical application. These metals may have either topical and/or systemic effects in humans [22]. At the site of application, heavy metals may accumulate in the stratum corneum causing local effects, that may be exhibited as allergic contact dermatitis associated with an excess of a metal (such as Ni, cobalt and chromium) at the site [23, 24] because of their binding to keratin [22]. In the case of Ni, this metal has a high affinity to the histidine component of the filaggrin in the stratum corneum [25]. Although, the mechanism by which heavy metals damage the skin is not yet fully understood, it is believed that this may be provoked by the formation of free radicals and/or by an inflammatory effect [26]. It was discovered that metals may accumulate with repeated application of contaminated cosmetics. Consequently, such areas would serve as reservoirs and hence lead to long-term exposure of the individual to the metal, even when such cosmetics are no longer applied [27]. The exposure of the skin to heavy metals may extend the dermal inflammation to the systemic system [26]. Ni can only penetrate and reach the general circulation, when damage is provoked by other metals [28]. Other metals such as Hg, Pb and Cd may enter the general circulation through the skin layers and transported to various organs within the body [29, 30]. This permeation can occur via sweat glands and hair follicles. Besides direct contact, oral ingestion of heavy metals may occur by the application of cosmetic products either to the lips (e.g., lipstick, lip gloss and lip balms) or within the buccal cavity (toothpastes, mouthwashes and breathe sprays) or by hand to mouth transfer of any cosmetic applied to any body part [31]. Thinner facial skin is more permeable than skin elsewhere [32]. Several studies show that certain heavy metals (such as Pb, Hg and Cd) found in topical cosmetics are found in high concentrations in the blood, urine and internal organs of individuals who use cosmetics when compared to individuals who do not [33–40]. Cosmetics that are applied directly to the skin may contain moisturizing agents that increase skin permeation that may allow the entry of xenobiotics, to which some heavy metals may be bound, into the general circulation [41]. Apart from the binding of heavy metals to exogenous substances [42], these may also bind to endogenous biological molecules, hence replacing the beneficial metals. These metals may bind to several functional groups such as the amine, carboxylic and thiol function groups present in several proteins, some of which have functional roles (such as enzymes), while others have structural roles (such as collagen, keratin, actin and myosin). These metals can also bind to nucleic acids which may lead to defective DNA and RNA synthesis that may result in carcinogenesis. The application of underarm products has been hypothetically associated with the possibility of breast cancer [43]. Therefore, heavy metals provoke several toxic effects at the cellular and molecular levels [44, 45].

Several authors reported the presence of heavy metals in cosmetic products. This review gives an insight into the presence and effects of the most notorious and underestimated heavy metals in cosmetics products. The metals under discussion are lead, cadmium, nickel, mercury and arsenic.

4. The presence of typical heavy metals in cosmetics and their effects

4.1 Lead

One of the most studied heavy metals is lead. Lead is not normally used for its potential properties but it is rather considered as a contaminant with serious effects on human health. When a lead comes in contact with vital organs, it is neurotoxic, nephrotoxic and hepatotoxic [46, 47] and may provoke effects also on the reproductive system [48]. Lead can also affect fetal development through its passage via the placenta [49, 50]. Some studies have shown that it is considered a potential carcinogen to humans [51]. It has been reported that the level of Pb in the blood of consumers who use eye cosmetics was threefold higher than that of non-consumers [52]. Lead is acquired from industrial dust and fumes, car emissions, industrial chemicals such as old paints and pesticides, and a burning of fossil fuels. Food contamination may occur from some of these sources. Authorities worldwide are in a continuous struggle to establish permissible limits for Pb. The World Health Organization established a limit of 10 ppm [53]. The permissible level according to [54] is 0.1 mg/l. The FDA established a maximum permissible content of 10 ppm for Pb in color additives for the manufacture of cosmetics using Good Manufacturing Practices [10]. However, in color additives, the Pb content should not exceed 20 ppm [55]. Lead and its salts are prohibited in any cosmetic product within the EU [5]. Health Canada established a limit of 10 ppm for lead in cosmetic products [11].

As shown in **Tables 1–3**, several researchers investigated the presence of lead in several cosmetic products. The lipstick group is one of the most widely investigated groups with over fifteen citations. Only one study reported the absence of Pb in lipsticks [67, 70] whereas three other studies reported negligible Pb content in lipsticks [56–58] as their lower limit. Four studies showed a significantly high lead content (73.1–3760 ppm) [14, 16, 18, 57]. Most studies reported a Pb content that is within the 20 ppm permissible limit established by the FDA [59]. Some studies also tried to establish any differences between the high- and low-priced lipsticks (0.06–0.106 ppm) [60]. Eyeshadows ranks second in terms of investigated groups for Pb content. Four studies reported negligible Pb content in eyeshadows [31, 71–73] as their lower end. Whereas some studies have reported low Pb content as their lower end, the

	Pb	Cd	Ni	Hg	As
Lipsticks	<DL-252.4 [14, 18, 45, 56–66] 0.27–3760 [16]	ND-60.20 [14, 18, 45, 56, 58, 60–69]	ND-22.8 [18, 45, 56, 60–62, 64, 66, 70]	<DL to 80.00 [58, 61–63]	0.01–6.931 [58, 62, 63]
Eye-shadows	<DL-81.5 [16, 19, 31, 58, 61, 62, 65, 71–74]	<DL-55.59 [19, 58, 61, 62, 65, 71, 72, 74]	< 0.5–359.4 [19, 31, 61, 62, 71, 72, 74]	<DL-181.00 [58, 62]	<DL-1630 [58, 62, 75]
Eyebrow pencils and eye liners	0.109–61.22 [58, 59, 62, 63]	ND-1.12 [58, 62, 63]	2.1–10.52 [62]	ND-67.42 [58, 62, 63]	ND-2.071 [58, 62, 63]
Mascaras	ND-12.51 [58, 59, 73]	ND-0.034 [58]	ND-0.028 [76, 77]	ND-0.002 [58]	0.050–1.656 [58]
Make-up foundation	<DL to 190 [60, 61, 63, 64, 78]	<DL to 17 [45, 60, 61, 63, 64]	<DL to 13.1 [60, 61, 64]	48.99–60.77 [63]	0.12–1.0 [63]
Face paints	0.02–370 [79]	0.01–19.2 [79]	7.6 ppm [79]	ND-0.004 [69]	0.125.0 [79]
Face cream	ND-1.9 [62]	ND-0.37 [14, 62]	ND [70]	ND-1.27 [62]	ND-0.171 [62]
Toothpaste	ND-18.092 [21, 80, 81]	ND-2.490 [80–82]	0.025–18.535 [80–82]	ND-13.14 [81, 83]	0.06–26.94 [83]

Table 1.
The content of heavy metals in face products; cosmetics and face care products.

	Pb	Cd	Ni	Hg	As
Body lotions	<DL to 47.5 [61, 62]	ND [62]	ND-0.003 [62]	<DL to 47.5 [61, 62]	ND-0.007 [62]
Hair shampoos and conditioners	0.66–54.56 [14, 62]	ND [14, 62]	0.01–0.06 [62]	ND-21.08 [62, 83]	0.002–0.2 [62, 84]
Cleansers	0.04–22.14 [14, 62]	ND [14]	ND-0.08 [62]	ND-0.72 [62]	ND-0.009 [62]
Lotions	0.068–8.29 [64]	0.007–2.13 [64]	0.012–6.29 [64]	18.98–19.02 [83]	1.537–1.543 [83]
Hair dyes and creams	0.402–17.70 [17, 63, 64]	0.001–1.11 [63, 64]	0.081–4.167 [64]	53.74–90.32 [63]	0.16–0.71 [63]
Tonic creams	0.35–0.55 [85]	0.35–0.55 [85]	3.40–4.70 [85]	—	—
Beauty cream	14.38–50.39 [14, 86]	2.40–6.27 [86]	0.0175–5.09 [87, 88]	47.17–124.8 [83]	5.08–10.74 [83]

Table 2.
The content of heavy metals in body products.

higher end exceeded the 20 ppm limit established by FDA [16, 19, 61, 71, 74]. Other eye products include eyebrow pencils (0.109–18.60 ppm) [58, 62, 63] and mascaras (ND-12.51 ppm) [58, 59, 73], all within the 20 ppm limit established by the FDA.

	Pb	Cd	Ni	Hg	As
Skin lightening creams	<DL-143 [21, 60, 61, 64, 89]	0.1–1.276 [60, 64, 89]	2.59–11.17 [60, 64]	<DL-126,000 [15, 53, 61, 70, 89–94]	0.7–12.30 [53, 89]
Sunblock cream	ND-6.889 [62, 64]	ND-0.155 [62, 64]	ND-12.37 [62, 64]	ND-1.62 [62]	ND-0.01 [62]

Table 3.
The content of heavy metals in face and body products.

There is only one study that reported Pb levels of 61.218 ppm in mascaras [59]. Face products vary significantly in their Pb content. Make-up foundation and face powders contain negligible Pb content up to 190 ppm as reported by [61]. Other studies did not exceed a content of 22.57 ppm [60, 63, 64], although one study reported a maximum of 41 ppm [78]. Face washes exceed the FDA limit (24.06–40.61 ppm) [14] and face creams contain minimal content of Pb (0.77 ± 1.13 ppm) [62]. Cosmetic face paint is a potential threat to frequent consumers as reported levels go up to 16.6 ppm [79]. Most hair products do not pose a potential problem as reported for hair shampoos, conditioners and dyes [17, 62–64]. Only one study reported a level of 54.56 ppm in hair products [14]. Several studies investigated body products that are applied over a larger surface area than those mentioned previously. Beauty creams contain a considerable amount of Pb with levels reaching 50.39 ppm [14, 86]. Although, somebody lotions contain low Pb levels [62], one study shows also high Pb content in such products [61]. This same study also reports a high Pb level in skin-lightening creams (up to 43.04 ppm) alongside another study showing a maximum level of 143 ppm [89]. On the other hand, two studies show low Pb contents (<4.015 ppm) [60, 64]. Cleansers and lotions also contain low Pb levels (< 22.14 ppm) [14, 62, 64]. Products that are applied to the buccal cavity include toothpaste. These may pose a problem in addition to mucosal absorption, Pb can be also ingested and absorbed via the gastrointestinal tract. In some studies, the level of Pb in toothpaste is minimal (0.036 ppm) [80], however, in other studies, the highest levels were 12.04 ppm [81] and 18.092 ppm [21].

Lead is considered as a contaminant that is present to different extents in various cosmetic preparations. Several authorities recognize Pb as a toxic metal by setting up limits for its presence in cosmetics.

4.2 Cadmium

Cadmium is one of the metals that has been used in cosmetics for its colored salts, ranging from deep yellow to orange [2]. It has been associated with several toxicities in humans, mainly attributed to its absorption after topical application of several cosmetics [49, 86, 95] though this is very low (0.5%). Topically, it may cause irritant dermatitis [96]. The main concern with Cd is that it tends to accumulate in human tissues and then release slowly into the general circulation. However, it normally binds to the keratin. Systemically, it mainly affects the skeletal, reproductive, metabolic [88], respiratory and renal systems [97, 98]. It has been associated with osteoporosis, diabetes, lung cancer and kidney damage [99]. It contributes also to skin ageing as it may provoke oxidative stress [25]. Despite of its presence in cosmetics, it may be found in several sources such as industrial wastes, agrochemicals (pesticides and fertilizers) and batteries. According to the WHO, the permissible limit for cadmium is 0.3 ppm [53]. The permissible level according to [54] is 0.06 mg/l. The oral limit for Cd is 0.09 µg/kg to 3 ppm as given by USP for nutritional supplements. Cadmium and its salts are prohibited in any cosmetic product within the EU [5]. Health Canada established a limit of 3 ppm for Cd in cosmetic products [11].

Cadmium is another metal, which is banned by several authorities but still found in several cosmetics, as reported by several researchers (**Tables 1–3**). One of the studies that reports the absence of Cd in lipsticks was conducted by [67]. Several other studies reported levels that were lower than 0.002 ppm and levels that reached content of 60.20 ppm. Despite this, only two studies showed levels of Cd higher than 5 ppm [65, 68]. Due to possible ingestion, lipstick use may be linked to systemic toxicity. A study established a difference between high-priced (0.34 ± 0.20 ppm) and low-priced (0.89 ± 0.58 ppm) lipsticks [60]. Although, eye shadows may contain the deep yellow to orange pigments more than lipsticks, the Cd content of eye shadows as reported by several researchers did not exceed 3 ppm [19, 58, 62, 63, 71, 72, 74]. However, levels of 8.89 ppm [71] and 55.59 ppm [65] were reported for eye shadows. Another eye make-up was reported to contain low Cd content by three studies. For eyebrow pencils, the maximum content of Cd was 1.12 ppm [63], whereas the content of Cd in mascaras was reported to be lower, i.e., 0.034 ppm [58]. Other facial cosmetics were also reported to contain a low amount of Cd, with levels less than 0.96 ppm [45, 60, 63, 64]. The highest level was expressed to be that of 17 ppm [61] in the make-up foundation. Similarly, Cd levels did not exceed 0.67 ppm and 0.37 ppm in face washes and face creams, respectively [14]. Tonic creams did not exceed a level of 0.55 ppm [85]. In face paints, the average level reported was 0.6 ppm [79]. Cadmium was also absent in hair shampoos and conditions, with levels being below the detection limit [14, 62]. Negligible Cd content was also present in hair dyes with levels not exceeding 1.11 ppm [63, 64]. Body cosmetics are also generally low in Cd with levels not exceeding 0.92 ppm in skin lightening creams [60, 64, 89], 0.121 ppm in sunblock creams [62, 64] and 2.13 ppm in lotions [64]. No cadmium was detected in body lotions [62] and cleansers [14]. The level of Cd in most toothpastes did not exceed 0.058 ppm [80, 81] but a maximum of 2.49 ppm was reported [82].

Although, Cd is considered one of the most common and noxious heavy metals, it seems that its use in cosmetics is very limited. It may be concluded that Cd presence in cosmetics is considered to be a contaminant rather than an ingredient, as a coloring agent.

4.3 Nickel

Nickel is one of the metal impurities which is inevitably found in several natural ingredients used in cosmetic products. Most of the salts containing nickel are green in color, hence also its potential use as a colorant. However, nickel is considered to be a contact allergen that may provoke dermal sensitization, allergies, and dermatitis [100] by direct and often prolonged exposure. Persons have been diagnosed with Ni allergy due to its presence in topical cosmetic products and jewelry [101]. Nickel may also affect the respiratory system which may result in nasal and lung cancer [98]. Despite of the potential use of cosmetics to maintain the skin in a rejuvenating state, Ni in such products may lead to oxidative stress and hence promote skin aging [25]. This may be due to the overexpression of collagenases in the skin leading to the weakening of the skin matrix and a subsequent loss in elasticity [102]. The International Agency for Research on Cancer (IARC) has classified metallic Ni as a potential carcinogen to humans (Group 2B) and its compounds as carcinogenic (Group 1) [103]. Nickel may be found occurring naturally in soil and volcanic dust. This may be acquired from industrial dust and fumes. Due to the potential skin sensitization, limits for Ni presence in products have been proposed. Limits of 5 ppm [88] and 1 ppm [104] were suggested for certain household products and detergents, respectively. Likewise, in cosmetics, a Ni limit was also proposed particularly aimed for sensitized persons. Most "nickel-free" products on the market, contain less than 1 ppm of Ni [31, 105, 106]. The permissible level is 0.20 ppm according to [107] for oral consumption. Nickel and a number of its salts are prohibited in any cosmetic product within the EU [5]. These include nickel monoxide, dinickel trioxide, nickel dioxide, trinickel disulphide, tetracarbonynickel, nickel sulphide, nickel dihydroxide, nickel carbonate and nickel sulphate.

Due to its possible implications in allergenic reactions, the content of Ni in a number of cosmetic products was reported by a number of research groups (**Tables 1–3**). The risks associated with Ni intoxication is more possible with cosmetics that are potentially ingested. Lipsticks and lip products are amongst these candidates. Most lipstick products investigated by research groups rarely contained less than 0.20 ppm of Ni for oral consumption. Due to the short-term duration of lipstick on the lips, these products are applied frequently by consumers. This may pose a further exacerbation if ingested. The maximum levels in most studies range from 1.61 to 22.8 ppm of Ni in lipsticks [18, 45, 56, 61, 64, 66, 70]. However, in one study it was reported that the mean Ni content was 0.10 ± 0.14 ppm [62]. It was demonstrated that the price has no impact on Ni content of lipsticks (high-priced 8.24 ± 3.29 ppm and low-priced 5.15 ± 4.19 ppm) [60]. Oral consumption may be due to the accidental swallowing of toothpaste. Studies have shown the range of Ni content in most toothpaste is between 0.02 and 2.54 ppm [80, 81] but another study reported maximum levels of 18.535 ppm [82]. Dermal sensitization has been associated with eye cosmetic products. In this situation the 1 ppm threshold is applicable. Several studies have reported levels of Ni which exceed 1 ppm. Only two studies show that the minimum level of Ni in eye-shadows was less than 1 ppm [19, 31]. In several studies, maxima for Ni levels ranged between 4.133 and 359.4 ppm [61, 62, 71, 72, 74]. Nickel has been found in a green eye liner which provoked a form of contact dermatitis in a 47-year-old woman [76] whereas another study reported contact allergy to a Ni-containing mascara [77]. A study reported a mean Ni content of 6.31 ± 4.21 ppm in eyebrow pencils [62]. Most make-up foundation products seem to contain high amounts of Ni, quoting the minimum values above 3 ppm and the maximum values to 13.01 ppm [60, 64]. Only one study reported values being less than the detection limit [61]. In a study on face paints, the average Ni content was 7.6 ppm [79]. However, in other facial formulations such as face washes and creams, the content was reported to be very low with a mean of 0.04 ± 0.11 ppm [14] or not detectable [70] for these formulations respectively. Hair products are of no major concern, as the levels in shampoos and conditions do not exceed 0.06 ppm whereas the highest content of Ni in hair dyes is 4.167 ppm [64]. Body products vary in Ni content. In general, Ni does not exceed 12.37 ppm, but the 1 ppm of Ni is exceeded for most products that include skin lightening creams, sun blocks, tonic creams and body creams [60, 64, 85, 87, 88]. Body lotions and cleansers seem to contain very low Ni contents (<0.08 ppm) [62].

Whereas Ni in mascaras and eye shadows has been implicated in its involvement in allergic chronic dermatitis, several studies have reported that Ni allergy cannot be considered as the main risk factor in patients reporting eye-lid dermatitis [108].

4.4 Mercury

Mercury is one of the heavy metals that is widely used in cosmetic formulations. Although, mercury is known as a shiny, silvery, dense liquid, it may occur in various inorganic and organic compounds. In the inorganic form, such as ammoniated Hg, it is used for its skin lightening properties, whereas in the organic form, such as phenyl mercuric and ethyl mercuric salts, it is used as a preservative in mascaras and eye makeup cleansing products [90, 109]. After dermal application, Hg penetrates through the skin via the hair follicles and sweat glands [29, 96]. During this process, part of the Hg is reduced to the metallic form that accumulates in the skin tissue. Hg blocks tyrosinase in situ, inhibiting the melanin-forming enzyme [110], hence its use in skin-lightening creams [22]. Only one study mentioned that cream had the abbreviation 'precip blanc' on the label, which should suggest that this product contains Hg [36]. Systemically, Hg may exhibit a range of signs which include vomiting, nausea and kidney damage, central nervous sytstem effect which include irritability, tremors, weakness, nervousness, fatigue and memory loss. It may affect also the sensorial systems, that is, loss in hearing, taste and vision. Finally, high Hg content may lead to death [111, 112]. In some instances, following dermal absorption and systemic uptake, Hg may induce autoimmune glomerulonephritis. Studies reveal significant accumulation of mercury in several organs and body fluids, such as hair (22.5 ppm, twice that in non-cosmetic users), blood (up to 233 nmol/l, more than four times than that in non-cosmetic users] and urine (up to 2531 nmol/day, fifty times more than that in non-cosmetic users) [39, 91, 113, 114].

Mercury is a metallic element that is naturally occurring in the environment and its compounds are the most common form that exists naturally in the environment. Due to its ubiquity, several authorities issued limitations for Hg use. For instance, the FDA restricts its use and is regulated in cosmetic products. The FDA allows a maximum level of 1 ppm of Hg in mercury-contaminated lead acetate when used as a colour in cosmetics [10]. Within the European Union, mercury and its compounds are not allowed in cosmetics, whereas phenyl mercuric salts are only allowed as preservatives in eye care products at a maximum allowable level of 70 ppm [5] whereas in the US it is allowed up to a level of 65 ppm by weight [10]. Health Canada allows a maximum Hg content of 1 ppm in cosmetics [11].

The main emphasis of analysis by researchers was conducted on skin-lightening creams due to their interaction with melanin metabolism. Very few studies report levels below the 3 ppm threshold [15, 53, 70] (**Tables 1–3**). Other studies reveal values up to 126,000 ppm [61, 89–93]. Although, in some lipstick products, the Hg content was below the detection limit [58, 61, 62], some of these same studies and others reveal contents up to 80 ppm [63]. The status of Hg contamination in eye cosmetic products varies significantly with levels of up to 181 pm in eye shadows [58, 62], 67.42 ppm in eyebrow pencils [58, 62, 63] but levels of up to 0.002 ppm in mascaras [58]. Apart from the make-up foundation where levels of Hg reach a maximum level of 60.77 ppm [63], other face products contain minimal amounts of Hg, such as face creams (0.09 ± 0.37 ppm [62]), sunblock creams (0.41 ± 1.21 ppm [62]) and face paints (<0.004 ppm [69]). In some studies, it was reported that the content of Hg in hair products is below the detection limit [62] whereas in other studies, it reached a maximum of 90.32 ppm [63, 83]. Body care products such as body lotions and cleansers contain varied amounts of Hg [62], with products from the first group with contents up to 47.5 ppm [61]. In some beauty creams, the Hg level reached a maximum of 124.8 ppm [83]. The presence of mercury in toothpaste has not been widely investigated, but in two studies, the Hg level was reported to reach a maximum of 13.14 ppm [81, 83].

As can be concluded from these studies, Hg is one of the least detected heavy metals in most cosmetics [22], but not in skin-lightening products. Mercury is found intentionally in face and skin care products rather than in products for purely cosmetic use. Apart from skin-lightening properties, Hg compounds are claimed to reduce and remove freckles, treat acne, and prevent and remove wrinkles [22].

4.5 Arsenic

Arsenic is a metalloid that is present ubiquitously as a major contaminant in the environment. Although, it is redox inactive, its target functional groups are sulfydryl groups on proteins which may lead to the depletion of glutathione [115], an essential antioxidant of an amino acid origin, which prevents damage of cellular components caused by radicals and heavy metals. On long-term dermal exposure, As can cause hyperpigmentation and keratosis *in situ*, but systemically it may lead to carcinogenesis and vascular diseases [115, 116]. Though considered as a less significant contaminant in cosmetics than other heavy metals, authorities sought to establish limits for its presence in cosmetic products due to long-term exposure. The problem with As contamination goes beyond the legal cosmetic products placed on the market. It has been found in high quantities in cosmetic products from the underground market [58]. As with Hg, the FDA established a limit for As (up to 3 ppm) for lead acetate, as a colorant, contaminated with this metalloid [10]. This acceptable maximum limit goes also under the general limits set by Health Canada for all cosmetics [11]. Arsenic and its salts are prohibited in any cosmetic product within the EU [5].

In general, several studies reveal that As is not a significant contaminant and levels of this metalloid rarely exceed the 3 ppm limit (**Tables 1–3**). In lipsticks, although most studies report a level of up to 0.34 ppm [62, 63], a study reported a maximum level of 6.931 ppm of As [58]. In eye cosmetics, the maximum permissible limit is rarely exceeded. In general, eye shadows, eyebrow pencils and mascaras do not contain As levels more than 3.704, 2.071 and 1.656 ppm, respectively [58, 62, 63]. However, in a study, it was reported that in kohl, there was an alarming presence of As (810–1630 ppm). Kohl is an ancient eye cosmetic still used nowadays [75]. The danger with kohl is not solely because of its use as a cosmetic but there are claims that falsely indicating its use for the treatment of eye conditions. Apart from the presence of As in kohl, other face cosmetics, such as foundations and creams contain minimal quantities of As, reported as up to 1.0 and 0.171 ppm, respectively [62, 63]. Less commonly used products include face paints particularly used by opera actors in China. Levels of As reach a maximum of 25 ppm [79]. Shampoos, conditioners and dyes used on the hair contain low As concentrations (<0.71 ppm) [62, 63, 94]. Likewise, in a study, the As content for cleansers and sunblocks does not exceed 0.010 ppm [62] but higher levels were reported for body lotions (1.543 ppm) [83]. There are some concerns with skin-lightening creams as some exceed the 3 ppm threshold [53, 89] and other creams with levels up to 10.74 ppm [83]. In some toothpaste, the As content was 26.94 ppm [83]. Most likely, As is present as a contaminant with other heavy metals used for this purpose. In spite of these findings, As is one of those elements that is rarely found in cosmetics [22]. However, its presence may raise concerns particularly in legal products that are used on a long-term basis and in illegal cosmetic products on the underground market.

5. General considerations

The studies discussed in the previous sections highlight the importance of specific metals as contaminants and additives in cosmetic products. To determine any particular relationships between formulations, a multi-variate meta-analysis was carried out using Spearman correlation and Principal Component Analysis, taking into account the maximum levels obtained for the various cosmetic formulations. Pearson correlation statistics (**Table 4**) reveal a relationship between all five metals ($r > 0.466$). Two latent factors had an eigenvalue greater than 1, which together explained 80.54% of the total variance. The factor loadings demonstrated the different groups of variables (**Figure 1**). Factor 1, displayed on the horizontal axis, weighed heavily on Pb, Cd, Ni and As with lipsticks, eye shadows, face paints, make-up foundation and skin

Variables	Cd	Ni	Hg	As
Pb	0.538	0.495	0.527	0.579
Cd		0.779	0.334	0.750
Ni			0.446	0.641
Hg				0.465

Table 4.
Spearman correlation matrix for the five metals.

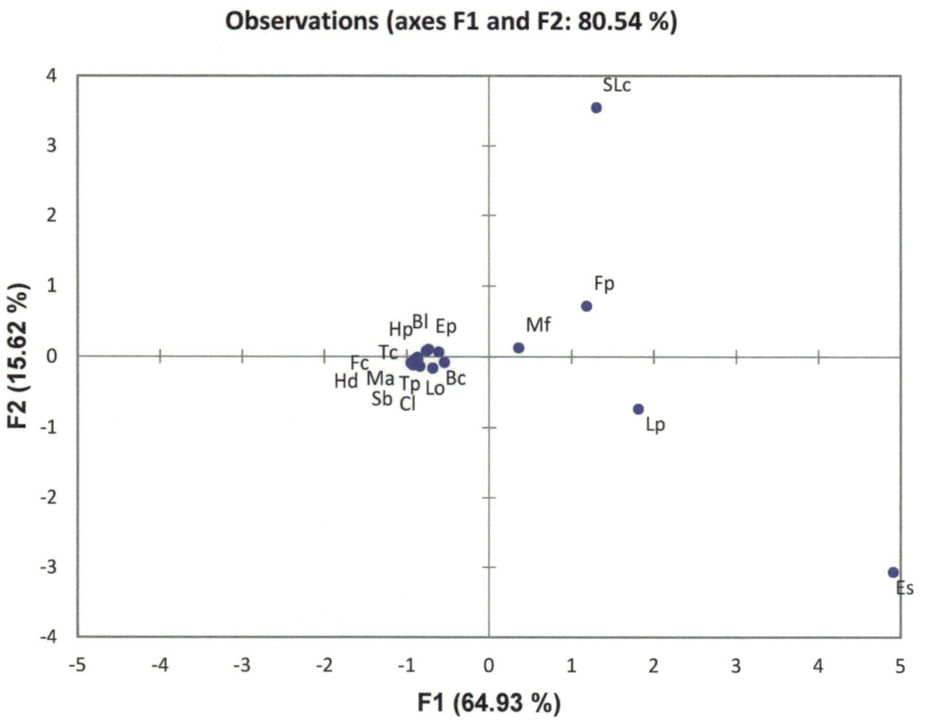

Figure 1.
Observations plot for the formulations. Legend: Lp = lipsticks; Es = eye shadows; Ep = eyepencils; Ma = mascaras; Mf = foundation; Fp = face paint; Fc = face cream; Tp = toothpaste; SLc = skin-lightening creams; Sb = sunblock; Hp = hair products; Hd = hair dyes; Bl = body lotion; Cl = cleansers; Lo = lotions; and Tc = tonic creams.

lightening creams exhibiting high levels of these metals. These formulations were discriminately different from the rest. On the other hand, F2, displayed on the vertical axis, weighed heavily on Hg with skin lightening creams having superior quantities of this metal for the other formulations. This multi-variate analysis consolidates the findings from previous studies. The findings of such research works are at the disposition of authorities and policy makers for the formulation of high-quality cosmetic products.

Author details

Tamara Attard[1] and Everaldo Attard[2]*

1 Department of Pharmacy, Faculty of Medicine and Surgery, University of Malta, Msida, Malta

2 Division of Rural Sciences and Food Systems, Institute of Earth Systems, University of Malta, Msida, Malta

*Address all correspondence to: everaldo.attard@um.edu.mt

References

[1] Oyedeji FO, Hassan GO, Adeleke BB. Hydroquinone and heavy metals levels in cosmetics marketed in Nigeria. Trends in Applied Sciences Research. 2011;**6**:622-639. DOI: 10.3923/tasr.2011.622.639

[2] Safavi SO, Najarian RA, Rasouli-Azad M, Masoumzadeh SH, Ghaderi AM, Eghtesadi RA. A narrative review of heavy metals in cosmetics; health risks. International Journal of Pharmaceutical Research. 2019;**11**:182-190. DOI: 10.31838/ijpr/2019.11.04.031

[3] Council of the European Union. Council Directive 93/35/EEC of 14 June 1993 amending for the sixth time Directive 76/768/EEC on the approximation of the laws of the Member States relating to cosmetic products. Official Journal of the European Communities. 1993;**151**:33-36

[4] European Parliament and Council of the European. Directive 2001/83/EC of 6 November 2001 on the Community code relating to medicinal products for human use. Official Journal of the European Communities. 2001;**L311**:67-128

[5] European Parliament and Council of the European. Regulation (EC) No 1223/2009 of 30 November 2009 on cosmetic products. Official Journal of the European Union. 2009;**L342**:59-209

[6] Roman A, Madras-Majewska B, Popiela-Pleban E. Comparative study of selected toxic elements in propolis and honey. Journal of Apicultural Science. 2011;**55**(2):97-106

[7] Mohammed FA, Bchitou R, Boulmane M, Bouhaouss A, Guillaume D. Modeling of the distribution of heavy metals and trace elements in argan forest soil and parts of argan tree. Natural Product Communications. 2013;**8**:21-23. DOI: 10.1177/1934578X1300800105

[8] Brkljača M, Giljanović J, Prkić A. Determination of metals in olive oil by electrothermal atomic absorption spectrometry: Validation and uncertainty measurements. Analytical Letters. 2013;**46**:2912-2926. DOI: 10.1080/00032719.2013.814056

[9] La Pera L, Saitta M, Di Bella G, Dugo G. Simultaneous determination of Cd (II), Cu (II), Pb (II), and Zn (II) in citrus essential oils by derivative potentiometric stripping analysis. Journal of Agricultural and Food Chemistry. 2003;**51**:1125-1129. DOI: 10.1021/jf020804t

[10] United States Food and Drug Authorities, FDA's Testing of Cosmetics for Arsenic, Cadmium, Chromium, Cobalt, Lead, Mercury, and Nickel Content [Internet] 2020. Available from: https://www.fda.gov/cosmetics/potential-contaminants-cosmetics/fdas-testing-cosmetics-arsenic-cadmium-chromium-cobalt-lead-mercury-and-nickel-content [Accessed: December 13, 2021]

[11] Health Canada. Guidance on Heavy Metal Impurities in Cosmetics [Internet] 2012. Available from: https://www.canada.ca/en/health-canada/services/consumer-product-safety/reports-publications/industry-professionals/guidance-heavy-metal-impurities-cosmetics.html [Accessed: December 13, 2021]

[12] Al-Trabulsy HA, Khater AE, Habbani FI. Heavy elements concentrations, physiochemical characteristics and natural radionuclides

levels along the Saudi coastline of the Gulf of Aqaba. Arabian Journal of Chemistry. 2013;**6**:183-189. DOI: 10.1016/j.arabjc.2010.10.001

[13] Borg D, Attard E. Honeybees and their products as bioindicators for heavy metal pollution in Malta. Acta Brasiliensis. 2020;**4**:60-69. DOI: 10.22571/2526-4338282

[14] Salve KS, Sonwane NS. Detection of heavy metals in cosmetics. World Journal of Pharmaceutical Research. 2015;**4**:1368-1372

[15] Peregrino CP, Moreno MV, Miranda SV, Rubio AD, Leal LO. Mercury levels in locally manufactured Mexican skin-lightening creams. International Journal of Environmental Research and Public Health. 2011;**8**:2516-2523. DOI: 10.3390/ijerph8062516

[16] Al-Saleh I, Al-Enazi S, Shinwari N. Assessment of lead in cosmetic products. Regulatory Toxicology and Pharmacology. 2009;**54**:105-113. DOI: 10.1016/j.yrtph.2009.02.005

[17] Amartey EO, Asumadu-Sakyi AB, Adjei CA, Quashie FK, Duodu GO, Bentil NO. Determination of heavy metals concentration in hair Pomades on the Ghanaian market using atomic absorption spectrometry technique. British Journal of Pharmacology and Toxicology. 2011;**2**:192-198

[18] Nnorom IC, Igwe JC, Oji-Nnorom CG. Trace metal contents of facial (make-up) cosmetics commonly used in Nigeria. African Journal of Biotechnology. 2005;**4**:1133-1138. DOI: 10.5897/AJB2005.000-3224

[19] Volpe MG, Nazzaro M, Coppola R, Rapuano F, Aquino RP. Determination and assessments of selected heavy metals in eye shadow cosmetics from China, Italy, and USA. Microchemical Journal. 2012;**101**:65-69. DOI: 10.1016/j.microc.2011.10.008

[20] Bocca B, Pino A, Alimonti A, Forte G. Toxic metals contained in cosmetics: A status report. Regulatory Toxicology and Pharmacology. 2014;**68**:447-467. DOI: 10.1016/j.yrtph.2014.02.003

[21] Iwegbue CM, Bassey FI, Tesi GO, Onyeloni SO, Obi G, Martincigh BS. Safety evaluation of metal exposure from commonly used moisturizing and skin-lightening creams in Nigeria. Regulatory Toxicology and Pharmacology. 2015;**71**:484-490. DOI: 10.1016/j.yrtph.2015.01.015

[22] Borowska S, Brzóska MM. Metals in cosmetics: Implications for human health. Journal of Applied Toxicology. 2015;**35**:551-572. DOI: 10.1002/jat.3129

[23] Larese F, Gianpietro A, Venier M, Maina G, Renzi N. In vitro percutaneous absorption of metal compounds. Toxicology Letters. 2007;**170**:49-56. DOI: 10.1016/j.toxlet.2007.02.009

[24] Filon FL, D'Agostin F, Crosera M, Adami G, Bovenzi M, Maina G. In vitro absorption of metal powders through intact and damaged human skin. Toxicology in Vitro. 2009;**23**:574-579. DOI: 10.1016/j.tiv.2009.01.015

[25] Chavatte L, Juan M, Mounicou S, Leblanc Noblesse E, Pays K, Nizard C, et al. Elemental and molecular imaging of human full thickness skin after exposure to heavy metals. Metallomics. 2020;**12**:1555-1562. DOI: 10.1039/d0mt00121j

[26] Thyssen JP, Menné T. Metal allergy—A review on exposures, penetration, genetics, prevalence, and clinical implications. Chemical

Research in Toxicology. 2010;**23**:309-318. DOI: 10.1021/tx9002726

[27] Franken A, Eloff FC, Du Plessis J, Du Plessis JL. In vitro permeation of metals through human skin: A review and recommendations. Chemical Research in Toxicology. 2015;**28**:2237-2249. DOI: 10.1021/acs.chemrestox.5b00421

[28] Irvine AD, McLean WI, Leung DY. Filaggrin mutations associated with skin and allergic diseases. New England Journal of Medicine. 2011;**365**:1315-1327. DOI: 10.1056/NEJMra1011040

[29] Palmer RB, Godwin DA, McKinney PE. Transdermal kinetics of a mercurous chloride beauty cream: An in vitro human skin analysis. Journal of Toxicology: Clinical Toxicology. 2000;**38**(7):701-707. DOI: 10.1081/CLT-100102383

[30] Sun CC, Wong TT, Hwang YH, Chao KY, Jee SH, Wang JD. Percutaneous absorption of inorganic lead compounds. AIHA Journal. 2002;**6**:641-646. DOI: 10.1080/15428110208984751

[31] Sainio EL, Jolanki R, Hakala E, Kanerva L. Metals and arsenic in eye shadows. Contact Dermatitis. 2000;**42**:5-10. DOI: 10.1034/j.1600-0536.2000.042001005.x

[32] Feldman RJ. Regional variation in percutaneous penetration of 14C cortisol in man. Journal of Investigative Dermatology. 1967;**48**:181-183. DOI: 10.1038/jid.1967.29

[33] Warley MA, Blackledge P, O'gorman P. Lead poisoning from eye cosmetic. British Medical Journal. 1968;**1**:117

[34] Shaltout A, Yaish SA, Fernando N. Lead encephalopathy in infants in Kuwait: A study of 20 infants with particular reference to clinical presentation and source of lead poisoning. Annals of Tropical Paediatrics. 1981;**1**:209-215. DOI: 10.1080/02724936.1981.11748090

[35] Al MF. Lead encephalopathy in a Bahraini infant. Bahrain Medical Bulletin. 1993;**15**(3):109-111

[36] Özkaya E, Mirzoyeva L, Ötkür B. Mercury-induced systemic allergic dermatitis caused by 'white precipitate' in a skin lightening cream. Contact Dermatitis. 2009;**60**:61-63. DOI: 10.1111/j.1600-0536.2008.01460.x

[37] Li SJ, Zhang SH, Chen HP, Zeng CH, Zheng CX, Li LS, et al. Mercury-induced membranous nephropathy: Clinical and pathological features. Clinical Journal of the American Society of Nephrology. 2010;**5**:439-444. DOI: 10.2215/CJN.07571009

[38] Benz MR, Lee SH, Kellner L, Döhlemann C, Berweck S. Hyperintense lesions in brain MRI after exposure to a mercuric chloride-containing skin whitening cream. European Journal of Pediatrics. 2011;**170**:747-750. DOI: 10.1007/s00431-010-1333-1

[39] Chakera A, Lasserson D, Beck LH Jr, Roberts IS, Winearls CG. Membranous nephropathy after use of UK-manufactured skin creams containing mercury. QJM: An International Journal of Medicine. 2011;**104**:893-896. DOI: 10.1093/qjmed/hcq209

[40] Dickenson CA, Woodruff TJ, Stotland NE, Dobraca D, Das R. Elevated mercury levels in pregnant woman linked to skin cream from Mexico. American Journal of Obstetrics and Gynecology. 2013;**209**:e4-e5. DOI: 10.1016/j.ajog.2013.05.030

[41] Björklund S, Engblom J, Thuresson K, Sparr E. Glycerol

and urea can be used to increase skin permeability in reduced hydration conditions. European Journal of Pharmaceutical Sciences. 2013;**50**:638-645. DOI: 10.1016/j.ejps.2013.04.022

[42] Flora SJ, Pachauri V. Chelation in metal intoxication. International Journal of Environmental Research and Public Health. 2010;7:2745-2788. DOI: 10.3390/ijerph7072745

[43] Darbre PD. Underarm cosmetics and breast cancer. Journal of Applied Toxicology: An International Journal. 2003;**23**:89-95. DOI: 10.1002/jat.899

[44] Shanker AK. Mode of action and toxicity of trace elements. In: Prasad MNV, editor. Trace Elements: Nutritional Benefits, Environmental Contamination, and Health Implications. Chichester: Wiley; 2008. pp. 528-555. DOI: 10.1002/9780470370124.ch21

[45] Ullah H, Noreen S, Rehman A, Waseem A, Zubair S, Adnan M, et al. Comparative study of heavy metals content in cosmetic products of different countries marketed in Khyber Pakhtunkhwa, Pakistan. Arabian Journal of Chemistry. 2017;**10**:10-18. DOI: 10.1016/j.arabjc.2013.09.021

[46] Patrick L. Lead toxicity, a review of the literature. Part I: Exposure, evaluation, and treatment. Alternative Medicine Review. 2006;**11**:2-22

[47] Karri V, Schuhmacher M, Kumar V. Heavy metals (Pb, Cd, As and MeHg) as risk factors for cognitive dysfunction: A general review of metal mixture mechanism in brain. Environmental Toxicology and Pharmacology. 2016;**48**:203-213. DOI: 10.1016/j.etap.2016.09.016

[48] Chauhan AS, Bhadauria R, Singh AK, Lodhi SS, Chaturvedi DK, Tomar VS. Determination of lead and cadmium in cosmetic products. Journal of Chemical and Pharmaceutical Research. 2010;**2**:92-97

[49] Ayenimo JG, Yusuf AM, Adekunle AS, Makinde OW. Heavy metal exposure from personal care products. Bulletin of Environmental Contamination and Toxicology. 2010;**84**:8-14. DOI: 10.1007/s00128-009-9867-5

[50] Horowitz Y, Greenberg D, Ling G, Lifshitz M. Acrodynia: A case report of two siblings. Archives of Disease in Childhood. 2002;**86**:453. DOI: 10.1136/adc.86.6.453

[51] Sprinkle RV. Leaded eye cosmetics: A cultural cause of elevated lead levels in children. Journal of Family Practice. 1995;**40**:358-362

[52] Mohiuddin AK. Heavy metals in cosmetics: The notorious daredevils and burning health issues. American Journal of Biomedical Science and Research. 2019;**4**:332-337. DOI: 10.34297/AJBSR.2019.04.000829

[53] Sukender K, Jaspreet S, Sneha D, Munish G. AAS estimation of heavy metals and trace elements in Indian herbal cosmetic preparations. Research Journal of Chemical Sciences. 2012;**2**:46-51

[54] Singh R, Gautam N, Mishra A, Gupta R. Heavy metals and living systems: An overview. Indian Journal of Pharmacology. 2011;**43**:246. DOI: 10.4103/0253-7613.81505

[55] United States Food and Drug Authorities. Draft Guidance for Industry: Lead in Cosmetic Lip Products and Externally Applied Cosmetics: Recommended Maximum Level. FDA-2014-D-2275. [Internet]

2016. Available from: https://www.fda.gov/regulatory-information/search-fda-guidance-documents/draft-guidance-industry-lead-cosmetic-lip-products-and-externally-applied-cosmetics-recommended [Accessed: December 13, 2021]

[56] Liu S, Hammond SK, Rojas-Cheatham A. Concentrations and potential health risks of metals in lip products. Environmental Health Perspectives. 2013;**121**:705-710. DOI: 10.1289/ehp.1205518

[57] Brandão JDO, Okonkwo OJ, Sehkula M, Raselek ARM. Concentrations of lead in cosmetics commonly used in South Africa. Toxicological & Environmental Chemistry. 2012;**94**(1):70-77. DOI: 10.1080/02772248.2011.633911

[58] Saadatzadeh A, Afzalan S, Zadehdabagh R, Tishezan L, Najafi N, Seyedtabib M, et al. Determination of heavy metals (lead, cadmium, arsenic, and mercury) in authorized and unauthorized cosmetics. Cutaneous and Ocular Toxicology. 2019;**38**:207-211. DOI: 10.1080/15569527.2019.1590389

[59] Philip J, John S, Othman O. Levels of heavy metals in selected facial cosmetics marketed in Dar es Salaam, Tanzania. Tanzania Journal of Science. 2018;**44**:81-88

[60] Sani A, Gaya MB, Abubakar FA. Determination of some heavy metals in selected cosmetic products sold in kano metropolis, Nigeria. Toxicology Reports. 2016;**3**:866-869. DOI: 10.1016/j.toxrep.2016.11.001

[61] Ababneh FA, Al-Momani IF. Assessments of toxic heavy metals contamination in cosmetic products. Environmental Forensics. 2018;**19**:134-142. DOI: 10.1080/15275922.2018.1448908

[62] Lim DS, Roh TH, Kim MK, Kwon YC, Choi SM, Kwack SJ, et al. Non-cancer, cancer, and dermal sensitization risk assessment of heavy metals in cosmetics. Journal of Toxicology and Environmental Health, Part A. 2018;**81**:432-452. DOI: 10.1080/15287394.2018.1451191

[63] Nasirudeen MB, Amaechi AU. Spectrophotometric determination of heavy metals in cosmetics sourced from Kaduna Metropolis, Nigeria. Science World Journal. 2015;**10**:1-5

[64] Arshad H, Mehmood MZ, Shah MH, Abbasi AM. Evaluation of heavy metals in cosmetic products and their health risk assessment. Saudi Pharmaceutical Journal. 2020;**28**:779-790. DOI: 10.1016/j.jsps.2020.05.006

[65] Nourmoradi H, Foroghi M, Farhadkhani M, Vahid DM. Assessment of lead and cadmium levels in frequently used cosmetic products in Iran. Journal of Environmental and Public Health. 2013;**2013**:1-5. DOI: 10.1155/2013/96272

[66] Gao P, Liu S, Zhang Z, Meng P, Lin N, Lu B, et al. Health impact of bioaccessible metal in lip cosmetics to female college students and career women, northeast of China. Environmental Pollution. 2015;**197**:214-220. DOI: 10.1016/j.envpol.2014.11.006

[67] Vafaei F, Nouri G, Razi A. Spontaneous cholecystocutaneous fistulae: A case report. International Journal of Pharmaceutical Research. 2018;**10**:345-344

[68] Gondal MA, Seddigi ZS, Nasr MM, Gondal B. Spectroscopic detection of health hazardous contaminants in lipstick using laser induced breakdown spectroscopy. Journal of

Hazardous Materials. 2010;**175**:726-732. DOI: 10.1016/j.jhazmat.2009.10.069

[69] Hepp NM, Mindak WR, Cheng J. Determination of total lead in lipstick: Development and validation of a microwave-assisted digestion, inductively coupled plasma-mass spectrometric method. Journal of Cosmetic Science. 2009;**60**:405

[70] Sahu R, Saxena P, Johnson S, Mathur HB, Agarwal HC. Heavy metals in cosmetics. Centre for Science and Environment. 2014:1-28

[71] Omolaoye J, Uzairu A, Gimba C. Heavy metal assessment of some eye shadow products imported into Nigeria from China. Archives of Applied Science Research. 2010;**2**:76-84

[72] Faruruwa MD, Bartholomew SP. Study of heavy metals content in facial cosmetics obtained from open markets and superstores within Kaduna metropolis, Nigeria. American Journal of Chemistry and Application. 2014;**1**:27-33

[73] Al-Dayel O, Hefne J, Al-Ajyan T. Human exposure to heavy metals from cosmetics. Oriental Journal of Chemistry. 2011;**27**:1-11

[74] Świerczek L, Cieślik B, Matysiak A, Konieczka P. Determination of heavy metals in eyeshadows from China. Monatshefte für Chemie-Chemical Monthly. 2019;**150**:1675-1680. DOI: 10.1007/s00706-019-02467-7

[75] Zakaria A, Ho YB. Heavy metals contamination in lipsticks and their associated health risks to lipstick consumers. Regulatory Toxicology and Pharmacology. 2015;**73**:191-195. DOI: 10.1016/j.yrtph.2015.07.005

[76] Travassos AR, Bruze M, Dahlin J, Goossens A. Allergic contact dermatitis caused by nickel in a green eye pencil. Contact Dermatitis. 2011;**65**:307. DOI: 10.1111/j.1600-0536.2011.01960.x

[77] Karlberg AT, Liden C, Ehrin E. Colophony in mascara as a cause of eyelid dermatitis. Chemical analyses and patch testing. Acta Dermato-Venereologica. 1991;**71**:445-447

[78] Nnorom IC. Trace metals in cosmetic facial talcum powders marketed in Nigeria. Toxicological & Environmental Chemistry. 2011;**93**:1135-1148. DOI: 10.1080/02772248.2011.577075

[79] Wang B, Su Y, Tian L, Peng S, Ji R. Heavy metals in face paints: Assessment of the health risks to Chinese opera actors. Science of The Total Environment. 2020;**724**:138163. DOI: 10.1016/j.scitotenv.2020.138163

[80] Odukudu FB, Ayenimo JG, Adekunle AS, Yusuff AM, Mamba BB. Safety evaluation of heavy metals exposure from consumer products. International Journal of Consumer Studies. 2014;**38**:25-34. DOI: 10.1111/ijcs.12061

[81] Vella A, Attard E. Analysis of heavy metal content in conventional and herbal toothpastes available at maltese pharmacies. Cosmetics. 2019;**6**:28. DOI: 10.3390/cosmetics6020028

[82] Orisakwe OE, Okolo KO, Igweze ZN, Ajaezi GC, Udowelle NA. Potential hazards of toxic metals found in toothpastes commonly used in Nigeria. Roczniki Państwowego Zakładu Higieny. 2016;**67**:197-204

[83] Salama AK. Assessment of metals in cosmetics commonly used in Saudi Arabia. Environmental Monitoring and Assessment. 2016;**188**:1-1. DOI: 10.1007/s10661-016-5550-6

[84] Abdel-Fattah A, Pingitore NE. Low levels of toxic elements in Dead Sea black mud and mud-derived cosmetic products. Environmental Geochemistry and Health. 2009;**31**:487-492. DOI: 10.1007/s10653-008-9201-x

[85] Abata EO, Ogunkalu OD, Adeoba AA, Oluwasina OO. Evaluation of the heavy metals in tonic creams using the wet acid and dry ashing methods. Earthline Journal of Chemical Sciences. 2019;**1**:37-43. DOI: 10.34198/ejcs.1119.3743

[86] Alam MF, Akhter M, Mazumder B, Ferdous A, Hossain MD, Dafader NC, et al. Assessment of some heavy metals in selected cosmetics commonly used in Bangladesh and human health risk. Journal of Analytical Science and Technology. 2019;**10**:1-8. DOI: 10.1186/s40543-018-0162-0

[87] Theresa OC, Onebunne OC, Dorcas WA, Ajani OI. Potentially toxic metals exposure from body creams sold in Lagos, Nigeria. Researcher. 2011;**3**:30-37

[88] Bocca B, Forte G, Petrucci F, Cristaudo A. Levels of nickel and other potentially allergenic metals in Ni-tested commercial body creams. Journal of Pharmaceutical and Biomedical Analysis. 2007;**44**:1197-1202. DOI: 10.1016/j.jpba.2007.04.031

[89] Alqadami AA, Naushad M, Abdalla MA, Khan MR, Alothman ZA, Wabaidur SM, et al. Determination of heavy metals in skin-whitening cosmetics using microwave digestion and inductively coupled plasma atomic emission spectrometry. IET Nanobiotechnology. 2017;**11**:597-603. DOI: 10.1049/iet-nbt.2016.0212

[90] Al-Saleh I, Al-Doush I. Mercury content in skin-lightening creams and potential hazards to the health of Saudi women. Journal of Toxicology and Environmental Health. 1997;**51**:123-130. DOI: 10.1080/00984109708984016

[91] Chan MH, Cheung RC, Chan IH, Lam CW. An unusual case of mercury intoxication. British Journal of Dermatology. 2001;**144**:192-193. DOI: 10.1046/j.1365-2133.2001.03978

[92] Al-Ashban RM, Barratt DA, Shah AH. Mercury contents of skin-lightening creams marketed in Saudi Arabia. Journal of Saudi Chemical Society. 2006;**10**:383-388

[93] Sin KW, Tsang HF. Large-scale mercury exposure due to a cream cosmetic: Community-wide case series. Hong Kong Medical Journal. 2003;**9**:329-334

[94] Bogusz MJ, Al Tufail M, Hassan H. How natural are 'natural herbal remedies'? Adverse Drug Reactions and Toxicological Reviews. 2002;**21**:219-229. DOI: 10.1007/BF03256198

[95] Ahmadi-Jouibari T, Fattahi N, Mirzaei N, Sharafi K, Ghafari HR. Determination of cadmium in cosmetics from Kermanshah, Iran by graphite furnace atomic absorption spectrometry. New Journal of Chemistry. 2017;**41**:11948-11954. DOI: 10.1039/c7nj00406k

[96] Guy RH. Metals and the Skin: Topical Effects and Systemic Absorption. London: CRC Press; 1999

[97] Rebelo FM, Caldas ED. Arsenic, lead, mercury and cadmium: Toxicity, levels in breast milk and the risks for breastfed infants. Environmental Research. 2016;**151**:671-688. DOI: 10.1016/j.envres.2016.08.027

[98] Mahurpawar M. Effects of heavy metals on human health. International

Journal of Research-Granthaalayah. 2015;**3**:1-7. DOI: 10.29121/granthaalayah.v3.i9SE.2015.3282

[99] Alissa EM, Ferns GA. Heavy metal poisoning and cardiovascular disease. Journal of Toxicology. 2011;**2011**:1-21. DOI: 10.1155/2011/870125

[100] Tuchman M, Silverberg JI, Jacob SE, Silverberg N. Nickel contact dermatitis in children. Clinics in Dermatology. 2015;**33**:320-326. DOI: 10.1016/j.clindermatol.2014.12.008

[101] Halicz L, Portugal-Cohen M, Russo MZ, Robino F, Vanhaecke T, Rogiers V. Safety evaluation of traces of nickel and chrome in cosmetics: The case of Dead Sea mud. Regulatory Toxicology and Pharmacology. 2015;**73**:797-801. DOI: 10.1016/j.yrtph.2015.10.016

[102] Cammarota M, Lamberti M, Masella L, Galletti P, De Rosa M, Sannolo N, et al. Matrix metallo-proteinases and their inhibitors as biomarkers for metal toxicity in vitro. Toxicology in Vitro. 2006;**20**:1125-1132. DOI: 10.1016/j.tiv.2006.02.004

[103] Ahlström MG, Thyssen JP, Wennervaldt M, Menné T, Johansen JD. Nickel allergy and allergic contact dermatitis: A clinical review of immunology, epidemiology, exposure, and treatment. Contact Dermatitis. 2019;**81**:227-241. DOI: 10.1111/cod.13327

[104] Basketter DA, Briatico-Vangosa G, Kaestner W, Lally C, Bontinck WJ. Nickel, cobalt and chromium in consumer products: A role in allergic contact dermatitis? Contact Dermatitis. 1993;**28**:15-25. DOI: 10.1111/j.1600-0536.1993.tb03318.x

[105] Le Coz CJ, Leclere JM, Arnoult E, Raison-Peyron N, Pons-Guiraud A, Vigan M, et al. Allergic contact dermatitis from shellac in mascara. Contact Dermatitis. 2002;**46**:149-152. DOI: 10.1034/j.1600-0536.2002.460304.x

[106] Diepgen TL, Weisshaar E. Contact dermatitis: Epidemiology and frequent sensitizers to cosmetics. Journal of the European Academy of Dermatology and Venereology. 2007;**21**:9-13. DOI: 10.1111/j.1468-3083.2007.02381.x

[107] Amer MM, Sabry BA, Marrez DA, Hathout AS, Fouzy AS. Exposure assessment of heavy metal residues in some Egyptian fruits. Toxicology Reports. 2019;**6**:538-543. DOI: 10.1016/j.toxrep.2019.06.007

[108] Borghi A, Corazza M, Maietti E, Patruno C, Napolitano M, Schena D, et al. Eyelid dermatitis and contact sensitization to nickel: Results from an italian multi-centric observational study. Endocrine, Metabolic & Immune Disorders-Drug Targets (Formerly Current Drug Targets-Immune, Endocrine & Metabolic Disorders). 2019;**19**:38-45. DOI: 10.2174/1871530318666180731114418

[109] Ladizinski B, Mistry N, Kundu RV. Widespread use of toxic skin lightening compounds: Medical and psychosocial aspects. Dermatologic Clinics. 2011;**29**:111-123. DOI: 10.1016/j.det.2010.08.010

[110] Hostynek JJ. Factors determining percutaneous metal absorption. Food and Chemical Toxicology. 2003;**41**:327-345. DOI: 10.1016/S0278-6915(02)00257-0

[111] Agorku ES, Kwaansa-Ansah EE, Voegborlo RB, Amegbletor P, Opoku F. Mercury and hydroquinone content of skin toning creams and cosmetic soaps, and the potential risks to the health of Ghanaian women. Springerplus. 2016;**5**:1-5. DOI: 10.1186/s40064-016-1967-1

[112] Barr RD, Woodger BA, Rees PH. Levels of mercury in urine correlated with the use of skin lightening creams. American Journal of Clinical Pathology. 1973;**59**:36-40. DOI: 10.1093/ajcp/59.1.36

[113] Soo YO, Chow KM, Lam CW, Lai FM, Szeto CC, Chan MH, et al. A whitened face woman with nephrotic syndrome. American Journal of Kidney Diseases. 2003;**41**:250-253. DOI: 10.1053/ajkd.2003.50017

[114] Tang HL, Mak YF, Chu KH, Lee W, Fung SK, Chan TY, et al. Minimal change disease caused by exposure to mercury-containing skin lightening cream: A report of 4 cases. Clinical Nephrology. 2013;**79**:326-369

[115] Gibb H, Haver C, Gaylor D, Ramasamy S, Lee JS, Lobdell D, et al. Utility of recent studies to assess the National Research Council 2001 estimates of cancer risk from ingested arsenic. Environmental Health Perspectives. 2011;**119**:284-290. DOI: 10.1289/ehp.1002427

[116] Bhattacharjee P, Chatterjee D, Singh KK, Giri AK. Systems biology approaches to evaluate arsenic toxicity and carcinogenicity: An overview. International Journal of Hygiene and Environmental Health. 2013;**216**:574-586. DOI: 10.1016/j.ijheh.2012.12.008

Section 2

Evaluation of Heavy Metals and Their Risks to Irrigation Water

Chapter 8

Concentration Levels of Heavy Metals and Selected Ions in the Irrigation Water: The Case of Little Akaki River, Addis Ababa, Ethiopia

Mekonnen M. Tarekegn and Getaneh L. Weldekidan

Abstract

Irrigable water resources have been challenged by the contamination of heavy metals and unwanted ions that impair plant growth and human health. It impaired the quality of edible fruit & vegetables. The objective of this study was to determine the concentration of heavy metals (Pb, Cd, Cr, and Cu) and selected ions (chloride, Na, Mg, Ca), and to evaluate its suitability for irrigation use. Water samples were collected from three various locations (the upstream, middle stream, and downstream) of the river. Temperature (T), EC, pH, and total dissolved solids (TDS) were tested onsite using handheld multiparameter testing equipment, while the heavy metals (Pb, Cd, Cr, and Cu) and selected ions (Chloride, Na, Mg, and Ca) of the water sample were analyzed using (ICP-OES). ANOVA test was conducted to examine the concentration variations of heavy metals and selected ions between the sampling locations along the river. The concentrations of (Pb and Cd) were found (2.5–3.9), (0.03–0.4) mg/L respectively, and (Cr and Cu) were below the detectible limit of the (ICP-OES) equipment. Cadmium (Cd) was found to be higher than the permissible limit of FAO (0.01mg/L) for irrigation water. There was no significant variation of heavy metals and selected ions between the sampling locations.

Keywords: heavy metals, selected ions, sodium adsorption risks (SAR)COD, physic-chemical parameters, irrigation water quality

1. Introduction

Inappropriately managed urbanization and expansion of industrialization are the major causes of river water pollution in urban and pre-urban areas due to the introduction of undesirable materials into soils and irrigation water sources [1]. Contamination of heavy metals and other toxic ions in irrigation water sources is a worldwide problem and harmful for human health and the ecosystems. The excessive accumulation of heavy metals in irrigation water and soils resulted in contamination of human diets [2]. Heavy metals are entering into the river water and the environment primarily through anthropogenic activities. The main sources of heavy metals and other pollutants entering into the Little Akaki River basin can be industrial effluent, municipal solid waste, oily wastes from garages, and fuel stations. Industries like textile, dyeing, garment, pharmaceutical, ceramic, paint, packaging, etc. discharge their effluents into the rivers could be the causes of heavy metal contamination in the irrigation water sources [3]. Industry expansion has brought severe water pollution in Little Akaki catchment from domestic, commercial, and industrial effluents [4] and the waste management systems of industries and other commercial centers are very poor. According to Addis Ababa environmental pollution Authority 2007 report, 90% of all industries lack waste treatment facilities and subsequently dispose of their effluents into river streams. Lack of proper waste management system in the catchment areas, the irrigation water source in Little Akaki River is highly polluted with point and non-point waste sources.

The rapid urbanization and expansion of industries together with poor effluent management systems have a significant effect on the quality of irrigation water sources in the catchment areas. In the recent few decades, the social and economic structure of Addis Ababa city has changed radically. Rapid urbanization and industries expansion are observed and all other economic activities are also highly concentrated in Addis Ababa city, particularly in the Little Akaki River catchment. Besides the urbanization and industries expansion, the intensification of pre-urban and urban farming activities is also becoming one of the other social and economic features of the city. However, most of these rapid changes are brought without considering the negative environmental consequences. As a result, irrigation water pollution with heavy metals and other undesirable pollutants become an increasingly emphasized problem. Wastes generated from households, industries, fuel stations, hospitals, different business centers, and sewerages are getting into the river streams. Contamination of water bodies from various sources denies current and future generations of a birthright and puts at risk ecological integrity [5].

Little Akaki River is the primary irrigation water source for pre-urban and urban agriculture in the catchment area. The demand for irrigation water is markedly increasing in the study area for the production of fruits and vegetables. Many households are involved in urban farming activities to sustain their life. The use of industrial and municipal wastewater in urban agriculture is a common practice in many parts of the world including Ethiopia [6]. The shortages of safe irrigation water sources in the study area forced the farmers to look for to use contaminated river water for irrigated agriculture and access to quality irrigation water sources is becoming a serious concern these days in the study catchment.

The heavy metals and other pollutant elements are entering the soil because using severely contaminated irrigation water source for prolonged periods in the catchment area affect the physical and chemical characteristics of the soil. The contaminated soil of irrigated farm sites in the catchment area harms vegetables and fruit production. The heavy metals become highly concentrated in the edible parts of fruits and vegetables which alter human health. Heavy metal accumulation in soils, and subsequently, in vegetation by long-term wastewater irrigation has a potentially detrimental effect on humans via their transfer along the food chain [7]. In the existing situation, vegetables and other edible crop products produced in contaminated soil are distributed in the local market of Addis Ababa city. Residents are consuming the infected vegetables subsequently by purchasing from the local market and farmers also use the contaminated vegetables for their home consumption before going to market.

Few studies have been conducted so far in Addis Ababa city particularly in Little Akaki River to investigate the contamination levels of Little Akaki River irrigation water with heavy metals [8–12] but it was still inadequately researched. Because the heavy metal contamination and irrigation water pollution is a very dynamic problem and become progressively increasing. The intensification of industrialization in the Little Akaki River catchment aggravated the progression of river water contamination with heavy metals and toxic pollutants over time. Thus, the dynamism of the problem and the need for updated information about heavy metal contamination are the triggering points for the initiative of this research work. Determination of the existing heavy metals and selected ions is a relatively newer issue or insight to provide possible suggestions. Therefore, this research has focused to investigate the concentration level of heavy metals (Pb, Cd, Cr, and Cu), selected ions (Chloride, Na, Ca, and Mg), and other physic-chemical parameters of the Little Akaki River and to evaluate the suitability of the river water for irrigation uses.

2. Materials and methods

The area description, materials and methods, water sample collection and laboratory testing procedures, and method of data analysis are discussed in detail in this chapter.

2.1 Description of the study area

This study was conducted in Addis Ababa city, particularly in the Little Akaki River basin. The Little Akaki River basin is located in the western part of Addis Ababa and the river flows starting from northwest upstream of the city about 40 km before it reaches the downstream of the Aba Samuel reservoir which is indicated in **Figure 1**.

Little Akaki River is highly contaminated with industry effluents and other different anthropogenic activities. According to the Addis Ababa city administration industry development commission report, more than 667 different sized industries are found in Addis Ababa city and the distributions of manufacturing industries are relatively higher in the Little Akaki River catchment area than in other parts of the city. Most of the streams/tributaries flow from the Northwestern side of the catchment area meets Little Akaki River at Gullele area where different industries are found. Gullele Soap and Marble factories, Awash Winery, National Alcohol and liquor factory are found in Little Akaki River catchment around, Lideta and Mekanisa areas [8]. He also explained that tributaries that come from the Northwestern direction also receive wastes from abattoirs. These different industries release their effluents into the river stream thereby adversely influencing the irrigation water quality.

2.2 Water samples collection and sample preparation

Three water samples were collected from three purposively selected sampling locations from the upstream, middle, and downstream of Little Akaki River on March 2021. All three water samples are collected on a similar day of local time (at morning 4:00 am). The sampling points are selected by considering the different variation factors along the river stream and collecting representative water samples. The collected

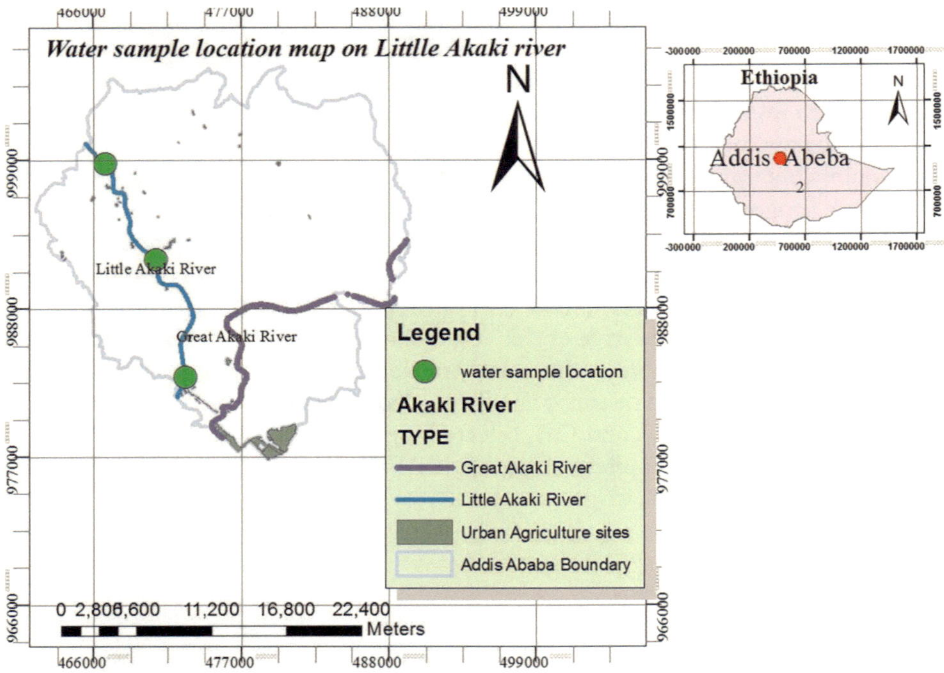

Figure 1.
Water sampling location map in little Akaki River.

river water samples were tested for the analysis of major heavy metals (Pb, Cd, Cr, and Cu), selected ions (Chloride, Na, Ca, and Mg), and other physic-chemical quality parameters of the irrigation water. Samples were collected with 500 mL plastic bottles from the representative flowing river water of medium velocity or free from any turbulence by dipping the bottles in the deeper mid-stream of the river flow to fill it to overflowing. The current weather conditions during sample collection were cleared the sky and sunny condition and the air temperature was ranges between 20 and 28°C.

2.3 Water sample laboratory testing procedures

Inductively Coupled Plasma Optical Emission Spectrometry (ICP-OES) was used to analyze the concentration level of heavy metals (Pb, Cd, Cr, Cu) and selected ions (chloride, Na, Ca, and Mg). Temperature (T), EC, pH, and TDS were tested onsite using handheld multiparameter testing equipment. After sample collection, the water samples were acidified with 10 ml of concentrated nitric acid and preserved in the refrigerator. The acidified water samples were digested to dissolve the organic matter and then the digested wastewater samples were analyzed for concentration levels of heavy metals (Pb, Cd, Cr, and Cu) and selected ions (chloride, Na, Ca, and Mg) by ICP-OES with ES ISO 11885:2007 testing standard procedures. In ICP-OES the samples are exposed to a high energy source such as inductively coupled plasma (ICP) at a temperature of 5000 to 10,000 K [13] and the elements emit light of a spectrum being characteristics of each element. The emitted light is collected by a photomultiplier and the intensity of emitted light is directly proportional to the concentration of elements within a sample.

Chemical oxygen demand (COD) was tested with American Public Health Association (APHA) 5220-B open reflux testing methods. Chemical oxygen demand (COD) is defined as the amount of a specified oxidant that reacts with the samples under controlled conditions [14]. Organic and inorganic components of samples were subject to oxidation [15]. The dichromate ion ($Cr_2O_7^{-2}$) is the specified oxidant in 5220-B testing methods. Wastewater samples were refluxed in a strongly acid solution for a minimum of two hours with a known excess of potassium dichromate ($K_2Cr_2O_7$). After digestion, the remaining un-reduced $K_2Cr_2O_7$ is titrated with ferrous ammonium sulfate (FAS) to determine the amount of $K_2Cr_2O_7$ consumed and the oxidized matter is calculated in terms of oxygen equivalence. The very important apparatus that has been used in COD testing was 150 ml Erlenmeyer flasks with ground-glass 24/40 neck and jacket Liebig or equivalent condenser with 24/40 ground glass joint and a hot plate having sufficient power to produce at least 1.4 W/cm^2 of heating surface or equivalence.

COD was can be calculated by Eq. (1)

$$\text{COD as mg } O_2/L = \frac{(A - B) * M * 8000}{\text{mL sample}} \tag{1}$$

Where:

A = mL FAS used for blank (volume of blank titrant).

B = mL FAS used for sample (volume of sample titrant).

M = molarity of FAS (Molarity of titrant).

8000 = mill equivalent weight of oxygen * 1000 mL/L.

The alkalinity of the wastewater samples was tested with APHA 2320-B titration methods. The alkalinity of water is explained by its acid-neutralizing capacity [16]. Bicarbonate, carbonate and hydroxide ions are the primary contributors to the alkalinity of water, other constituents such as borate, phosphates, or silicates may also contribute to alkalinity [17]. The alkalinity of the wastewater samples was determined from the volume of standard acid required to titrate a portion to a designated pH value. It was titrated at room temperature with a properly calibrated pH meter or electrically operated titration. The most important apparatus used for the alkalinity test were pH meter and electrode, magnetic stirrer, stir bar, Beaker, titration vessel, Burette, standard sulfuric acid titrant, Pipets volumetric, flasks volumetric.

Procedures: standardized sulfuric acid titrant solutions were prepared as required. The clean burette was filled with the standard acid titrant. Samples temperature is equilibrated with the room's temperature and transferred volumetrically enough samples to 100/150 mL beaker to provide the titrant good volumetric precision. The stir bar is placed in the beaker and the beaker is placed on the magnetic stirrer and inserts the pH electrode into the beaker. The sample pH is measured and the initial burette reading was recorded when sample pH is measured above 8.3. And standard acid titrant is added until the pH endpoint of 4.5 is reached. The endpoint pH value of 4.5 is used for routine measurement of alkalinity in most environmental water and wastewater samples. Then the final burette reading is recorded.

Calculation and reporting of alkalinity were done by Eq. (2).

$$\text{Alkalinity as } (T_{Alk})\text{mg/LCaCO}_3 = \frac{B * N * 50000}{\text{mL sample (S)}} \tag{2}$$

Where:
B = mL of sulfuric acid titrant used to reach end point pH.
N = normality of the standardized acid titrant.
S = mL of sample volume.
Finally, total alkalinity is reported as "total alkalinity to endpoint pH mg/L as $CaCO_3$".

2.4 Data analyses

Data were analyzed by the statistical tools (SPSS software version 21) and Microsoft excels software. The analysis results of water samples were presented in descriptive texts, tables, and respective graphs for each heavy metal element and other irrigation water quality parameters. The relationship of the variables or heavy metals and other irrigation water quality parameters were tested with correlation analysis. Variations of heavy metals and selected ions between water sampling locations along the river stream were tested with ANOVA. SAR was computed to test the level of existing sodium hazard in the Little Akaki River for irrigation uses.

3. Result and discussion

3.1 Descriptive statistics of the irrigation water quality parameters

The descriptive statistics illustrated the analysis result of heavy metals, selected toxic ions, and other physic-chemical parameters of Little Akaki River irrigation water (**Table 1**).

3.2 Levels of heavy metals in little Akaki River water

As presented in **Table 2**, the concentration levels of heavy metals (Pb and Cd) were ranged between (2.5–3.9 mg/L) and (0.03–0.4 mg/L) respectively. But Cr and Cu were found below the detectible concentration limits of the laboratory instrument (ICP-OES) in all three sampling locations along Little Akaki River. The detectible concentration limits of ICP-OES for both heavy metals (Cr and Cu) are 0.005 mg/L [18]. The result has revealed that the contamination level of the river water with Chromium and Copper was very low. Woldetsadik et al. has reported Cr (0.02–0.029 mg/L) and Cu (0.028–0.039 mg/L) around Lekuanda and Mekanisa respectively and (Aschale, 2015) also reported Cr (0.0074 mg/L) and Cu (0.0056 mg/L)

	COD	Chloride	Alkalinity	Pb	Cd	pH	TDS	ECw	Ca	Na	Mg	SAR
Max	295.6	394	400	3.9	0.4	8.42	1036	1584	24.46	87.9	5.25	4.25
Min	168.9	284	366.2	2.5	0.03	7.8	198	288	16.27	66.32	4.95	3.68
Mean	252.37	334.67	382.4	3.23	0.16	8.01	519.3	791	21.54	78.52	5.11	3.94
Range	126.7	110	33.8	1.4	0.37	0.62	838	1296	8.19	21.58	0.3	0.57
S. D.	72.3	55.51	16.943	0.70	0.21	0.35	451.86	694.97	4.57	11.06	0.15	0.29

Table 1.
The descriptive statics.

Sample ID	Sampling location	Pb	Cd	Cr	Cu
01	Gelan	2.5	0.03	<0.005	<0.005
02	Gofa	3.9	0.04	<0.005	<0.005
03	Lekuanda	3.3	0.4	<0.005	<0.005
	MRL (mg/L)	5	0.01	0.1	0.2

Table 2.
Concentration of heavy metals in (mg/L).

around Kera [9]. This shows that previous studies also confirmed that the level of chromium and copper in the Little Akaki River was very low this is because the possible reasons are it could be the presence of hydrological process of the river water. In the natural aquatic environments or surface water, chromium does not persist for long in the dissolved states and is precipitated as a suspension in the river water and the soluble species of chromium are readily adsorbed by Phyto and zooplankton.

- RML-Recommended maximum limit for irrigation water set by FAO (Ayers and Westcot 1985)
- <0.005 represents the detectible limit of Cr & Cu. These two heavy metals were found below the detectible limits in all of the three sampling locations.

Cadmium ion found in the ranges between (0.03 and 0.4 mg/L) with an average value of 0.1566 mg/L (**Figure 2**). The study shows that cadmium was found extremely higher than the maximum recommended permissible limit of FAO (0.01 mg/L) guidelines for irrigation uses in the catchment irrigated areas.

In addition, the concentration of cadmium was very higher upstream around lekuanda (0.4 mg/L) and it has decreased downstream of the river around Gelan (**Figure 3**). This is because the possible reason is that the concentration of cadmium in the upper catchment is influenced by naturally occurring sources like weathering of parent materials, sources of soils, and rocks than the anthropogenic effects. Cadmium is also governed by the types and extent of land use in the catchment areas. However, cadmium is decreasing downstream of the river. This is due to the presence of

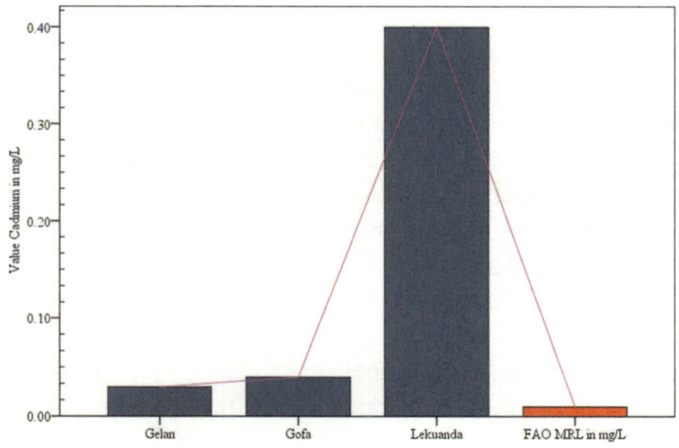

Figure 2.
Concentration levels of cadmium ions between the sampling points across little Akaki River streams.

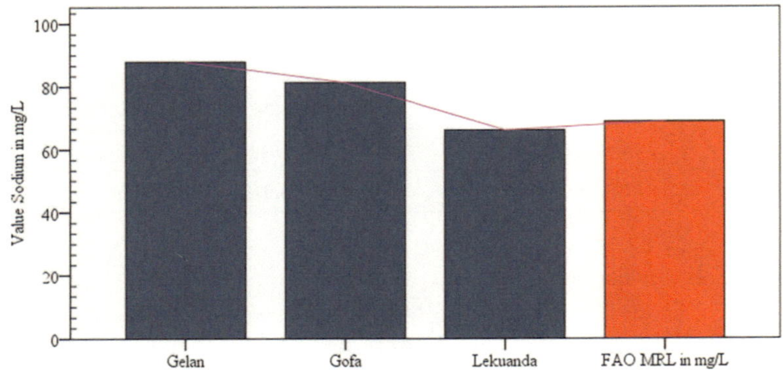

Figure 3.
Concentration level of sodium ion in mg/L between sampling locations of little Akaki River stream.

complex physic-chemical interaction and hydrological processes of the river water. The river flow rate also can determine the concentration of heavy metals, in a lower rate of river flow intensity, heavy metals become deposited at the bottom of the rivers and can be adsorbed with different suspended particulate matter that could be deposited in the bottom of the river flow while the heavy metals are transported in the long-distance along with the river flow. The other point is Cadmium concentration is highly influenced by the pH and other physic-chemical parameters of the river water. Because heavy metals like Cadmium are strongly adsorbent with the organic and inorganic matter in alkaline conditions. So, the analysis result also supported this assumption that pH, selected ions (chloride, Na, Ca, and Mg), TDS, and ECw are higher downstream of the river and they can influence the dissolution rate of cadmium downstream. Woldetsadik et al. has reported a similar trend of Cadmium concentration along the river streams that cadmium was higher in Kera (0.00282 mg/L) and Lafto (0.00148) but the very lower value of cadmium in Akaki (0.00033 mg/L) and it was also extremely lower than the current study result (0.03–0.4 mg/L) which indicated that the problem of cadmium concentration is progressively increasing in Little Akaki River water [9].

Lead (Pb) was also found in ranges between (2.5–3.9 mg/L) with an average value of (3.23 mg/L). The concentration level of lead was higher (3.9 mg/L) in the middle catchment around Gofa followed by the upper catchment in the lekuanda sampling location and it was lower downstream of the river (**Figure 4**).

This is because in the middle catchment the anthropogenic activities such as industries and other commercial activities are higher and lead is accumulated in the river water through various sources such as industrial emission; burning of lead-containing gasoline etc. the concentration of leads has been reduced to the down catchment it is because of different hydrological process and interaction of other physicochemical properties in the river water. The low-intensity water flow downstream can reduce its carrying energy causing lead has to be adsorbed on suspended particles and become deposited in the bottom of the river and river banks. The downstream river water is also diluted with different small tributaries which are joining to the main river stream at the down catchment. Factors such as pH, Alkalinity of water, TDS, and ECw affect the concentration of leads in the down catchment. In the downstream, the pH is relatively higher and has a slightly alkaline nature than the upper catchment. So, this slightly alkaline water content tends to the heavy metal ions converted to poorly soluble forms and to adsorb on the suspended matter in the river water. And heavy

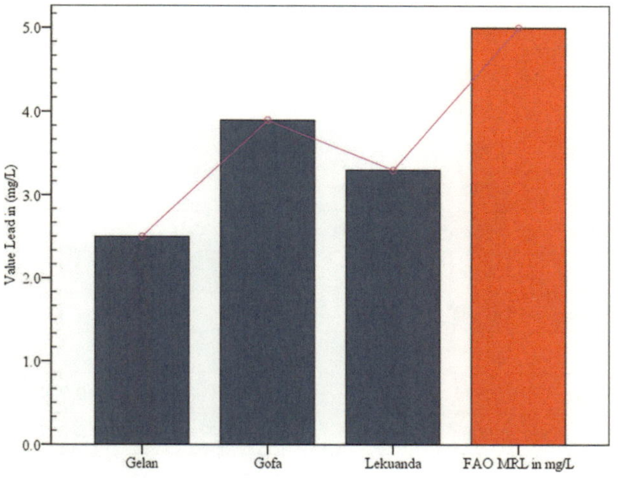

Figure 4.
Concentration levels of Lead ions between the sampling points across the little Akaki River streams.

metals like leads have strongly adsorbent properties and can be retained at river banks, aquatic vegetation, hydro-engineering structures, suspended particles, and other solid bodies in the river water. pH above 7 in irrigation water sources inactivate the heavy metals and reduce their mobility and availability to crops (Office, F A O Regional and Cairo, 2003). However, under acidic conditions (pH < 7) heavy metals could be a problem. The correlation analysis (**Table 3**) also supported these assumptions because Pb is perfectly (100%) correlated with Alkalinity and has a negative association. Woldetsadik et al. have reported a similar trend along the river stream that the level of lead was higher in Lafto (0.0369 mg/L) and kera (0.0477 mg/L) whereas the value was lower in 0.0168 mg/L) in Akaki [9]. The result of previous studies has revealed similar concentration trends along the catchment of the Little Akaki River and their result was extremely lower than the current study result. This shows that the concentration of heavy metals especially Pb and Cd are progressively increasing in Little Akaki River irrigation water. And the current study shows that the existing concentration level of lead somehow seems to be hazardous for irrigation agriculture to produce vegetables because the value was found nearly lower than the maximum recommendation limits of FAO (5 mg/L) guidelines for irrigation water. So, however, currently, it was found below the permissible limits of FAO for irrigation water, the possible hazardous condition should not be overlooked because it is persistent and highly toxicant heavy metals to plants and human health even at lower concentrations.

3.3 Heavy metals variations between sampling locations

Variation of heavy metals concentration levels between the three water sampling locations along Little Akaki River streams were tested with ANOVA and the result is illustrated in **Table 4**.

The research Hypothesis was that Ha: the concentration of heavy metals has a significant concentration variation between the sampling locations at 0.05. According to the ANOVA test, the alternative hypothesis is rejected because the F calculated value (1.079) is less than the F tabulated value (5.14325). So, it is confirmed that there is no statistically significant concentration variation of heavy metals (Pb and Cd) at 0.05 and 0.01 between the sampling locations along the river stream.

		COD	Alkalinity	Pb	Cd	pH	TDS	ECw
COD	P. Correlation	1						
	Sig.							
Pb	P. correlation	0.895	−1.000**	1				
	Sig.	0.294	0.007					
Cd	P. Correlation	0.538	−0.095	0.106	1			
	Sig.	0.638	0.939	0.933				
pH	P. Correlation	−1.000**	0.887	−0.892	−0.544	1		
	Sig.	0.005	0.306	0.299	0.633			
TDS	P. Correlation	−0.993	0.83	−0.836	−0.634	0.994	1	
	Sig.	0.076	0.377	0.37	0.563	0.071		
EC_w	P. Correlation	−0.991	0.822	−0.828	−0.645	0.992	1.000**	1
	Sig.	0.085	0.386	0.379	0.554	0.08	0.009	
Ca	P. Correlation	−0.463	0.008	−0.019	−0.996	0.469	0.565	0.576
	Sig.	0.694	0.995	0.988	0.056	0.689	0.618	0.609
Na	P.	−0.748	0.364	−0.374	−0.962	0.753	0.822	0.83
	Sig.	0.462	0.763	0.756	0.177	0.457	0.386	0.377
Mg	P. Correlation	−0.128	−0.338	0.328	−0.905	0.135	0.245	0.258
	Sig.	0.918	0.781	0.787	0.28	0.914	0.843	0.834
SAR	P. Correlation	−0.945	0.693	−0.701	−0.784	0.948	0.977	0.98
	Sig.	0.212	0.513	0.506	0.427	0.207	0.136	0.127

**Correlation is significant at the 0.05 level (2-tailed).*
***Correlation is significant at the 0.01 level (2-tailed).*

Table 3.
Correlation coefficient.

Source of Variation	SS	df	MS	F	P-value	F crit
Sampling location	0.285	2	0.142	1.079	0.39777	5.14325
Heavy metals	22.767	3	7.589	57.567	8.1750E-05	4.75706
Error	0.791	6	0.132			
Total	23.843	11				

Table 4.
The analysis of variation (ANOVA) for heavy metals.

3.4 Selected ions of water sample

The concentration levels of selected ions such as chloride, calcium, Sodium, and Magnesium were presented in **Table 5**.

The concentration of chloride was varying between 284 to 394 mg/L with an average value of 334.66 mg/L in Little Akaki River irrigation water. The result revealed that the value of chloride was surpassed the maximum permissible limit

Sample ID	Sample location	Chloride	Ca	Na	Mg	SAR
01	Gelan	326	23.88	87.90	5.13	4.25
02	Gofa	394	24.46	81.34	5.25	3.88
03	Lekuanda	284	16.27	66.32	4.95	3.68
	MRL (FAO in (mg/l)	350		69		6

Table 5.
Selected ions in little Akaki River water (mg/L).

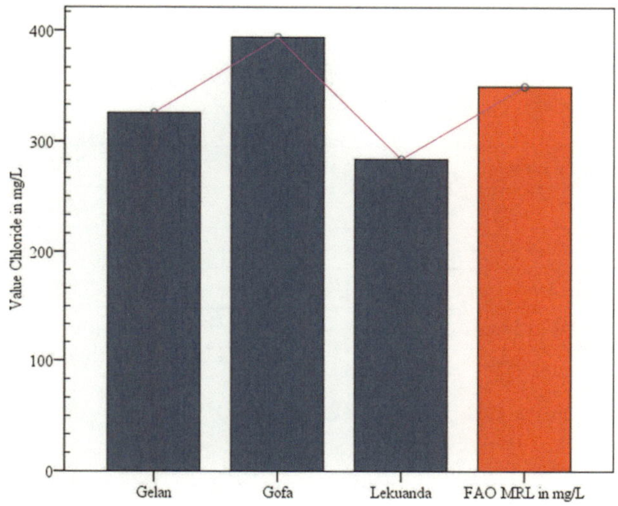

Figure 5.
Levels of chloride concentrations between sampling points across the downstream of Little Akaki River.

of FAO for irrigation water (350 mg/L) in the middle catchment (**Figure 5**). The Excess amount of chloride in the middle catchment is due to domestic and industrial wastes have been discharged into the river streams. Chloride is originating from natural resources, sewage, and industrial effluents, excessive chloride concentrations increase rates of corrosion of metals in the irrigation structure system, The excessive chloride ions in irrigation water have great impacts on the accumulation of chloride ions in soil solution through long time irrigation uses and can affect the vegetable production since excessive chloride in soil solution is very toxic to plants.

3.5 SAR and Sodium hazard

The computed SAR value of the water samples ranged between 3.68 to 4.25 and it is found below the Maximum limits of FAO (6) for irrigation water (**Figure 6**). SAR of the irrigation water has explained the impact of sodium in the destruction of soil structure and water infiltration problems through the application of contaminated irrigation water for long periods. The computed value of the Na/Ca ratio from (**Table 6**) was 3.64. In this regard, however, the computed SAR value is lower than the maximum limit of FAO, sodium ions can cause toxicities to sodium-sensitive crops at a lower SAR value in sodium-dominated irrigation water (ratio of Na/Ca > 3:1). At a

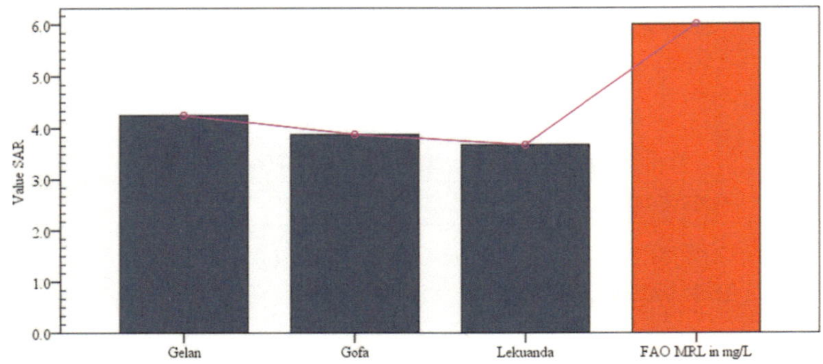

Figure 6.
Computed SAR between sampling locations across Little Akaki River stream.

Sample ID	Sample location	COD	Alkalinity	PH	TDS	EC_w (µS/cm)
01	Gelan	168.9	400	8.42	1036	1584
02	Gofa	292.6	366.2	7.82	324	501
03	Lekuanda	295.6	381	7.80	198	288
	MRL (FAO) (mg/l)			6.5–8.4	1400–2000	2000–3000

Table 6.
Physic-chemical characteristics of Little Akaki River water (mg/L).

given SAR value, the potential effect of sodium toxicity and soil water infiltration problems increases in sodium-dominated irrigation water (Na/ca >3:1).

Sodium-ion concentration was found in ranges between 66.32 and 87.9 mg/L with an average of 78.52 mg/L which is surpassed the maximum limit of FAO (69 mg/L) in the middle and downstream of the river for irrigation and the concentration is increasing to the downstream of the river (**Figure 3**). Sodium-ion concentration is told us the extent of its toxicity for plants. Therefore, according to the analysis result, sodium ion concentration is reached at the middle to slight restriction level for vegetable production in the study area because it has a significant toxicity effect on plants at higher concentration levels. The concentration of Sodium ions in the aquatic system is mainly derived from atmospheric deposition and silicate weathering [19].

3.6 Variation of selected ions between sampling locations

Variation of selected ions (Chloride, Sodium, Calcium, and Magnesium) concentration levels between the three sampling locations along the river streams were tested with ANOVA and the result is illustrated in **Table 7**.

The research Hypothesis was that Ha: the selected ions have a significant concentration variation between the sampling locations at 0.05. According to the ANOVA test, the alternative hypothesis is rejected because the F calculated value (1.538) is less than the F tabulated value (4.45897). So, it is confirmed that there is no statistically significant concentration variation of selected ions (Chloride, Na, Ca, and Mg) at 0.05 and 0.01 between the sampling locations along the Little Akaki River stream.

Source of variation	SS	df	MS	F	P-value	F crit
Sample location	1791.28	2	895.642	1.538194	0.272124	4.45897
selected ions	237858.70	4	59464.68	102.1259	6.7E-07	3.837853
Error	4658.14	8	582.2684			
Total	244308.20	14				

Table 7.
Analysis of variation (ANOVA) test for selected ions.

3.7 The physic-chemical characteristics of Little Akaki River water

As indicated in **Table 6**, COD has been decreased downstream of the rivers from 295.6 mg/L at Lekuanda to 168.9 mg/L at Gelan. Whereas, the other parameters such as Alkalinity, pH, TDS, and EC were increased across the downstream (**Table 3**). This is because the COD content was diluted and attenuated across the path of the river course from upstream to downstream. The physical and chemical properties of the river water are characterized by several interdependent interactions and their relationships are extremely complex. The hydrological process is also the most determinant factor that influence the concentration of the physic-chemical parameters of the river water.

The pH value of water samples ranged between7.8 to 8.4 and it has a mean value of 8.01. The value of pH also increases downstream of the river this is because the alkalinity and ECw of river water also increase downstream of the river and have a significant contribution to raising the pH of the river water downstream because they have a positive correlation. Hence the result indicated that Little Akaki River irrigation water is slightly alkaline and it lies in the acceptable ranges of FAO guidelines (6.5–8.4). pH is the most determinant factor for the quality of irrigation water and it can greatly influence the toxicity of heavy metals and other impairing selected ions. Alkaline irrigation water prohibits the solubility and bioavailability of heavy metals.

Electrical conductivity (EC_w) of the river water was varied between (288–1584 μS/cm) with an average of 791 μS/cm and the value was increased to the downstream of the river from 288 to 1584 μs/cm and it is found below the permissible limits of FAO (3000 μS/cm) for irrigation uses (**Figure 7**).

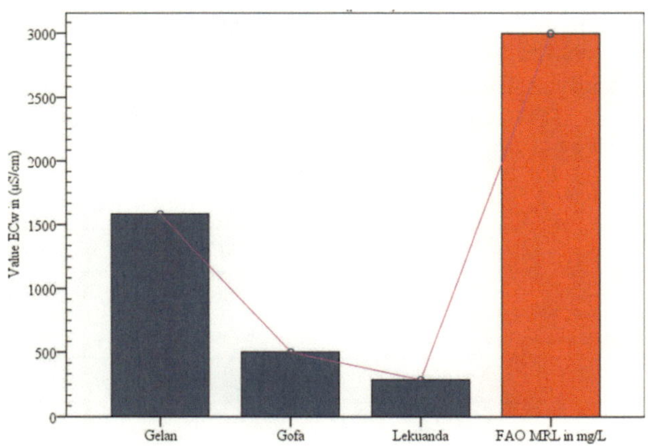

Figure 7.
Level of ECw in (μS/cm) between sampling locations across the Little Akaki River streams.

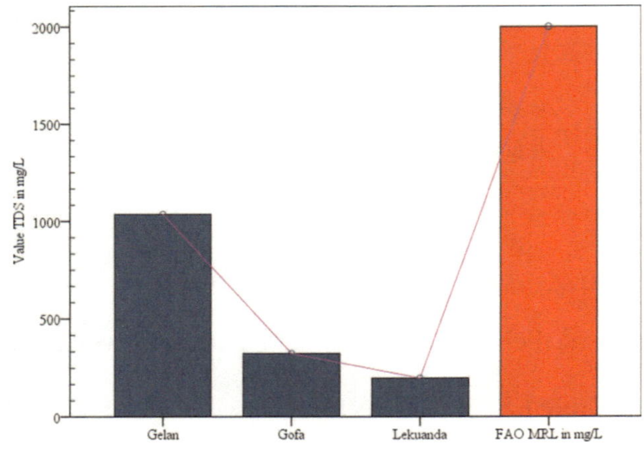

Figure 8.
Levels of TDS in mg/L between sampling locations across Little Akaki River streams.

The total dissolved solids (TDS) of water samples were varied between 198 and 1036 ppm with an average of 519 mg/L and it has a higher value in the downstream (1036 mg/L) and the concentration is decreased to the upstream of the river (**Figure 8**). TDS is found below the maximum recommended limit of FAO (1400–2000 mg/L) for irrigation water. But relatively the higher value is obtained downstream of the river and it indicated the presence of a higher amount of basic or alkaline compounds like bicarbonates, sulfates, chlorides, etc. In general, according to the result in (**Figures 7** and **8**), the value of EC and TDS illustrated that salinity is not a serious problem in the existing condition in Little Akaki River irrigation water.

3.8 The variation of physic-chemical parameters

Variation of physic-chemical parameters (COD, Alkalinity, pH, TDS, and EC_w) between the sampling locations along the river streams were tested with ANOVA and the result is illustrated in **Table 8**.

The research Hypothesis was that Ha: the physic-chemical parameters have a significant concentration variation between the sampling locations at 0.05. According to the ANOVA test, the alternative hypothesis is rejected because the F calculated value (2.085346) is less than the F tabulated value (4.45897). So, it is confirmed that there are no statistically significant variations in the physic-chemical parameters (COD, Alkalinity, pH, TDS, and ECw) at 0.05 and 0.01 between the sampling locations along the Little Akaki River stream.

Source of Variation	SS	Df	MS	F	P-value	F crit
Sample location	474733.3	2	237366.6	2.09	0.18668	4.45897
Physic-chemical parameters	1,027,322	4	256830.5	2.25	0.152122	3.837853
Error	910608.30	8	113,826			
Total	2,412,664	14				

Table 8.
Analysis of variation (ANOVA) for physic-chemical parameters.

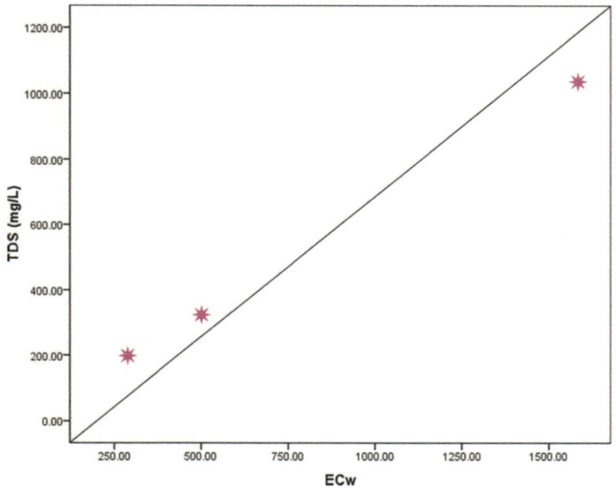

Figure 9.
Correlations between TDS and ECw.

3.9 Correlation of heavy metals and other physic-chemical parameters

The correlation analysis is conducted to show the relationship and interaction of heavy metals and other physic-chemical properties of the river water. The interaction between the heavy metals and other properties are the major factor for their concentration variations in the river water. Therefore, the correlation result is presented in **Table 3**.

The concentration of heavy metals and other selected ions were correlated with some physic-chemical interactions and different hydrological processes of the river water. The scatter plot analysis also depicted that ECw and TDS have positive associations and are strongly correlated with each other (**Figure 9**).

TDS concentration describes the presence of inorganic salts and organic matter in the irrigation water and EC is the measure of irrigation water capacity to conduct electric current. Both EC and TDS are very determinant irrigation water quality parameters, which are used to describe the salinity level of the irrigation water [20]. These two parameters are correlated and usually expressed by a simple equation: TDS (mg/L) = k*EC (μS/cm in 25°C. The value of k will increase along with the increase of ions in water. However, the relationship between conductivity and TDS is not always directly linear; it depends on the activity of specific dissolved ions in the liquids and ionic strength [21].

Accordingly, the TDS/ECw ratio of the water samples in the Little Akaki River was = 519/791 = 0.656 or it can be written as equation TDS = 0.656*EC. This indicates that the correlation of both parameters is strongly influenced by the EC values. Unlike freshwater, the correlation between TDS and EC in wastewater cannot be described well because the water is heavily influenced by many contaminants [21].

4. Conclusions

The study was aimed to determine the concentration levels of heavy metals (Pb, Cd, Cr, and Cu) and selected ions (chloride, Calcium, Sodium, and Magnesium) of

Little Akaki River irrigation water and to test the concentration variation of heavy metals and selected ions between the sampling locations along the river stream. To this end, the study was brought out the following observations and conclusions. The concentration level of Cadmium (Cd) ranged between (0.03–0.4 mg/L) and it was extremely higher than the permissible limits of FAO (0.01 mg/L) for irrigation water and the values of Pb was also varying in (2.5–3.9 mg/L) and it was found in approaching the maximum permissible limit of irrigation water set by FAO (5 mg/L). Both heavy metals (Cd & Pb) have higher concentrations in the middle and upstream than the downstream of the river and their concentration level reached the hazardous condition for irrigation water in Little Akaki River water. The concentration level of Pb and Cd were extremely higher than the previous study findings and this indicated heavy metal contamination problem is progressively increasing in the river stream. Whereas, heavy metals such as Cr and Cu were found below the detectible limits of the laboratory equipment (ICP-OES). Selected ions such as chloride and sodium were reached at the maximum permissible limits of FAO in the Little Akaki River and this can inhibit the growth of vegetables in the irrigation sites of the study catchment area. Other physic-chemical parameters (TDS, pH, and ECw) were found in optimum conditions for irrigation water in all three-sampling locations. The other main point that has been observed is a variation of heavy metals and selected ions between the sampling locations were not statically significant at 0.5 and 0.1. In general, the most important water quality parameters such as heavy metals (Pb and Cd), selected ions like (chloride and sodium) were exceeded the maximum recommendation limits of FAO guidelines for irrigation water in the Little Akaki River. Therefore, the study revealed that irrigation water quality is reached at a great concern for vegetable production and it could be a potential risk for human health through the food chain of vegetable consumption.

Acknowledgements

We would like to acknowledge Ethiopian Civil Service University for financing this study.

Conflict of interest

"The authors declare no conflict of interest."

Author details

Mekonnen M. Tarekegn[1*] and Getaneh L. Weldekidan[2]

1 Department of Environment and Climate Change, Ethiopian Civil Service University, Addis Ababa, Ethiopia

2 Department of Agricultural Investment Support Directorate, Ministry of Agriculture, Addis Ababa, Ethiopia

*Address all correspondence to: maschalm12@gmail.com

References

[1] Maleki A et al. Concentration levels of heavy metals in irrigation water and vegetables grown in peri-urban areas of Sanandaj, Iran. Journal of Advance Environment Health Research. 2002; **1**(2):81-88

[2] Fay DL. Heavy metals problems and solutions. Angewandte Chemie International Edition. 1967;**6**(11):951-952

[3] Gebre G, Van Rooijen D. Institutional repository urban water pollution and irrigated vegetable farming urban water pollution and irrigated vegetable farming in Addis Ababa. In: 34th WEDC International Conference. Vol. 166. Ethiopia: Addis Ababa; 2009

[4] Sayato Y. WHO guidelines for drinking-water quality. Eisei kagaku. 1989;**35**(5):307-312. DOI: 10.1248/jhs1956.35.307

[5] Sorial GA and Hong J. Proceedings from the 8th International Conference on Environmental Science and Technology. Yıldız Technical University, Dept. of Environmental Engineering Istanbul, Turkey. 2016

[6] Gashaye D. Wastewater-irrigated urban vegetable farming in Ethiopia: A review on their potential contamination and health effects. Cogent Food and Agriculture. 2020;**6**(1):1-17. DOI: 10.1080/23311932.2020.1772629

[7] Mohanty B et al. Heavy metals in soils and vegetation from wastewater irrigated croplands near Ahmedabad, Gujarat: Risk to human health. Nature Environment and Pollution Technology. 2021;**20**(1):163-175. DOI: 10.46488/NEPT.2021.V20I01.017

[8] Sinshaw TA. Understanding the Situation of Wastewater Irrigation in Community-Based Irrigation Schemes the Case of Akaki Catchment. Ethiopia: Wageningen University; 2011. p. 74 https://edepot.wur.nl/176803

[9] Woldetsadik D, Drechsel P, Keraita B, Itanna F, Gebrekidan H. Heavy metal accumulation and health risk assessment in wastewater-irrigated urban vegetable farming sites of Addis Ababa, Ethiopia. International Journal of Food Contamination. 2017;**4**(1). DOI: 10.1186/s40550-017-0053-y

[10] Tegegn FE. Physico-Chemical Pollution Pattern in Akaki River Basin. Ethiopia: Addis Ababa; 2012. p. 50. Available at: http://www.diva-portal.org/smash/get/diva2:555414/fulltext02

[11] Tamiru SM. Assessment of the Ecological Impacts of Floriculture Industries Using Physico-Chemical Parameters and Benthic Macroinvertebretes Metric Index along Wedecha River, Debrezeit, Ethiopia (July) 2007

[12] Weldesilassie AB et al. Wastewater use in crop production in peri-urban areas of Addis Ababa: Impacts on health in farm households. Environment and Development Economics. 2011;**16**(1): 25-49. DOI: 10.1017/S1355770X1000029X

[13] Zeiner M, Rezić I, Steffan I. Analytical methods for the determination of heavy metals in the textile industry. Kemija u industriji/Journal of Chemists and Chemical Engineers. 2007;**56**(11):587-595

[14] Abdulla HJ, Al-Quraeshi NKB, Al-Awadi FNJ. Study of chemical oxygen demand (COD) in relation to biochemical oxygen demand (BOD).

Journal of kerbala university. 2012;**10**(3): 8-11 https://iraqjournals.com/article_65514_0.html

[15] Yadvika et al. A modified method for estimation of chemical oxygen demand for samples having high suspended solids. Bioresource Technology. 2006; **97**(5):721-726. DOI: 10.1016/j.biortech.2005.04.013

[16] Rounds RSA. Alkalinity and acid neutralizing capacity. U.S. Geological Survey TWRI Book 9 Chapter. 2012;**A6**: 1-45

[17] Deshpande L. Water quality analysis laboratory methods. Council of Scientific & Industrial Research, New Delhi, Govt. of India. 2012:68

[18] Sarojam P. APP_MetalsinWastewater. Perkin Elmer Instruments. 2010:11

[19] Zakir HM, Sattar MA, Quadir QF. Cadmium pollution and irrigation water quality assessment of an urban river: A case study of the Mayur river, Khulna, Bangladesh. Journal of Chemical Biology Physics Science Section D. 2015;**5**(2): 2133-2149

[20] Namieśnik J, Rabajczyk A. The speciation and Physico-chemical forms of metals in surface waters and sediments. Chemical Speciation and Bioavailability. 2010;**22**(1):1-24. DOI: 10.3184/095422910X12632119406391

[21] Rusydi AF. Correlation between conductivity and total dissolved solid in various type of water: A review. IOP Conference Series: Earth and Environmental Science. 2018;**118**(1):1-6. DOI: 10.1088/1755-1315/118/1/012019

Chapter 9

Toxicants in Water: Hydrochemical Appraisal of Toxic Metals Concentration and Seasonal Variation in Drinking Water Quality in Oil and Gas Field Area of Rivers State, Nigeria

Morufu Olalekan Raimi, Henry Olawale Sawyerr, Ifeanyichukwu Clinton Ezekwe and Salako Gabriel

Abstract

Groundwater pollution is a major issue in many tropical countries. Governments at all levels are doing little or nothing to supply inhabitants with clean and accessible water, particularly in Nigeria's Niger Delta region. This study compares differences in water quality parameters in the study area (determine the level of pollutions in the different sites). The investigation made use of standard analytical methodologies. All sampling, conservation, transportation as well as analysis followed the usual APHA procedures (2012). To prevent degradation of the organic substances, all obtained samples were transferred to the laboratory, while keeping in an icebox. Result shows that during wet season, the mean values obtained for water quality parameters were significantly lower in site 9 compared with that obtained in other sites ($p < 0.05$) with the exemptions of temperature, DO, BOD, COD, acidity, TH, TDS, K, Mg, Zn, Mn, Cd, Pb, Cu, Cr, NH_3, NO_2, NO_3, Ni though slightly lower in most cases in site 9 were not significantly different ($p > 0.05$) and both alkalinity and SO_4 which were significantly higher in site **9** than site 1 ($p < 0.05$). Result obtained during dry season reveals that there is no remarkable difference in pH, acidity, Pb and Ni between the nine sites ($p > 0.05$) while other water quality parameters were significantly lower in site 9 than other sites excluding Cl and Mg which were both significantly higher in site 9 than site 8 ($p < 0.05$). Extra efforts must be made to completely understand the hydrogeochemical properties and appropriateness of groundwater in Nigeria's core Niger Delta region in order to ensure quality groundwater supply for varied applications. As a result, this research will contribute to the establishment of a quantitative understanding of the effects of many causes on groundwater level changes in every aquifer worldwide. This analysis also reinforces a useful resource for scholars, activists, and public officials looking to improve community awareness, planning, and performance. The verdicts will serve as a valuable guideline for policymakers, the Ministry of Water Resources, and development practitioners, as they highlight the need for appropriate approaches to mitigating toxic elements of water resource contamination in the core Niger Delta in order to protect public health from carcinogenic and non-carcinogenic risks.

Keywords: reproductive health, human health risk, toxicants, community awareness, extractive industry, Core Niger Delta, Nigeria

1. Introduction

Approximately 80% of houses in Niger Delta rely on this domestic water supply. Concern about safety of groundwater supplies have centered on pollution induced by human activities, with natural contamination receiving less attention [1–4]. This is linked to a lack of safe water, which exacerbates health issues and reduces productivity. According to the UNDP [5] only about a quarter (24%) of the indigenous people and half of the urban population in the Niger Delta have access to safe drinking water. This is consistent with the findings of a Bayelsa State Micro Credit Administration Agency poverty baseline survey, which revealed that only a small percentage of the indigenous populace has access to safe drinkable water [6]. However, several studies (e.g., [6-11]) have shown and document (scientifically) that the increasing presence of geogenic contaminants in the Niger Delta can have serious health effects as well as wellbeing on the indigenous population, thus leading to both environmental and community concerns, resulting in the prohibition of oil and gas companies in some locations. Access to safe drinking-water is a key health as well as development concern at the local, regional and national levels [6–11]. As groundwater becomes an important source of freshwater for residential use in the Niger Delta and most Nigerian cities, it is necessary to analyze its quality, particularly in terms of geogenic contaminants. This is due to the fact that people rely on groundwater from shallow aquifers, putting a significant number of people at risk of contamination. While, trace elements are among the few compounds that have been shown to cause severe health concerns in humans as a result of excessive drinking-water exposure [6, 12–15]. The study aims to compare water quality parameters in the study region (determine the level of pollutions in the different sites) in the vicinity of "Gas Flaring Area of Ebocha-Obrikom of Rivers State, Nigeria". This research will provide valuable information and add to our understanding on the physico-chemical examination of drinking water associated with the contamination of the ground waters by petroleum products. Hence, the study will help in integrating the health needs of the populace into the state health scheme, in recognition of the fact that health is required for national development. The study will also bring to the awareness of the local people the type of water that is good for them as drinking water according to recommended standards. It will provide a structural framework for effective management of groundwater and provide an available reference source and base line data for researchers involved in water resources assessment.

2. Material and methods

2.1 The study area

Ebocha-Obrikom is located among latitude 5°20 N–5°27 N as well as longitude 6°40 E–6°4 6E (**Figure 1**). It includes the towns of Obor, Obie, Obrikom, Agip New

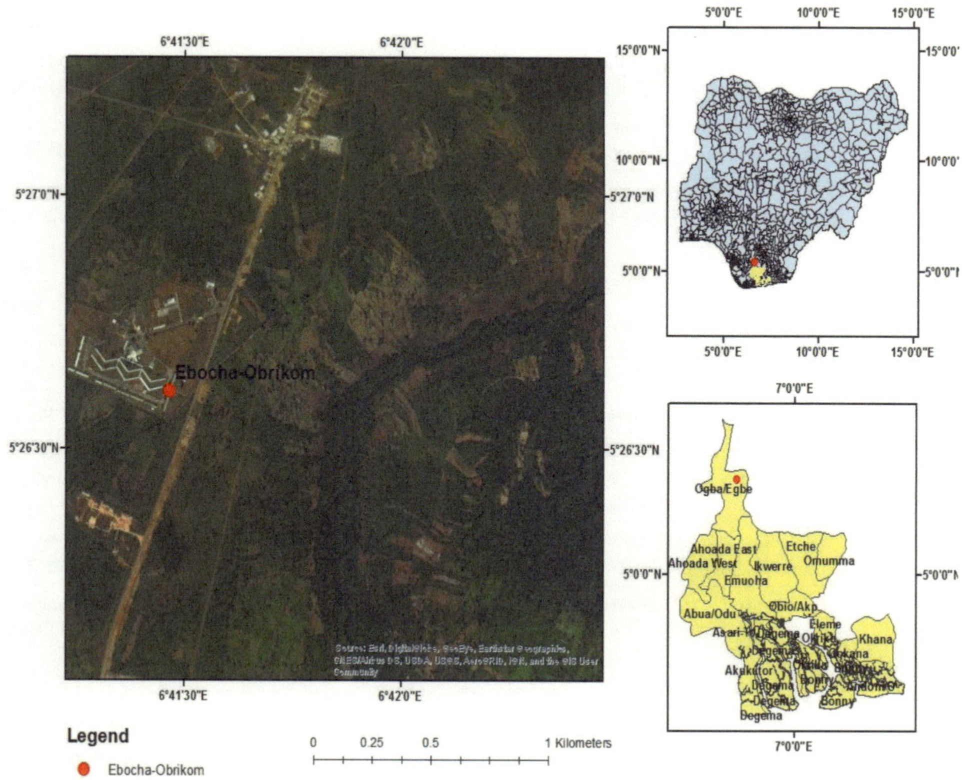

Figure 1.
Map showing the study area with Nigeria and River State insert. Sources: Adapted from Olalekan et al. [14].

Base and Ebocha, all of which are positioned in Ogba/Egbema/Ndoni Area (**Figure 1**) of Rivers State. The research study area is bordered to the North by the Nkissa River, to the West by the Orashi River, to the East by the Sombrero River, and to the South by Omoku town [1, 2].

2.2 Sample collection

The current research inquiry used a sample method similar to that used by Morufu and Clinton [13], Raimi and Sabinus [12], and Olalekan et al. [14], in which sampling was targeted in selected vulnerable quarters in a highly populated environment. These areas are prone to pollution not just due to their physical location, but also due to the existence of crude petroleum exploration and exploitation. Water samples were taken from groundwater sources near the sampling location (see **Table 1** below) and used mostly for drinking and domestic purposes. Only groundwater from dug wells or shallow pumping wells built specifically for residential use was sampled. The wells range in depth from 10 to 28 meters, indicating that they are in a phreatic aquifer. Portable GPS devices were used to record the sampling locations. Ground water sources in the neighborhood of the depot were chosen at random but at varying distances from one another for the purposes of this experiment. Furthermore, samples were manually collected from nine (9) strategic locations in the study area for ground water (boreholes and wells) into previously washed clean plastic sampling bottles after approximately 20 min of continuous water flow to ensure adequate aquifer that can remain suitably represented.

S/N	Locations	Altitude (m)	Latitude	Longitude
Site—1	(Borehole) (opposite Ijeoma Quarters. 750 m away from Agip Gas Flaring Center Ebocha)	10	Lat N05°27′068″	Long E006°41′480″
Site—2	(Borehole) (200 m opposite Agip Gas Flaring Centre Ebocha and 50 m from Agip Waste Pit)	—	Lat N05°27′28.7″	Long E006°41′58.1″
Site—3	(Well) (The Apple Hotel 500 m from Waste Pit and 150 m away from Mgbede Field Oil Well 7 Ebocha)	16	Lat N05°27′37.5″	Long E006°42′05.3″
Site—4	(Well) (1000 m away from the Agip Flare Stack Ebocha)	22	Lat N05°26′51.5″	Long E006°41′38.8″
Site—5	(Borehole) (Abacha Road Obrikom, 800 m away from Agip Gas Plant)	—	Lat N05°23′48.6″	Long E006°40′36.8″
Site—6	(Borehole) (Eagle Base Obor. 2500 m away from Agip Gas Plant)	28	Lat N05°23′00.9″	Long E006°41′07.4″
Sites—7	(Well) (Obor Road Obie. 2000 m away from Agip Gas Plant)	24	Lat N05°23′22.5″	Long E006°40′ 49.1″
Sites—8	(Borehole) (Green River Plant Propagation Centre Naoc 3000 m away from Agip Gas Plant)	17	Lat N05°24′18.9″	Long E006°40′55.0″
Sites—9	(Control) (35,000 m from Ebocha)	—	Lat N5°4′58.1412″	Long E6°39′30.4806″

Table 1.
Geographical coordinates of the nine (9) sampling sites (samples).

All samples were collected during the day, from 9:00 am to 4:00 pm. As a result of flooding, insecurity as well as lockdown from COVID-19. Night samples were not taken, and the sampling took place between the month of September 2019 and August 2020.

2.3 Sampling, preservation and analysis

Water sampling, conservation, transportation, and analysis have all been carried out in accordance with the standard methods specified in APHA [16]; Morufu and Clinton [13]; Raimi and Sabinus [12]; Olalekan et al. [14]; Morufu et al. [2].

2.4 Ground water collection

Ground water samples were obtained in pre-rinsed 1litre plastic containers for analyses of physico-chemical characteristics. Prior to storage, pre-rinsed ground water samples for trace metal analyses remained obtained in 1litre containers with nitric acid and treated with 2 ml nitric acid (assaying 100%, Fisher Scientific, Trace Metal Grade). These steps were taken to keep the metals oxidation settings stable. For Dissolved Oxygen (DO) and Biological Oxygen Demand (BOD) assays, groundwater samples remained obtained in two groups of 250 ml glass-stoppered-reagent bottles per sampling site. The BOD samples were carefully filled without air trapping and the bottles were wrapped in black polythene bags. This was done to exclude the presence of light in the samples, which was capable of creating DO by autotrophes (algae). The BOD samples were cultured for 5 days before being added to 2 ml of each sample. Winkler solutions I and II apply various dropping pipettes to each sample to slow

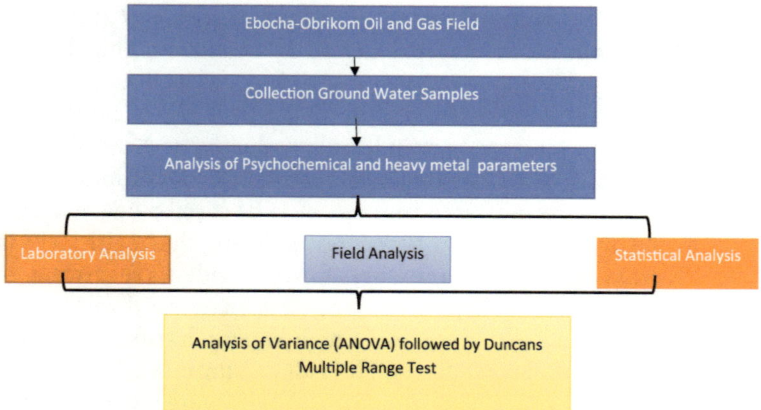

Figure 2.
A schematic illustration of quantification methodology adopted for the current study.

down additional biological activity. To precipitate the floc that was at the bottom of the bottles, the bottles were thoroughly shaken. Furthermore, Winkler solution I is a manganese sulphate solution, whereas solution II is a mixture of sodium or potassium iodide, sodium azide (sodium nitride), sodium or potassium hydroxide as well as sodium hydroxide. The samples of DO were collected in transparent bottles with tight-fitting stoppers. With dissolved oxygen samples kept on the spot using Winkler I and II solutions identical to those used for BOD samples [16]. For simple identification, all samples had remained carefully identified and kept at 4°C. On-site measurements were taken to determine the unstable concentrations and sensitive water quality indicators such as pH, electrical conductivity (EC), total dissolved solids (TDS), alkalinity (Alka.) as well as temperature (Temp). As a result, **Figure 2** depicts the key methodologies for researching groundwater composition.

2.5 Quality assurance and quality control (QA/QC)

Furthermore, using high purity analytical reagents and solvents, all analytical methods remained closely monitored using quality assurance and control methodologies. The instruments were calibrated using calibration standards. The analytical technique validation included the use of triplicate analysis, procedure blanks and the examination of certified reference materials (CRM). The limit of detection (LoD), precision, reproducibility, repeatability and accuracy of each organic pollutant from the groundwater samples were determined.

3. Results and discussion

3.1 Compare differences in water quality parameters in the study area (determine the level of pollutions in the different sites)

Trace elements exist in a variety of forms, including oxides, salts, organometallic complexes, sulphides, and dissolved ions in groundwater and soil. Chemical processes are driven by the partitioning of water, air as well as soil through particles adsorption or pH-dependent water dissolution [1–3, 6, 13, 17]. Thus, **Tables 2** and **3** provide the statistical analysis findings for the physicochemical parameters. The Ebocha-Obrikom area of Rivers State is significant for irrigation, drinking, as well as industrial uses [1–3, 6, 12–14]. Over the previous three decades it has been significantly altered by population expansion and increased agricultural productivity. A detailed analysis of groundwater geochemistry as well as associated estimation of community's health risk that are visible to the groundwater, remain yet to be carried out. A clear understanding of the utmost significant aspects regulating the health risks is vital toward taking effective management measures for the residents regarding drinking water. Thirty-four (34) water quality parameters were analyzed during rainy and dry season respectively. Eighteen (18) parameters such as temperature, pH, conductivity, turbidity, DO, BOD, Acidity, TSS, Salinity, Fluoride, Aluminum, Potassium, Magnesium, Iron, Zinc, Manganese, Cadmium and TPH were lowest at location nine (9) during the rainy season. For dry season, twenty-two (22) parameters, which include: turbidity, BOD, Acidity, TH, TDS, TSS, Salinity, Fluoride, Aluminum, Potassium, Calcium, Iron, Zinc, Manganese, Cadmium, Copper, Chromium, Sulphate, Ammonia, Phosphate, Nickel and TPH recorded minimum values below limits of detection in sampling locations nine (9). Thus, the results showed a significant disparity between the various sampling locations. As it is evident that samples from location 1 to location 9 must remain adequately observed since, there may be a remarkable increase in these heavy metal level in the future, which could eventually cause health-related threats for indigenous residents. While, there is high temperature variation in this region. Temperature was found to be highest with an average value of (28.77–32.46)°C at location 4 and 2 during the rainy and dry seasons. While it was lowest (26.01–29.00)°C at location 9 and 6 during the rainy and dry season. The maximum permissible limit for temperature has not been stated but ambient in nature. Groundwater temperature tend to be influenced more by dry season than rainy season. Thus, Morufu and Clinton [13]; Raimi and Sabinus, [12] and Olalekan et al. [14] indicated that temperature of groundwater in areas prone to pollution and gas flaring typically is higher than that of the surrounding environment, a study has shown. Unarguably, an increase in air temperature at a gas flare site must have led to a rise in groundwater temperature. Hence, rising temperature may adversely impact agriculture, thereby increasing the vulnerability of marginalized agriculture-dependent rural populations. This is particularly true for communities in Ogba/Egbema/Ndoni Local Government Area of Rivers State, which is home to Agip's Ebocha, Obrikom oil and gas facilities. The people of Ogba are predominantly fishermen and farmers who rely on groundwater and small streams for their water supply. Ogba is abundant in natural gas as well as is home to the state-owned gas turbine facility. There are several large and functioning gas stacks in the community. Water pH represents hydrogen ion concentration as well as is affected by the interaction of several compounds dissolved in water. The pH range in which aquatic creatures thrive varies [6, 14]. It is in itself poisonous at a certain level and can influence the toxicity of heavy metals and cyanides. The pH of water is determined by the carbonate cycle, which is composed of CO_2, H_2CO_3, HCO_3^- and CO_3^{-2}. The pH has no direct impact on human health but can change water taste as well as exhibit linked to other water quality characteristics [1–3, 6, 14, 18]. The alkalinity remarkability lies in its role for carbon dioxide chemistry, trace metal speciation and buffer capacity of the groundwater. Hydrogen-ion-concentration (pH) is a master control measure in the environment that demonstrates the chemical as well as biological features of water. pH ranges from 5.82 to 7.98, depending on the environment, with values

	Location								
Parameters	**1**	**2**	**3**	**4**	**5**	**6**	**7**	**8**	**9**
Temperature (°C)	28.63 ± 2.46^{a}	28.24 ± 2.06^{a}	27.83 ± 2.62^{a}	28.77 ± 1.73^{a}	27.96 ± 1.50^{a}	28.01 ± 1.18^{a}	28.27 ± 1.61^{a}	26.79 ± 1.80^{a}	26.01 ± 2.10^{a}
Hydrogen potential (pH)	7.20 ± 1.37^{c}	7.98 ± 0.73^{c}	7.06 ± 1.05^{c}	6.27 ± 0.59^{b}	6.49 ± 0.95^{b}	5.92 ± 0.27^{a}	6.44 ± 1.74^{b}	6.13 ± 0.52^{b}	5.82 ± 0.28^{a}
Conductivity (ЧSCM^{-1})	48.09 ± 24.23^{d}	52.29 ± 28.76^{d}	30.99 ± 18.42^{c}	18.57 ± 9.24^{a}	20.96 ± 9.27^{a}	36.88 ± 25.03^{c}	27.48 ± 14.18^{b}	26.93 ± 11.24^{b}	24.22 ± 7.61^{a}
Turbidity (NTU)	7.41 ± 2.97^{a}	7.36 ± 6.50^{a}	42.43 ± 14.40^{d}	31.43 ± 11.72^{c}	11.14 ± 5.05^{b}	3.35 ± 1.18^{a}	48.24 ± 20.57^{d}	4.99 ± 1.56^{a}	1.78 ± 0.66^{a}
Dissolved oxygen (DO) (mg/l)	17.27 ± 0.81^{a}	17.21 ± 1.38^{a}	17.84 ± 1.29^{a}	16.91 ± 1.19^{a}	16.37 ± 0.43^{a}	16.42 ± 0.45^{a}	17.01 ± 0.89^{a}	16.99 ± 0.73^{a}	16.29 ± 0.37^{a}
(BOD) (mg/l)	5.35 ± 0.29^{a}	5.31 ± 0.44^{a}	5.05 ± 0.10^{a}	5.48 ± 0.37^{a}	5.21 ± 0.40^{a}	5.42 ± 0.44^{a}	5.26 ± 0.21^{a}	5.47 ± 0.39^{a}	4.98 ± 0.10^{a}
(COD) (mg/l)	40.06 ± 12.15^{a}	37.27 ± 8.79^{a}	39.61 ± 11.84^{a}	41.78 ± 13.07^{a}	39.72 ± 6.80^{a}	32.03 ± 9.65^{a}	33.50 ± 7.15^{a}	32.75 ± 6.61^{a}	32.64 ± 6.57^{a}
Acidity (mg/l)	90.11 ± 48.55^{a}	90.96 ± 45.54^{a}	101.18 ± 46.68^{a}	90.01 ± 42.52^{a}	87.11 ± 42.97^{a}	89.92 ± 42.26^{a}	99.56 ± 46.44^{a}	86.18 ± 35.99^{a}	85.53 ± 45.14^{a}
Alkalinity (mg/l)	103.73 ± 60.46^{b}	119.56 ± 58.6^{b}	18.17 ± 7.42^{a}	134.96 ± 50.03^{b}	119.23 ± 65.28^{b}	119.15 ± 65.15^{b}	130.93 ± 43.65^{b}	134.76 ± 47.90^{b}	117.73 ± 63.97^{b}
(TH) (mg/l)	41.06 ± 2.27^{a}	41.49 ± 4.18^{a}	42.22 ± 2.26^{a}	37.71 ± 9.49^{a}	35.02 ± 7.45^{a}	37.64 ± 9.47^{a}	39.48 ± 3.08^{a}	40.30 ± 2.58^{a}	37.64 ± 4.64^{a}
TDS (mg/l)	11.12 ± 3.77^{a}	11.50 ± 3.89 ^{a}	8.24 ± 3.98 ^{a}	9.31 ± 4.62 ^{a}	9.27 ± 4.71 ^{a}	10.11 ± 5.14 ^{a}	8.06 ± 4.50 ^{a}	7.88 ± 4.27 ^{a}	10.61 ± 2.06 ^{a}
TSS (mg(l)	39.80 ± 3.66^{b}	34.72 ± 2.65 ^{a}	35.96 ± 1.95^{a}	37.49 ± 2.30 ^{b}	37.28 ± 1.85^{b}	37.11 ± 1.91 ^{b}	38.40 ± 3.05^{b}	38.73 ± 3.12^{b}	34.00 ± 2.47^{a}
Salinity (mg/l)	0.12 ± 0.08^{a}	0.14 ± 0.10^{a}	11.50 ± 10.60^{c}	15.74 ± 11.29 ^{c}	0.07 ± 0.01^{a}	1.71 ± 0.76^{b}	25.71 ± 5.35^{c}	0.09 ± 0.08 ^{a}	0.01 ± 0.02^{a}
Chloride (mg/l)	30.61 ± 2.11^{b}	31.43 ± 1.50 ^{b}	32.16 ± 1.81 ^{b}	32.10 ± 0.51 ^{b}	31.66 ± 0.24 ^{b}	31.13 ± 0.70 ^{b}	28.33 ± 1.96^{a}	29.05 ± 2.34^{a}	28.97 ± 2.31^{a}

	Location								
Parameters	**1**	**2**	**3**	**4**	**5**	**6**	**7**	**8**	**9**
Fluoride (mg/l)	0.00 ± 0.00 [a]	0.00 ± 0.00 [a]	0.00 ± 0.00[a]	0.29 ± 0.27[b]	0.53 ± 0.43 [b]	0.82 ± 0.25 [c]	0.85 ± 0.34[c]	0.79 ± 0.31[s]	0.00 ± 0.00 [a]
Aluminum (mg/l)	0.00 ± 0.00 [a]	0.00 ± 0.00[a]	0.01 ± 0.01[a]	0.01 ± 0.01[a]	0.02 ± 0.00 [b]	0.02 ± 0.01[b]	0.03 ± 0.01[b]	0.02 ± 0.01[b]	0.00 ± 0.00[a]
Sodium (mg/l)	12.32 ± 2.38[a]	16.06 ± 1.99	12.22 ± 2.72 [a]	14.54 ± 0.80[b]	14.45 ± 1.07[b]	14.65 ± 0.90[b]	16.39 ± 0.20[b]	15.48 ± 2.40[b]	12.85 ± 2.63 [a]
Potassium (K) (mg/l)	2.65 ± 0.76 [a]	2.49 ± 0.61 [a]	3.14 ± 0.03 [a]	3.23 ± 0.56 [a]	2.97 ± 0.64 [a]	3.07 ± 0.52 [a]	3.10 ± 0.04 [a]	3.29 ± 0.17 [a]	2.42 ± 0.59 [a]
Calcium (Ca) (mg/l)	52.22 ± 7.67 [a]	52.58 ± 8.57 [a]	56.20 ± 8.18 [b]	59.76 ± 6.15[b]	59.72 ± 6.47[b]	50.31 ± 6.60[a]	59.98 ± 6.29[a]	51.99 ± 7.61[a]	50.47 ± 6.77[a]
Magnesium (mg/l)	137.23 ± 11.45 [a]	136.11 ± 12.02 [a]	132.71 ± 48.02 [a]	129.99 ± 37.38 [a]	129.26 ± 37.07 [a]	129.66 ± 37.10 [a]	146.67 ± 21.78 [a]	145.74 ± 19.58 [a]	129.26 ± 37.07 [a]
Iron (mg/l)	2.23 ± 0.42 [a]	2.29 ± 0.73 [a]	3.27 ± 0.98 [b]	2.96 ± 2.18[b]	4.01 ± 0.12[c]	2.06 ± 0.02 [a]	5.16 ± 1.74 [c]	3.37 ± 1.60 [b]	1.21 ± 0.20 [a]
Zinc (mg/l)	0.67 ± 0.16 [a]	0.72 ± 0.15 [a]	0.77 ± 0.09 [a]	0.66 ± 0.24 [a]	0.67 ± 0.23 [a]	0.64 ± 0.17 [a]	0.67 ± 0.23 [a]	0.57 ± 0.04 [a]	0.56 ± 0.04 [a]
Manganese (mg/l)	0.02 ± 0.01 [a]	0.03 ± 0.02 [a]	0.02 ± 0.02 [a]	0.03 ± 0.02 [a]	0.04 ± 0.02 [a]	0.03 ± 0.02 [a]	0.02 ± 0.02 [a]	0.03 ± 0.03 [a]	0.00 ± 0.00 [a]
Cadmium (mg/l)	0.01 ± 0.02 [a]	0.02 ± 0.02 [a]	0.00 ± 0.00 [a]	0.01 ± 0.03 [a]	0.00 ± 0.00 [a]	0.00 ± 0.00 [a]	0.02 ± 0.03 [a]	0.02 ± 0.03 [a]	0.00 ± 0.00 [a]
Lead (mg/l)	0.12 ± 0.04 [a]	0.11 ± 0.07 [a]	0.09 ± 0.08 [a]	0.11 ± 0.05 [a]	0.09 ± 0.06 [a]	0.13 ± 0.07 [a]	0.14 ± 0.06 [a]	0.12 ± 0.08 [a]	0.11 ± 0.04 [a]
Copper (mg/l)	0.03 ± 0.03 [a]	0.04 ± 0.03 [a]	0.03 ± 0.03 [a]	0.04 ± 0.04 [a]	0.03 ± 0.04 [a]	0.05 ± 0.04 [a]	0.03 ± 0.04 [a]	0.03 ± 0.04 [a]	0.03 ± 0.03 [a]
Chromium (mg/l)	0.76 ± 1.21 [a]	0.60 ± 0.28 [a]	0.60 ± 0.28 [a]	0.56 ± 0.32 [a]	1.29 ± 1.05 [a]	1.17 ± 0.86 [a]	0.81 ± 1.06 [a]	1.27 ± 0.99 [a]	0.82 ± 1.30 [a]
Sulphate (mg/l)	0.87 ± 0.20 [a]	0.86 ± 0.17 [a]	0.86 ± 0.17 [a]	0.99 ± 0.22 [b]	0.92 ± 0.02 [b]	0.85 ± 0.04 [a]	0.84 ± 0.05 [a]	0.85 ± 0.11 [a]	0.94 ± 0.12 [b]

	Location								
Parameters	**1**	**2**	**3**	**4**	**5**	**6**	**7**	**8**	**9**
Ammonia (mg/l)	2.63 ± 1.01 ^{a}	2.79 ± 1.11 ^{a}	2.79 ± 1.11 ^{a}	2.79 ± 1.20 ^{a}	2.76 ± 1.08 ^{a}	2.80 ± 1.10 ^{a}	2.75 ± 1.06 ^{a}	2.22 ± 1.00 ^{a}	2.38 ± 1.03 ^{a}
Phosphate (mg/l)	0.15 ± 0.16 ^{a}	0.38 ± 0.02 ^{a}	0.38 ± 0.02 ^{a}	0.18 ± 0.15 ^{a}	0.24 ± 0.14 ^{a}	0.21 ± 0.14 ^{a}	0.20 ± 0.13 ^{a}	0.23 ± 0.20 ^{a}	0.24 ± 0.23 ^{a}
Nitrite (mg/l)	1.90 ± 1.12 ^{a}	1.70 ± 1.06 ^{a}	1.70 ± 1.06 ^{a}	2.33 ± 1.60 ^{a}	2.10 ± 1.62 ^{a}	1.66 ± 1.08 ^{a}	1.90 ± 1.09 ^{a}	1.95 ± 1.15 ^{a}	1.61 ± 0.48 ^{a}
Nitrate (mg/l)	2.87 ± 1.28 ^{a}	2.34 ± 0.80 ^{a}	2.34 ± 0.80 ^{a}	3.23 ± 0.71 ^{b}	3.36 ± 1.11 ^{b}	2.33 ± 0.76 ^{a}	1.90 ± 0.17 ^{a}	2.23 ± 0.67 ^{a}	2.01 ± 0.16 ^{a}
Nickel (mg/l)	0.97 ± 0.61 ^{a}	0.91 ± 0.26 ^{a}	0.91 ± 0.26 ^{a}	1.00 ± 0.42 ^{a}	0.90 ± 0.25 ^{a}	0.84 ± 0.22 ^{a}	0.94 ± 0.15 ^{a}	0.94 ± 0.17 ^{a}	0.95 ± 0.15 ^{a}
TPH (mg/l)	14.86 ± 0.38 ^{d}	10.41 ± 4.55 ^{c}	13.861 ± 1.21 ^{d}	13.00 ± 1.52 ^{d}	2.81 ± 1.58 ^{b}	11.57 ± 1.62 ^{c}	4.07 ± 0.94 ^{b}	3.84 ± 0.45 ^{b}	0.001 ± 0.00 ^{a}

Similar superscript means not significantly different ($p > 0.05$) while different superscripts indicates significantly difference in means ($p < 0.05$).

Table 2.
Comparison of the parameters in the different locations during rainy season.

	Location								
Parameters	**1**	**2**	**3**	**4**	**5**	**6**	**7**	**8**	**9**
Temperature (°C)	32.38 ± 0.58^{b}	32.46 ± 1.33^{b}	30.98 ± 0.92^{b}	29.62 ± 1.23^{a}	30.50 ± 0.96^{b}	29.00 ± 1.05^{a}	29.70 ± 0.64^{a}	29.74 ± 1.17^{a}	29.52 ± 1.17^{a}
pH	7.23 ± 0.90^{a}	5.99 ± 0.48^{a}	6.43 ± 0.70^{a}	6.99 ± 1.03^{a}	6.54 ± 0.57^{a}	7.22 ± 1.17^{a}	6.16 ± 1.01^{a}	6.56 ± 0.60^{a}	6.02 ± 0.19^{a}
Conductivity (ЧSCM^{-1})	35.66 ± 8.36^{c}	11.93 ± 1.27^{a}	32.26 ± 0.21^{c}	46.76 ± 15.34^{d}	39.35 ± 17.98^{c}	14.44 ± 2.08^{a}	24.58 ± 6.55^{b}	18.97 ± 0.3^{a}	17.34 ± 6.80^{a}
Turbidity (NTU)	5.27 ± 1.36^{a}	4.39 ± 4.26^{a}	11.40 ± 12.64^{b}	11.56 ± 8.23^{b}	2.84 ± 1.55^{a}	3.09 ± 1.19^{a}	16.75 ± 2.54^{c}	1.89 ± 0.19^{a}	1.82 ± 0.15^{a}
DO (mg/l)	18.69 ± 0.04^{b}	18.46 ± 1.17^{b}	19.10 ± 0.9^{b}	18.48 ± 0.54^{b}	17.09 ± 0.04^{a}	17.03 ± 0.14^{a}	17.90 ± 0.03^{a}	17.86 ± 0.28^{a}	17.09 ± 0.05^{a}
BOD) (mg/l)	5.60 ± 0.09^{b}	5.33 ± 0.44^{b}	4.95 ± 0.16^{a}	5.75 ± 0.11^{b}	6.03 ± 0.02^{c}	6.09 ± 0.05^{c}	5.54 ± 0.04^{b}	5.98 ± 0.11^{c}	4.86 ± 0.08^{a}
COD (mg/l)	21.87 ± 0.61^{a}	24.58 ± 0.26^{a}	22.64 ± 0.27^{a}	32.49 ± 12.21^{b}	24.53 ± 0.27^{a}	24.20 ± 0.06^{a}	23.19 ± 0.28^{a}	23.79 ± 0.19^{a}	22.53 ± 0.34^{a}
Acidity (mg/l)	156.72 ± 4.79^{a}	158.18 ± 11.08^{a}	168.82 ± 1.22^{a}	151.82 ± 0.44^{a}	152.18 ± 2.90^{a}	150.53 ± 0.88^{a}	167.09 ± 0.46^{a}	167.51 ± 0.49^{a}	138.08 ± 49.31^{a}
Alkalinity (mg/l)	53.52 ± 41.47^{b}	52.43 ± +46.56^{b}	30.57 ± 8.94^{a}	62.47 ± 0.47^{b}	23.66 ± 0.09^{a}	23.88 ± 0.26^{a}	64.25 ± 0.14^{b}	64.58 ± 0.47^{b}	24.22 ± 0.66^{a}
TH (mg/l)	44.79 ± 1.18^{b}	36.44 ± 1.52^{a}	37.35 ± +1.27^{a}	50.95 ± 0.68^{c}	51.66 ± 0.11^{c}	51.37 ± +0.30^{c}	43.38 ± 0.60^{b}	44.42 ± 0.52^{b}	35.78 ± 3.07^{a}
TDS (mg/l)	6.37 ± 1.11^{a}	7.56 ± 3.32^{a}	14.84 ± 0.17^{b}	16.25 ± 0.39^{c}	16.15 ± 0.12^{c}	15.72 ± 0.18^{c}	14.15 ± 0.42^{b}	14.01 ± 0.14^{b}	6.03 ± 0.07^{a}
TSS (mg(l)	34.66 ± 0.44^{d}	31.30 ± 0.76^{b}	32.20 ± 1.04^{c}	34.52 ± 0.27^{d}	34.54 ± 0.23^{d}	34.14 ± 0.20^{d}	33.77 ± 0.11^{d}	34.76 ± 0.43^{d}	29.56 ± 0.99^{a}
Salinity (mg/l)	0.11 ± 0.08^{a}	0.09 ± 0.04^{a}	16.00 ± 8.94^{b}	12.02 ± 13.02^{b}	0.11 ± 0.11^{a}	0.71 ± 0.39^{a}	8.00 ± 8.37^{b}	0.18 ± 0.06^{a}	0.04 ± 0.03^{a}
Chloride (mg/l)	27.10 ± 1.62^{b}	27.62 ± 0.88^{b}	29.22 ± 0.37^{d}	31.16 ± 0.03^{e}	31.27 ± 0.19^{e}	30.36 ± 0.35^{e}	25.26 ± 0.39^{a}	25.49 ± 0.54^{a}	26.44 ± 0.51^{b}
Fluoride (mg/l)	0.00 ± 0.00^{a}	0.00 ± 0.00^{a}	0.00 ± 0.00^{a}	0.02 ± 0.04^{a}	0.68 ± 0.41^{b}	0.45 ± 0.00^{b}	1.02 ± 0.22^{c}	0.32 ± 0.29^{b}	0.00 ± 0.00^{a}
Aluminum (mg/l)	0.00 ± 0.00^{a}	0.00 ± 0.00^{a}	0.00 ± 0.00^{a}	0.01 ± 0.00^{a}	0.01 ± 0.01^{b}	0.00 ± 0.01^{a}	0.02 ± 0.01^{b}	0.01 ± 0.00^{b}	0.00 ± 0.00^{a}

	Location								
Parameters	**1**	**2**	**3**	**4**	**5**	**6**	**7**	**8**	**9**
Sodium (mg/l)	9.47 ± +2.00^{a}	11.77 ± 0.12^{b}	9.22 ± 1.95^{a}	13.33 ± 0.16^{c}	12.66 ± 0.12^{b}	13.11 ± 0.21^{c}	11.70 ± 0.15^{b}	11.81 ± 0.19^{b}	9.41 ± 1.19^{a}
Potassium (K) (mg/l)	1.52 ± 0.05^{a}	1.57 ± 0.38^{a}	3.03 ± 0.03^{b}	4.11 ± 0.03^{c}	3.79 ± 0.13^{b}	3.81 ± 0.12^{b}	3.04 ± 0.02^{b}	3.11 ± 0.04^{b}	1.37 ± 0.36^{a}
Calcium (Ca) (mg/l)	44.38 ± 2.21^{b}	46.32 ± 6.36^{b}	44.33 ± 0.41^{b}	50.73 ± 0.11^{c}	50.58 ± 0.09^{c}	40.56 ± 0.29^{a}	50.44 ± 0.30^{c}	40.59 ± 0.21^{a}	40.37 ± 0.47^{a}
Magnesium (mg/l)	149.44 ± 10.30^{a}	151.87 ± 1.82^{a}	178.57 ± 0.25^{c}	184.68 ± 0.25^{d}	184.35 ± 0.57^{d}	184.28 ± 0.89^{d}	178.13 ± 0.71^{c}	175.85 ± 4.68^{c}	184.35 ± 0.57^{d}
Iron (mg/l)	1.74 ± 0.45^{a}	2.47 ± 1.27^{b}	3.82 ± 1.00^{c}	1.07 ± 0.52^{a}	2.71 ± 1.62^{b}	1.85 ± 0.45^{a}	4.42 ± 1.56^{c}	2.88 ± 1.64^{b}	0.95 ± 0.02^{a}
Zinc (mg/l)	0.86 ± 0.10^{b}	0.83 ± 0.12^{b}	0.91 ± 0.04^{c}	1.01 ± 0.00^{c}	1.00 ± 0.00^{c}	0.85 ± 0.04^{b}	0.80 ± 0.45^{b}	0.63 ± 0.02^{a}	0.61 ± 0.00 ^{a}
Manganese (mg/l)	0.04 ± 0.02^{a}	0.04 ± 0.01^{a}	0.07 ± 0.01^{c}	0.08 ± 0.00^{c}	0.08 ± 0.00^{c}	0.06 ± 0.01^{b}	0.06 ± 0.00^{b}	0.07 ± 0.00^{c}	0.03 ± 0.01^{a}
Cadmium (mg/l)	0.04 ± 0.00^{b}	0.05 ± 0.01^{c}	0.01 ± 0.00 ^{a}	0.06 ± 0.00^{d}	0.01 ± 0.00^{a}	0.01 ± 0.00^{a}	0.06 ± 0.00^{d}	0.05 ± 0.00^{c}	0.01 ± 0.00^{a}
Lead (mg/l)	0.01 ± 0.02^{a}	0.01 ± 0.00^{a}	0.03 ± 0.06^{b}	0.01 ± 0.00^{a}	0.01 ± 0.00^{a}	0.01 ± 0.01^{a}	0.01 ± 0.00^{a}	0.00 ± 0.00^{a}	0.01 ± 0.00^{a}
Copper (mg/l)	0.07 ± 0.00^{a}	0.08 ± 0.01^{b}	0.07 ± 0.01^{a}	0.09 ± 0.01^{c}	0.08 ± 0.00^{b}	0.08 ± 0.00^{b}	0.08 ± 0.01^{a}	0.09 ± 0.01^{c}	0.07 ± 0.01^{a}
Chromium (mg/l)	2.59 ± 0.03^{c}	1.01 ± 0.00^{b}	0.81 ± 0.28^{b}	2.81 ± 0.04^{c}	2.76 ± 0.06^{c}	2.38 ± 0.30^{c}	2.63 ± 0.08^{c}	2.76 ± 0.06^{c}	0.49 ± 0.49^{a}
Sulphate (mg/l)	1.08 ± 0.18^{b}	1.21 ± 0.01^{a}	1.17 ± 0.12^{b}	0.96 ± 0.01^{a}	0.93 ± 0.02^{a}	0.88 ± 0.05^{a}	0.99 ± 0.02^{a}	1.01 ± 0.01^{b}	0.88 ± 0.05^{a}
Ammonia (mg/l)	3.79 ± 0.86^{b}	4.30 ± 0.13^{c}	4.39 ± 0.26^{c}	3.97 ± 0.98^{c}	4.35 ± 0.12^{c}	4.35 ± 0.05^{c}	3.61 ± 0.05^{b}	3.67 ± 0.41^{b}	1.00 ± 0.00^{a}
Phosphate (mg/l)	0.50 ± 0.01^{c}	0.41 ± 0.01^{b}	0.37 ± 0.01^{a}	0.44 ± 0.00^{b}	0.43 ± 0.01^{b}	0.41 ± 0.02^{b}	0.53 ± 0.02^{c}	0.55 ± 0.05^{c}	0.36 ± 0.04^{a}
Nitrite (mg/l)	2.71 ± 0.98^{a}	2.95 ± 0.49^{a}	4.57 ± 0.07^{c}	4.53 ± 0.08^{c}	3.06 ± 0.13^{b}	3.55 ± 0.18^{b}	3.35 ± 0.66^{b}	2.31 ± 0.02^{a}	2.64 ± 0.05^{a}

Parameters	Location								
	1	2	3	4	5	6	7	8	9
Nitrate (mg/l)	3.98 ± 1.16^{b}	3.46 ± 0.04^{b}	4.57 ± 0.33^{c}	4.32 ± 0.84^{c}	3.46 ± 0.04^{b}	2.23 ± 0.03^{a}	3.39 ± 0.04^{b}	2.05 ± 0.49^{a}	2.14 ± 0.05^{a}
Nickel (mg/l)	1.32 ± 0.72^{a}	1.24 ± 0.16^{a}	1.40 ± 0.27^{a}	1.17 ± 0.00^{a}	1.16 ± 0.01^{a}	1.16 ± 0.01^{a}	1.17 ± 0.01^{a}	1.17 ± 0.04^{a}	1.14 ± 0.01^{a}
TPH (mg/l)	3.92 ± 1.43^{b}	13.40 ± 1.67^{d}	11.60 ± 0.89^{d}	12.80 ± 1.48^{d}	1.44 ± 0.96^{a}	3.10 ± 0.34^{b}	9.76 ± 3.36^{c}	1.04 ± 0.19^{a}	0.00 ± 0.00^{a}

Table 3.
Comparison of the parameters in the different locations during dry season.

ranging from acidic to alkaline. The main variables affecting pH in any milieu are: dissolved oxygen, water temperature, land runoff, decomposition of organic matter and processes such as precipitation and oxidation reduction. The highest pH was found in location 2, possibly due to more intensified human activity. The lowest was detected in location 9, indicating that gas flaring at Ebocha-Obrikom area of Rivers State affected groundwater acidity [1–3, 6, 12–14, 19]. Similarly, the analytical results during the dry season discovered that pH ranged from 5.99 to 7.23, which was within the WHO [20] acceptable pH range of 6.5-8.5 for drinking water, except at sampling location 1, which had the highest pH value. Thus, water with a pH of 7 denotes neutral water, and a value less than 7 denotes acidic water. Increasing pH values could lead to increase in the rate of corrosion. In any of the groundwater tests, nonetheless, no location was determined to remain beyond the maximum permissible limit. The pH ground water variance in the Ebocha-Obrikom area was below the allowable range and thus not dangerous for drinking [1–3, 6, 14]. Electric conductivity (EC) is a measurement of an ion's ability to carry electric current in a solution [1, 2, 14, 21]. The higher the conductivity, the warmer the water. Oil, alcohol, phenol, as well as sugar are organic compounds that do not transmit electrical current well. The EC is often used to calculate the ionic concentration of groundwater, which fluctuates based on the concentration, ions type and temperature of water. The most important test that reveals the total concentration of soluble salts is the conductivity test, according to Kanga et al., [22]. Thus, electrical conductivity (EC) diverges from (24.22–52.29) (11.93–46.76) μs/cm, with an average of (31.83–26.81) μs/cm (**Tables 2** and 3). Electrical conductivity (EC) remains a measure of a material's capacity toward conducting an electric current, and suggests a wide range of salts existing in groundwater. The percolation of agrochemicals and natural groundwater recharge processes increase the EC value [1, 2, 14, 21, 23]. Hence, higher values for conductivity at location 2&4 could be attributed to excessive accumulation of dissolved salts, spilled oil through possible emission of flared gases or salinization of groundwater [1, 2, 12, 14, 21, 24]. The findings were supported by Ehirim and Nwankwo, [25]; Morufu and Clinton, [13]; Olalekan et al., [14] and Morufu et al., [2] which established that electrical conductivity values of the ground water samples collected from the studied location are observed to be low throughout the sampling locations, as the variations in their mean concentrations at different distances. According to Okafor and Opuene [26]; Morufu and Clinton, [13] and Olalekan et al., [14], electrical conductivity reflects the degree of salinity; thus, it has a significant impact on water taste as well as user approval. The American Public Health Association (APHA) [16] and Morufu and Clinton [13] defines turbidity as "the optical quality of water sample that causes light to remain dispersed rather than diffused". The capacity of light to flow through water is related to the suspended particles volume. The more suspended particles there are in the water, the cloudier it becomes. An electronic turbidity meter is used to measure turbidity. APHA recommends that the turbidity of drinking water not exceed 5 NTUs. If turbidity is mostly caused by organic particles, depletion of dissolved oxygen in the water may occur [1, 2, 7, 8, 13, 27–29]. As a result, high turbidity levels may raise the risk of waterborne disease [1, 2, 6, 14, 21, 27–29]. Highest value of turbidity was noticed at location 7 (48.24–16.75) NTU during the rainy and dry season. Groundwater turbidities remained below the typical maximum tolerable limit of 5NTU for drinking water at location 6 & 9 for rainy season and location 5, 6, 8 & 9 for dry seasons. Despite this, location 8 (rainy season) and 1 & 2 (dry seasons) were close to the maximum permitted limit for drinking water. Thus, turbidity levels were higher during rainy season than the dry season. Hence, the wet season has a greater influence on turbidity than the dry season. This could be due to the research area's consistent and large tendency to receiving massive volumes of organic and inorganic material produced by gas flaring and oil spillage contaminating ground water. Dissolved Oxygen studies in water continue to be important since it is recognized as one of the most critical limiting elements for aquatic species' survival. It is an important metric in measuring pollution levels since sewage pollution is an organic pollutant that affects fish and other aquatic life. Bacteria devour oxygen as organic matter decomposes. As a result of this, an oxygen deficient milieu can emerge in lakes and rivers. The highest value of Dissolved Oxygen (DO) in groundwater was observed at location 3 (17.84–19.10) mg/l during both the rainy and dry seasons. These conditions can eventually lead to fish kills, restricted growth, disturbance of life cycles, migration to avoid unfavorable condition and mortality of benthic animals' creatures [1, 2, 6, 12–14, 21, 30, 31]. The BOD is the amount of oxygen required by bacteria during the breakdown of organic materials. It contains the oxygen required for the oxidation of numerous compounds found in water, such as sulfides, ferrous iron, and ammonia [1, 2, 7, 8, 9, 32]. Meanwhile, research by Chapman and Kimstach [32]; Morufu and Clinton, [13]; Raimi and Sabinus, [12]; Olalekan et al., [14] and Morufu et al., [2] discovered that DO concentration below 5 mg/l have a deleterious impact on the survival of biological communities. This measure represents the ability of microbial respiration to break down organic material in water, which results in low DO and may be a cause of hypoxia [1, 2, 12–14]. Despite this, biological oxygen demand (BOD) reflects the amount of oxygen needed by bacteria. It is used to determine any receiver environment pollution potential as well as assimilation capacity. The present study for BOD had its highest value at location 4 & 6 (5.48–6.09) mg/l during rainy and dry seasons. The values were higher during the dry season than during the rainy season. As a result, it is possible to extrapolate those anthropogenic activities may affect greater BOD during the dry season while supporting higher metabolic activity. Regardless of seasonal changes, both seasons influenced BOD. This pattern could have been caused by gas flaring. The chemical oxygen demand (COD) remain the amount of oxygen required to oxidize organic compounds in waste water using a powerful oxidant and convert them to carbon dioxide and water. The COD test is used to measure the pollution level in a certain location. COD readings are always greater than BOD_5 values because numerous organic molecules can be chemically oxidized but not physiologically [1, 2, 21, 33, 34]. As a result, the chemical oxygen demand (COD) is used to determine the level of pollution in water. When the COD level in the water surpasses 25 mg/l, it indicates that there is a larger concentration of contaminants. While, COD values were found to be highest at location 4 (41.78–32.49) mg/l during rainy and dry season. This indicates that organic pollution of water is more severe during rainy season than dry seasons. COD is used to calculate the amount of oxygen required by organic and inorganic substances. All reported values in this investigation were above the maximum acceptable limit of 10 mg/l for COD [1, 2, 21, 33, 34]. In unpolluted surface and ground waters, the existence of dissolved carbon dioxide is commonly the dominant acidifying agent. Apart from a palatability problem in very acidic waters, there is no specific implication [1, 2, 35]. The water acidity affects its corrosiveness as well as its speciation of other components. Thus, acidity values range from highest at location 3 (101.18–168.82) mg/l for both rainy season and dry season (**Tables 2** and **3**). There is currently no maximum value set for acidity according to WHO/SON/NAFDAC standards of potability. Alkalinity is a measure of water's ability toward neutralizing acids as well as indicates its buffer capacity [6, 12–14, 21, 35]. Also, the existence of bicarbonates generated in soil reactions via which water infiltrates contributes to the alkalinity of the natural water body. The eutrophication effects on water are also influenced by their pH. Thus, alkalinity was highest at location 4&8 (134.96–64.58) during the rainy and dry season and lowest at location 3&5 (18.17–23.66). Water hardness is a measure of water's ability toward reacting with soap as well as characterizes its ability toward binding soap to form scum or lather which is a reaction that is chemically harmful toward the process of washing [1, 2, 6, 12–14, 21]. The high rate might remain attributed toward constant discharge of acidic and chemicalized substances through oil spillage and gas flaring which latter find their way into the groundwater bodies and adjoining environment. Because calcium as well as magnesium remain the only remarkable minerals that are known to induce hardness. The sources of the metallic ions remain often found in sedimentary rocks, the most prevalent of which remain limestone ($CaCO_3$) as well as dolomite ($CaMg(CO_3)_2$) [1, 2, 6, 12–14, 21]. The concentration of TH levels varied from 39.17 mg/l to 44.02 mg/l [both rainy and dry season]. More than half of the sample's groundwater in Ebocha-Obrikom area of Rivers State possess TH below 100 mg/l. Thus, total hardness (TH) of the aquifers fluctuated on average from 39.17 mg/l for rainy season to 44.02 mg/l for dry seasons, with the lowest confined groundwater (mean = 35.02 at location 5) (35.78 at location 9), and highest confined groundwater (mean = 42.22 mg/l at location 3) (51.66 at location 5). Groundwater hardness varied from 198.5 to 409.5 mg/l, with a mean of 289.1 in the Upper Tigris River Basin, Diyarbakır-Batman, Turkey, Koffi et al. [36] discovered that the increased groundwater hardness was caused by carbonate sources. The research results contradicted Disli [37] who had found that the TH level varied "from 198 to 400 mg/l" in this region, but were mostly caused by mineral dissolution like carbonates and gypsum [1, 2, 6, 12–14, 21, 38, 39]. In the crystalline basement complex rock of India, Adimalla et al. [40] obtained TH values ranging from 60 to 750 mg/l, with approximately 18% of the samples falling into the moderately hard category, whereas Koffi et al. [36] recorded TH values ranging from 50.8 to 272 mg/l, with 60.6% of samples falling into the moderately hard category. Despite this, the maximum allowable level of TH for drinking purposes is 500 mg/l, with a suggested limit of less than 100 mg/l [20]. The groundwater in the Ebocha-Obrikom oil and gas area was found to be 100% safe, with all samples falling within the maximum permitted 500 mg/l limit. Conversely, Ezekwe et al., [41] claims that subsurface waters remain often tougher than surface waters. Total Dissolved Solids (TDS) refers to the various minerals that remain existent in water in dissolved form and is a pointer of water salinity as well as signifies dissolved salts in water [1, 2, 12–14, 42]. Consuming water of high TDS for an extended period of time can result in kidney stones as well as promote gastrointestinal discomfort in consumers [1, 2, 6, 27]. TDS levels above a certain threshold impair the palatability of water and promote gastrointestinal discomfort in consumers. It is also, an important metric for determining the appropriateness of irrigation as well as drinking water. WHO [20] claims that groundwater taste with a TDS level of less than 600 mg/l is regarded good for aquatic lives and residential water supply protection? High (TDS) levels in groundwater may cause unpleasant taste as well as gastrointestinal complications, according to the World Health Organization (WHO) [1, 2, 12–14, 43]. High TDS maybe derived from intensive or massive usage of agrochemical, dissolution of salts, ion exchange, organic materials, and sediment dissolution, aquifer percolation and allied substances emanating from oil related activities such as gas flaring. Thus, groundwater contamination in this wise could be due to the continuous contamination of groundwater by industrial pollutants as suggested by Olalekan et al., [14] and Olalekan et al., [6]. The total dissolved solids (TDS) show a very weak variability as seen by their low standard deviation (SD). The overall hydro chemical groundwater characteristics are regulated by major ions [1, 2, 12–14, 21, 44]. Hence, the groundwater samples were desirable as well as allowed for purposes of drinking based on the TDS categorization. WHO [20] recommends a concentration of 1500 mg/l for fisheries and aquatic life protection, and for household water supply? Because all values remained below the tolerable limit, they remain safe for drinking on TDS basis as supported by researches from Dami et al. [45]; Morufu and Clinton [13]; Raimi and Sabinus [12] and Olalekan et al. [14]. The difference in TDS indicates a wide variation in the geochemical processes. Concentrations of TDS in Ebocha-Obrikom area remain below the optimal threshold in all locations, according to earlier research, Besides, TDS concentrations remained sufficient in quality for drinking in all areas. According to Adimalla and Qian [46], about 95% of the total samples remained below ideal drinking threshold. The mean values for total suspended solids (TSS), demonstrate that the greatest value in groundwater was witnessed at location 1&8 (39.80–34.76) mg/l for rainy and dry seasons. The least value of (34.00–29.56) mg/l at location 9 for rainy or dry seasons respectively. All of the values noted in this investigation were above the maximum allowable limit. Indicating that gas flaring and oil spillage releases persistent non-combustible chemicals and less dense volatile chemicals into the environment. All groundwater comprises salts solution; and documented salt contents extend from less than 25 mg/l in a quartzite spring to above 300,000 mg/l in brines [1, 47]. Because of the larger exposure toward soluble elements in geologic strata, groundwater often has a larger concentration of dissolved components than surface water. Bicarbonate, which is typically the predominant groundwater anion, is produced from the released of carbon dioxide in the soil by organic breakdown. Salinity values range from highest at location 7 (25.71) mg/l for rainy season to location 3 (16.00) mg/l (**Tables 2** and **3**). The maximum value is set at 600 mg/l according to WHO/SON/NAFDAC standards of potability. All of the readings obtained in this study were less than the maximum allowable limit of 600 mg/l for drinking water. Chloride can be present in a variety of chemical and non-chemical components in the body. It is an essential component of the salt found in many foods and used in cooking. Even in small children, too much chloride from salted meals can raise blood pressure [1, 2, 12–14, 48, 49]. Excessive use of drinking water containing sodium chloride at concentrations greater than 2.5 g/l linked to hypertension [1, 2, 14]. Thus, a number of studies alleged that Cl excess in groundwater is an indicator pollution index and has a harmful influence on human health [1, 2, 12–14, 40, 50]. Though, chloride is also one of the prominent anions in Rivers State oil and gas producing area of Ebocha-Obrokom, ranging from (28.33–32.16) (25.26–31.27) mg/l for both rainy and dry season with a mean of (30.60–28.21) mg/l (**Tables 2** and **3**). It has been noted that while water with low chloride ions is not dangerous, but chloride ions at large concentrations can kill floras when used for horticultural or agricultural applications. It may also be to blame for the unpleasant taste of water consumed [51]. While samples at location (3 & 5) for rainy and dry season were high in the Ebocha-Obrikom region. High levels of chloride may perhaps remain linked to domestic waste effluents, septic tanks leakage, as well as chloride bearing rocks dissolution [1, 2, 6, 12–14, 21, 36, 52, 53]. In spite of the fact that no health dangers have been established, residents of Ebocha-Obrikom areas remain hesitant toward drinking water due to texture and taste issues. High Cl^- groundwater concentrations remain seen as a symptom of pollution from a number of sources, and they impart a salty flavor to the water [1, 2, 12–14, 54]. Furthermore, chloride concentration in drinking water above 200 mg/l has been linked to heart disease, asthma and possibly cancer. It is usually safe to drink water within the limits of 0.5–1.5 mg/l according to the suggested guidelines [1, 2, 6, 12–14, 20, 21, 55]. Fluoride becomes harmful to health at quantities above/below this recommendation, and is denoted as a double-edged sword [6]. Water consumers remain prone to dental carries at lower concentration, while at larger concentrations, it can induce skeletal fluorosis, debilitating fluorosis, dental fluorosis, as well as kidney damage [1, 2, 12–14]. Fluorine remains the 13th most prevalent element in the earth's crust but it is essential to human life. Excess consumption of fluoride has been associated to infertility, abortion, fertility, as well as hypertension [56]. Water ingestion and skin absorption remain the primary sources of trace elements intake in the milieu [1, 2, 12–14, 57, 58]. Heavy groundwater fluoride concentrations are a pervasive problem around the globe; particularly in the global south, where individuals remain disproportionately impacted by fluorosis due to high reliance on groundwater. Excessive levels of fluoride in drinking water might result in a decrease in total erythrocyte, hematocrit value and hemoglobin percentage, as well as protein content. In trace amounts, fluoride is advantageous to the human health as it can minimize dental caries risk even though encouraging strong bones formation [1, 2, 59, 60]. Fluorine ($F-$) in groundwater in this study either falls below or within the WHO/SON/NAFDAC limit. The lowest as well as highest values (0.00 and 0.85 mg/l) (0.00 and 1.02 mg/l) for rainy and dry season were observed in Ebocha-Obrikom area of Rivers State. Aside from increased hydrodynamics during gas flaring production, water mixing from diverse aquifers might also affect $F-$ in location 4, 5, 6, 7, and 8 [both rainy and dry season]. Thus, chronic exposure toward fluoride raises the risk of developing tooth decay, as well as other dental diseases [9]. Aluminum had highest recorded value of (0.03) (0.02) mg/l at location 7 during the rainy and dry season. However, its presence in other locations could be attributed to gas flaring and this of course calls for serious concern. Sodium [Na^+] is abundant in rocks and soils. It is always present in natural water and is used medicinally as a laxative. In the Ebocha-Obrikom oil and gas area of Nigeria, Na^+ varied from 14.33 mg/l in rainy season to 11.39 mg/l during dry season. Still, the highest [Na^+], 16.39 mg/l was observed in location 7 of the unconfined groundwater, while the lowest mean [Na^+], 12.22 mg/l, was observed in location 3 of the unconfined groundwater. This suggests that the region's groundwater is affected by more complex factors. Overall, Na^+ had the highest concentration in location 2, 4, 5, 6, 7 & 8. Na^+ ions in groundwater are largely regulated by weathering and water-rock interactions. Excess of Na^+ also indirectly indicates the process of ion exchange in water formation [61, 62]. In the Ebocha-Obrikom oil and gas area, principal lithology is occupied by crystalline rocks. Weathering of these rock forming minerals might likely be the chief source for elevated Na^+ concentration. For potassium (K^+), it usually exists at low concentrations in groundwater because of weak mobility [63]. The highest [K^+] (3.29 mg/l) was observed in location 8 of the shallow confined groundwater. Mg^+ possesses the highest SD value, indicating a very high spatial variability. Ca^{2+}, Mg^{2+}, and Na^+ range between (50.31–59.98) (40.37–50.73) (129.26–146.67) (149.44–184.68), and (12.22–16.39) (9.22–13.33) mg/l, respectively. As a result, He and Wu [64] reported that K^+ in groundwater is one of the essential trace elements for human health. K^+ occurs naturally in drinking water in amounts considerably below those deemed hazardous to human health; it is the most vital nutrient for humans, and too much of it can cause constipation [20]. However, high levels of K^+ in drinking water (beyond the regulatory limit) might cause hypertension, high blood pressure, hyperkalemia, and, in the worst-case scenario, a heart attack. Ca^{2+} is one of the dominant cations in the Ebocha-Obrikom oil and gas area of Rivers State groundwater, ranging from (50.31–59.98) (40.37–50.73) mg/l with a mean of (54.80–45.37) mg/l. There is some evidence that the incidence of heart disease is reduced in areas served by public water supply with a high degree of hardness, the primary constituent of which is calcium, so that the presence of the element in a water supply is beneficial to health. While, calcium and magnesium contribute to the formation and solidification of bones and teeth and play a role in the decrease of neuromuscular excitability, myocardial system, heart, and muscle contractility, intracellular information, transmission, and blood contractility [1, 2, 46]. They also play a major role in the metabolism of almost all cells of the body and interacts with many nutrients [1–3, 61]. However, inadequate, or excess intake of either nutrient can result in adverse health consequences [9]. Magnesium (Mg^{2+}) concentration in groundwater in the Ebocha-Obrikom oil and gas producing area of River State were between (129.26–146.67) and (149.44–184.68) mg/l throughout rainy and dry season. The main source of Mg^{2+} in groundwater was magnesium containing minerals in the host rocks and also animal, domestic, and industrial wastes [1, 2, 6, 14, 53, 65]. However, all groundwater samples collected above the maximum allowable limit of 150 mg/l. Thus, the presence of exchangeable Na^{+} in the soil may explain the high concentration of Mg^{2+} in groundwater [61, 66]. Although values higher or equal to 100 mg/l are stored particularly in cold climates [1, 2, 46, 67]. Low magnesium status has been implicated in hypertension, coronary heart disease, type 2 diabetes mellitus and metabolic syndrome, endothelial dysfunction, increased vascular reactions, elevated circulating levels of C-reactive protein (a proinflammatory marker that is a risk factor for coronary heart disease) and decreased insulin sensitivity [1, 2, 6, 14, 18, 67, 68]. The concentration of Fe ranges from (1.21–5.16) (0.95–4.42) mg/l both rainy and dry season, and four samples in rainy seasons have the Fe concentration higher than the permissible limit for drinking purpose. While, it is evident that trace metal can be toxic to human health if they are consumed in excess and accumulated in human bodies [1, 2, 6, 12–14, 21, 69]. High concentrations of iron could result in hemochromatosis which is characterized by tiredness, pains in the joints and abdomen [1, 2, 6, 12–14, 21, 70]. This condition is caused by rapid dissolution of iron species in anoxic groundwater. It is well-known that water-quality thresholds may be frequently breached for iron. While, the highest value for zinc was observed at location 3 (0.77) mg/l for rainy season and location 4 (1.01) mg/l for dry season. It was noticed that the maximum permissible limit of 3.00 mg/l for zinc was not exceeded by any of the locations. Zinc at these limit does not pose serious health and environmental effects though significant values were noticed at locations stated above between the seasons. Thus, zinc could be deposited in those locations due to oil related activities, especially during dry season. Zinc deficiency can cause nausea, lack of moisture, tiredness, pains in the abdomen, coordination of the muscles, and kidney failure. It can also cause malabsorption, Acrodermatitis enteropathica, liver damage, renal damage, sickle cell damage, diabetes, malignancy, and other chronic diseases. People most at risk are the elderly, children in rising nations, and individuals with renal deficiency. Signs of mild zinc insufficiency are varied but generally due to unsatisfactory dietary consumption [1, 2, 6, 7, 8, 9, 18, 21, 42]. Physiological consequences of eating a diet high in protein contain depressed development, diarrhea, weakness and late sexual development, alopecia, eye and skin abrasions, decreased appetite, changed perception, decreased host protection possessions, defects in carbohydrate utilization, and reproductive spermatogenesis [1, 2, 6, 12–14, 21]. Two (2) samples are not suitable for drinking (location 2&3–4&5) because of high Mn concentration in groundwater. Manganese is an essential nutrient but neurotoxic at high levels and evidence suggests infants could be uniquely vulnerable to its effects. Manganese exposure in drinking water has been associated with neurodevelopmental outcomes that include reduced IQ or poorer memory, hyperactivity, impulsivity and motor function in children [1, 2, 6, 12–14, 18, 21, 71–73]. Groundwater in Ebocha-Obrikom area of Rivers State contain widely varying amounts of manganese. Water that exceeds the state's reference dose (RfD) is likely to cause harmful effects over a lifetime of exposure. This finding should be seen as a wake-up call for many communities in the Niger Delta region to be aware of their groundwater levels. Cadmium (Cd) are known to increase the risks of lung cancer and renal carcinoma. The highest value for cadmium was observed at location 2, 7 & 8 (0.02) mg/l during the rainy season and location 4 &7 (0.06) mg/l during the dry season. All values recorded in this study area were above the maximum permissible limit of 0.003 mg/l for WHO/SON/NAFDAC. Thus, Cadmium (Cd) is known to cause damage to the kidney, bones in both young and old, also responsible for bronchitis, anemia [1, 2, 6, 12–14, 21]. Lead is classified as a prevalent toxic metal and a major environmental health hazard. Excessive lead causes problems in the synthesis of hemoglobin, kidney disease, mental retardation, anemia and acute or chronic damage to the nervous system. The primary cause of lead's toxicity is its interference with a variety of enzymes since it binds to sulfhydryl groups found in many enzymes. Lead also interferes with the activity of an essential enzyme called delta-aminolevulinic acid dehydrates, or ALAD and ferrochelatase which are important in the biosynthesis of heme, the cofactor found in hemoglobin. Extreme level of lead absorption in the human body can cause death or perpetual harm to the brain, central nervous system and kidneys [1, 2, 6, 12–14, 21, 74]. During the wet season, the greatest value for lead was found at location 7 (0.14) mg/l, whereas during the dry season, the highest value was observed at location 3 (0.03) mg/l. All levels obtained in this research region were either within or above the WHO/SON/NAFDAC maximum acceptable limit of 0.01 mg/l. Long-term lead exposure can be damaging to the circulatory and nervous systems. Lead is found in the human body mostly through water and food. It can be inhaled as lead particles in paints or as excess gases from leaded petroleum products. It is originated in minor quantities in several water bodies and food, particularly fish, which remain seriously focus to industrialized toxic waste. The capability of lead to permit above the barrier blood and brain is mostly due to its capability to extra for calcium ions. Major toxicity of lead causing the brain prefrontal hippocampus, cerebellum and cerebral cortex can lead to a variability of neurological disorder, such as brain injury, psychological delay and nerve injury [1, 2, 6, 12–14, 21, 41, 75]. Long-term exposure to copper can cause irritation of the nose, mouth and eyes and it causes headaches, stomachaches, dizziness, vomiting and diarrhea. Intentionally high uptakes of copper may cause liver and kidney damage even death [76–80]. Copper is a ductile metal with very high thermal and electrical conductivity. The metal and its alloys have been used for thousands of years. Copper had its highest of (0.05) mg/l at location 6 for rainy season and (2.81 mg/l at locations 4 during the dry season. Contamination of drinking water by copper could be by directly polluting water sources or through rusting of copper pipes and materials. High values of copper could lead to the development of chronic anemia [1, 2, 6, 12–14, 21, 81]. One of the most prevalent contaminants detected in industrial effluents is copper. Excessive copper consumption causes gastrointestinal issues, kidney damage, anemia, and lung cancer. Copper is deadly to humans in concentrations ranging from 4 to 400 mg/kg body weight. Lower concentrations of copper ions might elicit food poisoning symptoms (headache, nausea, vomiting, diarrhea) [1, 2, 6, 12–14, 21]. The liver is the major organ affected by copper poisoning in humans. Finally, copper intoxication produces Wilson's disease in humans. In this study, chromium had its highest of (1.29) mg/l at location 5 for rainy season and (2.81) mg/l at locations 4 throughout the dry season. All values remained above the maximum permissible chromium level. Thus, inhaling hexavalent chromium compounds on a regular basis raises lung cancer risk. Chromium (VI) ingestion can potentially induce stomach and intestinal discomfort or ulcers. Although, chromium does not pose any known serious environmental and public health threat, its current concentration must be continuously monitored since it may perhaps be attributable to gas flaring. While, chromium is a highly toxic element due to its ability to penetrate cell membranes and at high exposure level can cause liver damage. Levels more than 0.05 mg/l of chromium (VI) in drinking water can result in convulsions, diarrhea, abdominal pain, vomiting, indigestion, as well as damage to liver and kidney. Chromium is essential for human nutrition and is considered non-toxic [1, 2, 6, 12–14, 21, 82–86] Chromium exposure in the environment involves compound mixtures identified to cause multi organ poisonousness like allergy, asthma, kidney damage and, in severe cases, cancer of the respiratory tract in humans [1, 2, 6, 14, 87]. Impatience as well as small intestine ulceration and anemia are some of the most serious health problems reported in humans after consuming chromium combinations. Despite the fact that evidence of chromium carcinogenicity in humans and other living things appears to be difficult to come by [1, 2, 6, 12–14, 21, 88]. The highest value of sulphate at location 4 & 2 could remain found in water body (**Tables 2** and **3**), indicating a significant sulphate sensitivity toward changes in geochemical characteristics within the aquifer system. High sulphate concentrations are widely recognized to be caused by minerals dissolution that govern its water natural abundance or by land use. It may be said that sulphate is particularly unstable in the atmosphere, where it is transformed into forms ideal for its long-term presence in groundwater's. Ammonia (NH_3^-) values range from highest at location 6 (2.80) mg/l for rainy season to location 3 (4.39) mg/l (**Tables 2** and **3**). The maximum value is set at 3.0 mg/l according to WHO/SON/NAFDAC standards of potability. Nutrient salts (nitrite, nitrate and ammonia) are vital to the metabolism and growth of aquatic life, and when their concentrations rise, the biological balance shifts. Human activity has caused a significant increase in the amount of nutrients and salts in aquatic ecosystems, causing an issue with water quality. Extensive use of mineral fertilizers in some areas has resulted in atmospheric pollution, greenhouse gas emissions and eutrophication of water [1, 2, 6, 12–14, 21, 76, 78, 79, 89]. Nickel intake is determined by its physicochemical technique, with water-soluble techniques (nitrate, sulphate, chloride) providing additional readily consumed nickel. Thus, the values for nickel was higher at location 4 (1.00) mg/l for rainy season and location 3 (1.40) mg/l for dry season respectively. The values were higher than the WHO/SON/NAFDAC tolerable limits of 0.02 mg/l. The nickel values differed remarkably. Even though nickel has been identified as a vital trace metal, it could also be highly poisonous at higher doses. Hair loss, lung fibrosis, allergies of the skin, eczema, and various degrees of kidney and heart poisoning have been associated with humans exposed to high concentrations. Nickel also has the propensity of replacing iron and zinc in the body, thus interfering in the normal biochemistry [1, 2, 41, 82, 90]. Exposure to highly polluted water is likely to cause a number of clinical consequences in humans. Among these are skin allergies, respiratory cancer, lung fibrosis, and iatrogenic nickel toxicity. It has been established that nickel exposure has hematological implications in both animals and humans. Even if no reproductive repercussions have remained reported with humans' exposure to nickel. Location 1, 2, 3, 4 & 6 have higher TPH concentrations, while location 5, 7 & 8 have lower TPH concentration and in location 9 TPH was not detected for rainy season. The content of TPH in groundwater, on the other hand indicated that locations 2, 3, 4, & 7 had higher concentration above WHO/SON/NAFDAC standards. But location 9 did not show any presence of T PH for dry season. The findings found that five (5) locations in the rainy season and four (4) locations in the dry season did not fulfill the WHO/SON/NAFDAC criteria. Accordingly, the result show that TPH concentrations in drinking water remain much higher, signaling that water quality may have a detrimental effect on fish survival, eggs and larvae production and ecosystem development. Because of the high tidal velocities, the pollution is dispersed over a large area. There is also concern about the lengthy period required for total biodegradation of the heavier components, which contain extremely dangerous aromatic compounds with low boiling points. The high TPH values in those sites are a cautionary sign that everything is not well, since some water company and vendors

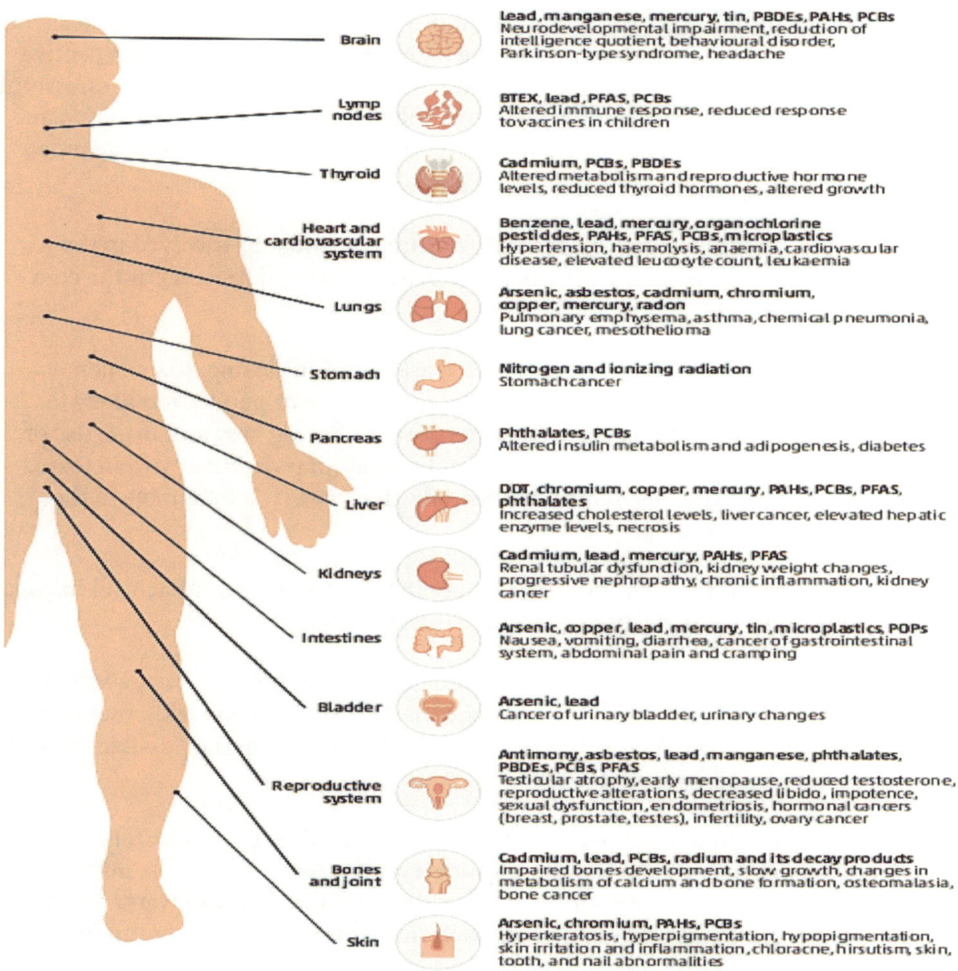

Figure 3.
Main effects of contaminants on human health, indicating the organs or systems affected and the contaminants causing them. Source: Adapted from Morufu et al. [1].

use ground water for production as well as sell it in places nearby or as far away as Yenagoa and Imo. Apart from its deadly effects, oil can induce death via producing narcosis, which causes animals to get detached from substrate. Total recovery may perhaps take close to 20 years. Oil coatings on the water surface in damaged areas impede light transmission and thus photosynthetic primary production. As a result, we must not forget that the general pollution effect on water bodies as well as ecosystem is significantly more problematic to anticipate. Summarily, trace metals cause respiratory irritation, kidney failure, neurological impairments, immunosuppression, anemia, gastrointestinal as well as cancer of liver, skeletal system abnormalities, liver inflammation, cardiovascular diseases following chronic exposure. The main contaminants effects on human health (see **Figure 3**) is represented thus:

4. Conclusion and recommendations

The Ebocha-Obrikom oil and gas producing area of River State, was chosen as the study area in order to gain awareness into the hydrochemistry as well as perspective of groundwater health. The aim of the research was to make available decision-useful information that could assist in taken action to solve the urgent threats facing societies across the Niger Delta. Thus, the following highlights are offered after the broad research findings:

- Development partners as well as local governments must be involved in the artificial recharge schemes implementation as well as maintenance at the community level.

- The relevant stakeholders have an urgent task ahead in closing down open wells in the Niger Delta region of Nigeria, for the sake of population likely to be affected through them, since they live near gas flaring area and make use of polluted groundwater, also because the indigenous population breath in toxins released from gas flaring. Groundwater pollution growth is inevitable in the Niger Delta, unless we act.

- This study found that groundwater contamination has a greater harmful impact on indigenous residents' life expectancy than COVID-19.

- Furthermore, open lines of communication among decision makers, academia, and society remain essential to guarantee that decision makers and other stakeholders have access to timely, science-based information on the possible hazards caused via contaminants.

- Eventually, public and school kids should be educated on groundwater quality and its management at the community level through a series of seminars, short videos, and other activities. Furthermore, seasonal groundwater quality monitoring, as well as other necessary actions to prevent further groundwater contamination, should continue.

- Appropriate management approaches, such as limiting human activities, implementing water treatments, developing public awareness, and establishing a groundwater quality monitoring network, are recommended to improve groundwater quality.

- Human interference (indiscriminate disposal of drainage wastes and unregulated use of agricultural pesticides) should be more effectively and rigidly monitored, as it is the most important technique of preventing groundwater contamination.

- It is strongly recommended that frequent monitoring and assessment of total water resource availability be encouraged. Waste management, land use, and agricultural practices that help to preserve the quality of water resources should all be implemented. The water should be thoroughly boiled before consumption.

- The first step in water pollution management is identifying and assessing risk at potentially polluted sites. If pollution at an assumed location remains at levels that may harm humans, evidence around that location should remain collected and made public, and appropriate remediation or risk-minimization actions should be taken, particularly if the location is utilized for water reservoir or production of food designed for human consumption.

- In light of the current global trend scenario of worsening groundwater pollution, stronger political, business, as well as social commitment is required to identify alternatives to the usage of extremely harmful pollutants as well as increased research investment in prevention as well as cleanup.

- Enhanced cooperation as well as partnership remain required to enable knowledge availability, the exchange of successful experiences, as well as worldwide access to safe and sustainable technologies, that leave no one behind.

- Agip should immediately begin replacing all old pipes in the Ebocha-Obrikom Oil Fields as soon as possible, and should collaborate with other agencies to complete a comprehensive Joint Investigation Visit (JIV) report. Furthermore, fair compensation should be provided to the impacted victims of Agip carelessness because their means of livelihood have been annihilated.

Author details

Morufu Olalekan Raimi[1*], Henry Olawale Sawyerr[2], Ifeanyichukwu Clinton Ezekwe[3] and Salako Gabriel[4]

1 Faculty of Clinical Sciences, Department of Community Medicine, Environmental Health Unit, Niger Delta University, Wilberforce Island, Bayelsa State, Nigeria

2 Department of Environmental Health Sciences, School of Health, Allied and Environmental Science, College of Pure and Applied Sciences, Kwara State University, Nigeria

3 Weather Forecasting Services, Nigerian Meteorological Agency Abuja/Department of Geography and Environmental Management, University of Port Harcourt, Choba Town, Rivers State, Nigeria

4 Department of Environmental Management and Toxicology, Kwara State University, Malete, Kwara State, Nigeria

*Address all correspondence to: morufuolalekan.raimi@gmail.com

References

[1] Raimi MO, Sawyerr HO, Ezekwe CI, Salako G. Many oil wells, one evil: Potentially toxic metals concentration, seasonal variation and human health risk assessment in drinking water quality in Ebocha-Obrikom Oil and Gas Area of Rivers State, Nigeria. medRxiv. 2021. DOI: 10.1101/2021.11.06.21266005 [Preprint]

[2] Raimi MO, Sawyerr OH, Ezekwe CI, et al. Quality Water not Everywhere: Exploratory Analysis of Water Quality Across Ebocha-Obrikom Oil and Gas Flaring Area in the Core Niger Delta Region of Nigeria, 04 October 2021, PREPRINT (Version 1) available at Research Square. DOI: 10.21203/rs.3.rs-953146/v1

[3] Olalekan RM, Ezekwe CI, Sawyerr OH. Problematic groundwater contaminants: Impact of surface and ground water quality on the environment in Ebocha-Obrikom oil and gas producing area of Rivers State, Nigeria. In: Oral Presentation Presented at the United Research Forum. 2nd International E-Conference on Geological and Environmental Sustainability during; July 29-30, 2021. Available from: https://assets.unitedresearchforum.com/documents/final-book-2021-91-309.pdf

[4] Raimi MO, Abdulraheem AF, Major I, Odipe OE, Isa HM, Onyeche C. The sources of water supply, sanitation facilities and hygiene practices in an island community: Amassoma, Bayelsa state, Nigeria. Public Health Open Access. 2019;**3**(1):000134. ISSN: 2578-5001. DOI: 10.23880/phoa-16000134

[5] United Nations Development Programme. Niger Delta Human Development Report. Abuja: UNDP; 2006

[6] Olalekan RM, Dodeye EO, Efegbere HA, Odipe OE, Deinkuro NS, Babatunde A, et al. Leaving No one behind? Drinking-water challenge on the rise in Niger Delta region of Nigeria: A review. Merit Research Journal of Environmental Science and Toxicology. 2020;**6**(1):031-049. DOI: 10.5281/zenodo.3779288. ISSN: 2350-2266

[7] Olalekan RM, Adedoyin OO, Ayibatonbira A, et al. "Digging deeper" evidence on water crisis and its solution in Nigeria for Bayelsa state: A study of current scenario. International Journal of Hydrology. 2019;**3**(4):244-257. DOI: 10.15406/ijh.2019.03.00187

[8] Olalekan RM, Omidiji Adedoyin O, Adedotun AT, Emmanuel OO, Anu B. An analysis of Bayelsa state water challenges on the rise and its possible solutions. Acta Scientific Agriculture. 2019;**3**(8):110-125. DOI: 10.31080/ASAG.2019.03.0572

[9] World Health Organization (WHO). Guidelines for drinking-water quality. WHO Chronicle. 2011;**38**(4):104-108

[10] Gift RA, Olalekan RM. Access to electricity and water in Nigeria: A panacea to slow the spread of Covid-19. Open Access Journal of Science. 2020;**4**(2):34. DOI: 10.15406/oajs.2020.04.00148. https://medcrave.com/index.php?/articles/det/21409/

[11] Gift RA, Olalekan RM, Owobi OE, Oluwakemi RM, Anu B, Funmilayo AA. Nigerians crying for availability of electricity and water: A key driver to life coping measures for deepening stay at home inclusion to slow covid-19 spread. Open Access Journal of Science. 2020;**4**(3):69-80. DOI: 10.15406/oajs.2020.04.00155

[12] Raimi MO, Sabinus CE. An assessment of trace elements in surface

and ground water quality in the Ebocha-Obrikom oil and gas producing area of Rivers state, Nigeria. International Journal for Scientific and Engineering Research (IJSER). 2017;**8**:6. ISSN: 2229-5518

[13] Morufu R, Clinton E. Assessment of Trace Elements in Surface and Ground Water Quality. Mauritius: LAP Lambert Academic Publishing; 2017. ISBN: 978-3-659-38813-2. Available from: www.omniscriptum.com

[14] Olalekan RM, Omidiji AO, Nimisngha D, Odipe OE, Olalekan AS. Health risk assessment on heavy metals ingestion through groundwater drinking pathway for residents in an oil and gas producing area of Rivers State, Nigeria. Open Journal of Yangtze Gas and Oil. 2018;**3**:191-206. DOI: 10.4236/ojogas.2018.33017

[15] Raimi MO. 21st century emerging issues in pollution control. In: 6th Global Summit and Expo on Pollution Control; May 06-07, 2019; Holiday Inn Amsterdam, Netherlands. 2019. Available from: https://pollutioncontrol.global-summit.com/2019

[16] American Public Health Association (APHA). Standard Methods for the Examination of Water and Waste Water. 22nd ed. American Public Health Association, American Water Works Association, Water Environment Federation; 2012

[17] Alloway BJ. Sources of heavy metals and metalloids in soils. In: Alloway BJ, editor. Heavy Metals in Soils: Trace Metals and Metalloids in Soils and Their Bioavailability, Environmental Pollution. Vol. 22. Dordrecht: Springer; 2012. pp. 11-50

[18] Ukah BU, Egbueri JC, Unigwe CO, Ubido OE. Extent of heavy metals pollution and health risk assessment of groundwater in a densely populated industrial area, Lagos, Nigeria. International Journal of Energy and Water Resources. 2019. DOI: 10.1007/s42108-019-00039-3

[19] Egbueri JC, Mgbenu CN, Chukwu CN. Investigating the hydrogeochemical processes and quality of water resources in Ojoto and environs using integrated classical methods. Modeling Earth Systems and Environment. 2019;**5**(4):1443-1461. DOI: 10.1007/s40808-01900613-y

[20] World Health Organization (WHO). Guidelines for Drinking-water Quality, Fourth Edition Incorporating the First Addendum. 2017. Available from: https://www.who.int/water_sanitation_health/publications/drinking-water-quality-guidelines-4-including-1st-addendum/en/ [Accessed: 29 June 2020]

[21] Afolabi AS, Raimi MO. When water turns deadly: Investigating source identification and quality of drjinking water in Piwoyi community of federal capital territory, Abuja Nigeria. Online Journal of Chemistry. 2021;**1**:38-58. DOI: 10.31586/ojc.2021.010105. Available from: www.scipublications.org/journal/index.php/ojc

[22] Kanga IS, Naimi M, Chikhaoui M. Groundwater quality assessment using water quality index and geographic information system based in Sebou River Basin in the North-west region of Morocco. International Journal of Energy Water Resources. 2020;**4**(4):347-355. DOI: 10.1007/s42108-020-00089-y

[23] Ako AA, Eyong GET, Shimada J, et al. Nitrate contamination of groundwater in two areas of the Cameroon volcanic line (Banana plain and Mount Cameroon area). Applied Water Science. 2014;**4**:99-113

[24] Olobaniyi SB, Owoyemi FB. Characterization by factor analysis of

the chemical facies of groundwater in the deltaic Plain Sands aquifer of Warri, Western Niger Delta. UNESCO/African journal of Science and Technology: Science and Engineering Series. 2006;**7**(1):73-81

[25] Ehirim CN, Nwankwo CN. Evaluation of aquifer characteristics and groundwater quality using Geoelectric method in Choba, Port Harcourt. Archives of Applied Sciences in Research. 2010;**2**(2):396-403

[26] Okafor EC, Opuene K. Preliminary assessment of trace metals and polycyclic aromatic hydrocarbons in the sediments. International Journal of Environmental Sciences and Technologies. 2007;**4**(2): 233-240

[27] Raimi MO, Pigha TK, Ochayi EO. Water-related problems and health conditions in the oil producing communities in central Senatorial District of Bayelsa state. Imperial Journal of Interdisciplinary Research (IJIR). 2017;**3**:6. ISSN: 2454-1362

[28] Suleiman RM, Raimi MO, Sawyerr HO. A deep dive into the review of national environmental standards and regulations enforcement agency (NESREA) act. International Research Journal of Applied Sciences. 2019. DOI: Irjas.2019.123.123. pISSN: 2663-5577, eISSN: 2663-5585. Available from: www.scirange.com. https://scirange.com/abstract/irjas.2019.108.125

[29] Raimi MO, Sabinus CE. Influence of organic amendment on microbial activities and growth of pepper cultured on crude oil contaminated Niger Delta soil. International Journal of Economy, Energy and Environment. 2017;**2**(4): 56-76. DOI: 10.11648/j.ijeee.20170204.12

[30] Amiri V, Rezaei M, Sohrabi N. Groundwater quality assessment using entropy weighted water quality index (EWQI) in Lenjanat, Iran. Environmental Earth Sciences. 2014;**72**:3479-3490

[31] Amiri V, Sohrabi N, Dadgar MA. Evaluation of groundwater chemistry and its suitability for drinking and agricultural uses in the Lenjanat plain, Central Iran. Environment and Earth Science. 2015;**74**(7):6163-6176

[32] Chapman D, Kimstach V. Selection of water quality variable. In: Chapman D, editor. Water Quality Assessments. London: Chapman and Hall; 1992. pp. 51-119

[33] Odipe OE, Raimi MO, Suleiman F. Assessment of heavy metals in effluent water discharges from textile industry and river water at close proximity: A comparison of two textile industries from Funtua and Zaria, North Western Nigeria. Madridge Journal of Agriculture and Environmental Sciences. 2018;**1**(1):1-6. DOI: 10.18689/mjaes-1000101. https://madridge.org/journal-of-agriculture-and-environmental-sciences/mjaes-1000101.php

[34] Henry OS, Odipe EO, Olawale SA, Raimi MO. Bacteriological assessment of selected hand dug wells in students' residential area: A case study of Osun State College of Health Technology, Ilesa, Nigeria. Global Scientific Journal. 2019;**7**:1. ISSN 2320-9186. Available from: www.globalscientificjournal.com

[35] USEPA. Regional Screening Levels (RSLs) Generic Tables. 2015. Available from: https://www.epa.gov/risk/regional-screening-levels-rsls-generictables-November-2015 [Accessed: 15 September 2016]

[36] Koffi KV, Obuobie E, Banning A, et al. Hydrochemical characteristics of groundwater and surface water for

domestic and irrigation purposes in Vea catchment, Northern Ghana. Environmental Earth Sciences. 2017;**76**:185. DOI: 10.1007/s12665-017-6490-3

[37] Disli E. Hydrochemical characteristics of surface and groundwater and suitability for drinking and agricultural use in the Upper Tigris River Basin, Diyarbakır–Batman, Turkey. Environmental Earth Sciences. 2017;**76**:500. DOI: 10.1007/s12665-017-6820-5

[38] Mahato MK, Singh PK, Singh AK, et al. Assessment of hydrogeochemical processes and mine water suitability for domestic, irrigation, and industrial purposes in East Bokaro Coalfield, India. Mine Water and the Environment. 2018;**37**:493-504

[39] Li P, Wu J, Qian H, Zhang Y, Yang N, Jing L, et al. Hydrogeochemical characterization of groundwater in and around a wastewater irrigated forest in the southeastern edge of the Tengger Desert, Northwest China. Exposure and Health. 2016;**8**(3):331-348. DOI: 10.1007/s12403-016-0193-y

[40] Adimalla N, Li P, Venkatayogi S. Hydrogeochemical evaluation of groundwater quality for drinking and irrigation purposes and integrated interpretation with water quality index studies. Environmental Processes. 2018;**5**(2):363-383. DOI: 10.1007/s40710-018-0297-4

[41] Ezekwe IC, Odubo E, Chima GN, Onwuchekwa IS. Groundwater occurrence and flow patterns in the Ishiagu mining area of southeastern Nigeria. Frontiers in Earth Science. 2012;**6**:18-28

[42] World Health Organisation. Guidelines for Drinking-Water Quality: Incorporating First Addendum, Recommendations. 3rd ed. Vol. 1. Geneva; 2006

[43] Chabukdhara M, Gupta SK, Kotecha Y, Nema AK. Groundwater quality in Ghaziabad district, Uttar Pradesh, India: Multivariate and health risk assessment. Chemosphere. 2017;**179**:167-178. DOI: 10.1016/j.chemosphere.2017.03.086

[44] Li P, He S, H, e X., Tian, R. Seasonal hydrochemical characterization and groundwater quality delineation based on matter element extension analysis in a paper wastewater irrigation area, Northwest China. Exposure and Health. 2018;**10**(4):241-258. DOI: 10.1007/s12403-17-0258-6

[45] Dami A, Ayuba HK, Amukali O. Effects of gas flaring and oil spillage on rainwater collected for drinking in Okpai and Beneku, Delta State, Nigeria. Global Journal of Human Social Sciences. 2012;**12**(13):7-10

[46] Adimalla N, Qian H. Groundwater chemistry, distribution and potential health risk appraisal of nitrate enriched groundwater: A case study from the semi-urban region of South India. Ecotoxicology and Environmental Safety. 2021;**207**:1-10. DOI: 10.1016/j.ecoenv.2020. 111277

[47] David KT. Groundwater Hydrology. 2nd ed, John Wiley and Sons, Inc; 2006

[48] Goldman LR et al. Low birth weight, prematurity and birth defects in children living near the hazardous waste site, Love Canal. Hazardous Waste and Hazardous Materials. 1985;**2**:209-223

[49] www.health.nytimes.com

[50] Li P, He S, Yang N, Xiang G. Groundwater quality assessment for

domestic and agricultural purposes in Yan'an City, Northwest China: Implications to sustainable groundwater quality management on the loess plateau. Environment and Earth Science. 2018;**77**(23). DOI: 10.1007/s12665-018-7968-3

[51] WHO. Guideline for Drinking-Water Quality, 3rd Edition (Recommendations). Geneva: World Health Organization; 2004

[52] Adimalla N, Li P. Occurrence, health risks, and geochemical mechanisms of fluoride and nitrate in groundwater of the rock-dominant semi-arid region, Telangana State, India. Human and Ecological Risk Assessment: An International Journal. 2018:1-23. DOI: 10.1080/10807039.2018.1480353

[53] Adimalla N, Venkatayogi S. Geochemical characterization and evaluation of groundwater suitability for domestic and agricultural utility in semi-arid region of Basara, Telangana State, South India. Applied Water Science. 2018;**8**:44. DOI: 10.1007/s13201-018-0682-1

[54] Marghade D, Malpe DB, Zade AB. Major ion chemistry of shallow groundwater of a fast growing city of Central India. Environmental Monitoring and Assessment. 2012;**184**:2405-2418

[55] Olalekan RM, Oluwatoyin OA, Olawale SH, Emmanuel OO, Olalekan AZ. A critical review of health impact assessment: Towards strengthening the knowledge of decision makers understand sustainable development goals in the twenty-first century: Necessity today; essentiality tomorrow. Research and Advances: Environmental Sciences. 2020;**1**:72-84. DOI: 10.33513/RAES/2001-13. Available from: https://ospopac.com/journal/environmental-sciences/early-online

[56] Olalekan RM, Oluwatoyin O, Olalekan A. Health impact assessment: A tool to advance the knowledge of policy makers understand sustainable development goals: A review. ES Journal of Public Health. 2020;**1**(1):1002. Available from: https://escientificlibrary.com/public-health/in-press.php

[57] Raimi MO, Ihuoma BA, Esther OU, Abdulraheem AF, Opufou T, Deinkuro NS, et al. Health impact assessment: Expanding public policy tools for promoting sustainable development goals (SDGs) in Nigeria. EC Emergency Medicine and Critical Care. 2020;**4**:9

[58] Zhai Y, Zhao X, Teng Y, et al. Groundwater nitrate pollution and human health risk assessment by using HHRA model in an agricultural area, NE China. Ecotoxicology and Environmental Safety. 2017;**137**:130-142

[59] Li P, Wu J, Qian H, Lyu X, Liu H. Origin and assessment of groundwater pollution and associated health risk: A case study in an industrial park, Northwest China. Environmental Geochemistry and Health. 2014;**36**:693-712

[60] Sun Y, Ling P, Li Y, Li Q, Sun Q, Wang J. Influences of coal mining water irrigation on the maize losses in the Xingdong mine area, China. Environmental Geochemistry and Health. 2014;**36**(1):99-106

[61] Olalekan MR, Abiola I, Ogah A, Dodeye EO. Exploring how human activities disturb the balance of biogeochemical cycles: Evidence from the carbon, nitrogen and hydrologic cycles. Research on World Agricultural Economy. 2021;**02**:03. DOI: 10.36956/rwae.v2i3.426. Available from: http://ojs.nassg.org/index.php/rwae

[62] Zhang Y, Wu J, Xu B. Human health risk assessment of groundwater nitrogen pollution in Jinghui canal irrigation area of the loess region, Northwest China. Environmental Earth Sciences. 2018;**77**(7):273. DOI: 10.1007/s12665-018-7456-9

[63] Wu J, Zhou H, He S, Zhang Y. Comprehensive understanding of groundwater quality for domestic and agricultural purposes in terms of health risks in a coal mine area of the Ordos basin, north of the Chinese Loess Plateau. Environment and Earth Science. 2019;**78**(15):446. DOI: 10.1007/s12665-019-8471-1

[64] He S, Wu J. Relationships of groundwater quality and associated health risks with land use/land cover patterns: A case study in a loess area, Northwest China. Human and Ecological Risk Assessment. 2019;**25**(1-2):354-373. DOI: 10.1080/10807039.2019.1570463

[65] Marghade D, Malpe DB, Duraisamy K, Patil PD, Li P. Hydrogeochemical evaluation, suitability, and health risk assessment of groundwater in the watershed of Godavari basin, Maharashtra, Central India. Environmental Science and Pollution Research. 2021;**28**(15):18471-18494. DOI: 10.1007/s11356-020-10032-7

[66] Henry OS, Morufu OR, Adedotun TA & Oluwaseun EO. Measures of harm from heavy metal pollution in battery technicians' workshop within Ilorin Metropolis, Kwara State, Nigeria. Scholink Communication, Society and Media 2019;**2**(2). ISSN 2576-5388 (Print) ISSN 2576-5396 (Online). Available from: www.scholink.org/ojs/index.php/csm. https://doi.org/10.22158/csm.v2n2p73

[67] Egbueri JC, Mgbenu CN. Chemometric analysis for pollution source identifcation and human health risk assessment of water resources in Ojoto Province, southeast. Applied Water Science. 2020:1-18. DOI: 10.1007/s13201-020-01180-9

[68] Edokpayi JN, Enitan AM, Mutileni N, Odiyo JO. Evaluation of water quality and human risk assessment due to heavy metals in groundwater around Muledane area of Vhembe District, Limpopo Province, South Africa. Chemistry Central journal. 2018;**12**(1):1-16. DOI: 10.1186/s13065-017-0369-y

[69] He X, Wu J, He S. Hydrochemical characteristics and quality evaluation of groundwater in terms of health risks in Luohe aquifer in Wuqi County of the Chinese Loess Plateau, Northwest China. Human and Ecological Risk Assessment: An International Journal. 2018:1-20. DOI: 10.1080/10807039.2018.1531693

[70] Standards Organization of Nigeria (SON). Nigerian Standard for Drinking Water Quality. Vol. 554. Abuja: Nigerian Industrial Standard; 2015. pp. 13-14

[71] Erikson KM, Thompson K, Aschner J, Aschner M. Manganese neurotoxicity: A focus on the neonate. Pharmacology & Therapeutics. 2007;**112**(2):369-377. DOI: 10.1016/j.pharmthera.2006.09.002

[72] Blanc PD. The early history of manganese and the recognition of its neurotoxicity, 1837-1936. Neurotoxicology. 2018;**64**:5-11. DOI: 10.1016/j.neuro.2017.04.006

[73] Oulhote Y, Mergler D, Barbeau B, Bellinger DC, Bouffard T, Brodeur MÈ, et al. Neurobehavioral function in school-age children exposed to manganese in drinking water. Environmental Health Perspectives. 2014;**122**(12):1343-1350. DOI: 10.1289/ehp.1307918

[74] United States Geological Survey (USGS). Arsenic Statistics and

Information. 2019. Available from: https://www.usgs.gov/centers/nmic/arsenic-statistics-and-information [Accessed: 15 June 2019]

[75] Khan AF, Srinivasamoorthy K, Rabina C. Hydrochemical characteristics and quality assessment of groundwater along the coastal tracts of Tamil Nadu and Puducherry, India. Applied Water Science. 2020;**10**(2):74. DOI: 10.1007/s13201-020-1158-7

[76] Raimi MO, Adeolu AT, Enabulele CE, Awogbami SO. Assessment of air quality indices and its health impacts in Ilorin Metropolis, Kwara State, Nigeria. Science Park Journals of Scientific Research and Impact. 2018;**4**(4):060-074. ISSN: 2315-5396. DOI: 10.14412/SRI2018.074. Available from: http://www.scienceparkjournals.org/sri/pdf/2018/September/Olalekan_et_al.pdf http://www.scienceparkjournals.org/sri/Content/2018/September/2018.htm

[77] Oluwaseun EO, Raimi MO, Nimisingha DS, Abdulraheem AF, Okolosi-Patainnocent E, Habeeb ML, et al. Assessment of environmental sanitation, food safety knowledge, handling practice among food handlers of Bukateria complexes in Iju Town, Akure North of Ondo-State, Nigeria. Acta Scientific Nutritional Health. 2019;**3**(6):186-200. DOI: 10.31080/ASNH.2019.03.0308

[78] Raimi MO, Adio ZO, Odipe OE, Timothy KS, Ajayi BS, Ogunleye TJ. Impact of sawmill industry on ambient air quality: A case study of Ilorin Metropolis, Kwara State, Nigeria. Energy and Earth Science. 2020;**3**(1):2020. DOI: 10.22158/ees.v3n1p1. Available from: www.scholink.org/ojs/index.php/ees

[79] Raimi OM, Samson TK, Sunday AB, Olalekan AZ, Emmanuel OO, Jide OT. Air of uncertainty from pollution profiteers: Status of ambient air quality of sawmill industry in Ilorin Metropolis, Kwara State, Nigeria. Research Journal of Ecology and Environmental Sciences. 2021;**1**(1):17-38. DOI: 10.31586/rjees.2021.010102. Available from: https://www.scipublications.com/journal/index.php/rjees/article/view/60

[80] Agency for Toxic Substances and Disease Registry (ATSDR). Toxicological Profiles, Toxic Substances Portal. 2015. Available from: https://www.atsdr.cdc.gov/toxprofiles/index [Accessed: Nov 2015]

[81] Iqbal J, Nazzal Y, Howari F, Xavier C, Yousef A. Hydrochemical processes determining the groundwater quality for irrigation use in an arid environment: The case of Liwa Aquifer, Abu Dhabi, United Arab Emirates. Groundwater for Sustainable Development. 2018. DOI: 10.1016/j.gsd.2018.06.004

[82] Olalekan RM, Vivien OT, Adedoyin OO, et al. The sources of water supply, sanitation facilities and hygiene practices in oil producing communities in central senatorial district of Bayelsa State, Nigeria. MedCrave Online Journal of Public Health. 2018;7(6):337-345. DOI: 10.15406/mojph.2018.07.00265

[83] Egbueri JC. Heavy metals pollution source identifcation and probabilistic health risk assessment of shallow groundwater in Onitsha, Nigeria. Analytical Letters. 2020;**53**:1620-1638. DOI: 10.1080/00032 719.2020.1712606

[84] Adeyeye OA, Xiao C, Zhang Z, Yawe AS, Liang X. Groundwater fluoride chemistry and health risk assessment of multi-aquifers in Jilin Qianan, Northeastern China. Ecotoxicology and Environmental Safety. 2021;**211**:111926-112014. DOI: 10.1016/j.ecoenv.2021.111926

[85] Wang J, Liu G, Liu H, Lam PKS. Multivariate statistical evaluation of dissolved trace elements and a water quality assessment in the middle reaches of Huaihe River, Anhui, China. Science of the Total Environment. 2017;**583**:421-431

[86] National Agency for Food and Drug Administration and Control (NAFDAC). National Agency for Food and Drug Administration and Control, Ministry Safety Bulletin. Recommendation. Vol. 2. Lagos, Nigeria: National Agency for Food, Drug, Administration and Control; 2008

[87] Goyal SK, Chaudhary BS, Singh O, Sethi GK, Thakur PK. GIS based spatial distribution mapping and suitability evaluation of groundwater quality for domestic and agricultural purpose in Kaithal district, Haryana state, India. Environmental Earth Sciences. 2010;**61**(8):1587-1597

[88] Chen J, Qian H, Wu H, Gao Y, Li X. Assessment of arsenic and fluoride pollution in groundwater in Dawukou area, Northwest China, and the associated health risk for inhabitants. Environmental Earth Sciences. 2017;**73**:314

[89] Premoboere EA, Raimi MO. Corporate civil liability and compensation regime for environmental pollution in the Niger Delta. International Journal of Recent Advances in Multidisciplinary Research. 2018;**05**(06):3870-3893

[90] Okoyen E, Raimi MO, Omidiji AO, Ebuete AW. Governing the environmental impact of dredging: Consequences for marine biodiversity in the Niger Delta region of Nigeria. Insights Mining Science and Technology. 2020;**2**(3):555586. DOI: 10.19080/IMST.2020.02.555586

Chapter 10

Effect of Mining on Heavy Metals Toxicity and Health Risk in Selected Rivers of Ghana

George Yaw Hadzi

Abstract

Heavy metal contamination of selected rivers in the mining areas of Ghana was studied. In the study, 44 composite water samples were collected, digested, and analyzed for selected metals using ICP-MS. The average concentrations (mg/L) of heavy metals from the pristine sites ranged from 0.003 (As) to 0.929 (Fe), and the mining sites ranged from 0.002 (Pb) to 20.355 (Fe). Generally, the metals were within the WHO and USEPA acceptable limits except Al, Fe, As, Cr, and Mn. Hazard quotients for ingestion (HQing) and dermal contact from pristine and mining sites ranged from 0.838 (Cr) to 3.00×10^{-4} (Cu) and from 0.181 (As) to 2.40×10^{-6} (Cu), respectively. The carcinogenic risks (CRs) for ingestion were within standard limit (10^{-6} to 10^{-4}). However, Arsenic showed high CRing above the acceptable limit (1.83×10^{-2}). The CRs for dermal contact range from 4.22×10^{-8} to 1.44×10^{-9} (Cr) and from 7.34×10^{-8} to 5.65×10^{-9} (Pb). Carcinogenic risk values for As in the mining areas raise carcinogenic concerns for the residents in the studied areas. PROMETHEE and GAIA indicate major contribution of the metals from the mining. Multivariate PCA and cluster analysis suggest anthropogenic activities as the major source of the metal toxicity of the mine rivers.

Keywords: surface water, pristine, mining, heavy metal, contamination, toxicity, health risk

1. Introduction

The issues of heavy metals contamination of local, regional, and global environment emanate directly from natural sources and indirectly from anthropogenic activities such as mining, rapid industrialization, urbanization, improper waste management, and other local and regional man-made activities [1].

Substantial quantities of heavy metals are released from different anthropogenic sources into the atmosphere from where they are deposited in soils and aquatic ecosystem through dry and wet deposition processes.

Anthropogenic inputs of heavy metals are currently getting higher and in some areas exceeding natural inputs where human activities are predominant [2, 3]. The metals accumulation and distribution in soil, water, and environment are increasing at a faster rate causing deposition and sedimentation in water reservoirs and affecting aquatic organisms [4, 5]. High levels of Cd, Cu, Pb, and Fe can act as ecological toxins in aquatic and terrestrial ecosystems [6].

Heavy metals are potentially harmful to humans and various ecological receptors due to their toxicity, persistence, bio-accumulative characteristics, and their nonbiodegradable nature. Toxic metals can cause different health problems depending on the type of the metal concerned, its concentration, and oxidation state. They are among the most toxic and persistent pollutants in freshwater systems [4, 5]. Certain heavy metals and metalloids are toxic and can cause adverse effects and severe problems such as oxidative stress by formation of free radicals even at low concentrations [7, 8].

Heavy metals contamination can result in several diseases and deformities; for instance, in the 1950s, an advanced country such as Japan was devastated by heavy metal poisoning known as the Fetal Minamata Disease, which resulted from contamination of fishes by organic mercury. The situation led to severe nerve damage of newborn babies from pregnant women [9].

In Iraq, babies walked at later age due to consumption of organic mercury contaminated grains by pregnant mothers. Similar incidence occurred in Faroe Islands where school children scored lower grades on brain function test due to consumption of mercury-contaminated whale meat by pregnant women [10].

A third world country such as Bangladesh in recent years has become vulnerable to heavy metal contamination of groundwater [11] and heavy metals contamination of drinking water sources by Cd, Pb, Cu, and Zn in Bolivia, Hong Kong, and Berlin [12, 13].

Efforts were made in both research and monitoring to establish sources, transport, and fate of these metals in the aquatic environment. However, studies have shown that contamination artifacts have seriously compromised the reliability of many past and current analyses and under certain circumstances, metal concentrations have been measured 100 times the true concentrations [14]. These errors are of great concern, since contaminant-free data are necessary to detect trends and to identify factors that control the transport and fate of toxic metals in water bodies.

Many mineral resources including gold represent significant material basis for socioeconomic development, justifying the exploitation and utilization of mineral resources essential to national development. Nonetheless, despite the importance of mineral resources, mineral extraction with its associated release of heavy metals has caused serious environmental damage in many developing and developed countries [15, 16].

As gold mineral is being mined actively in many developing countries, there are fears that the mining activity may be causing serious metal pollution to water resources. Disused and closed mines with huge mine waste materials including tailings were left from the extraction processes without adequate treatment, and as a result, soils, plants, water bodies, and sediments in the vicinity of mines were contaminated by potentially toxic metals from tailings through wind and Acid-Mine-Drainage [17, 18]. Reports from earlier studies have shown that metal levels of surface and groundwater exceeded World Health Organization (WHO)'s acceptable limits for drinking water around Tarkwa mining area [19]. Huge deposits of mine wastes as well as ore stockpiles and waste rocks are usually seen in large piles around both large- and small-scale mining areas. These deposits are gradually washed through weathering and leaching into far and near water bodies, thereby releasing toxic substances into water bodies [20].

Metals associated with gold mines, including Cd, Cu, Pb, and Zn, may be dispersed downstream due to the weathering process of tailings. Thus, the extent and degree of heavy metal contamination around mines may vary depending on geochemical characteristics and mineralization of tailings [21].

Mine tailings may result in the influx of metals and toxic chemicals into the environment. Waste rocks are known to contain arsenic (As), mercury (Hg), cadmium (Cd), lead (Pb), and other toxic metals, which are extensively dispersed into the environment [22]. According to the recent World Health Organization (WHO) report on arsenic, it was recognized that at least 140 million people in 50 countries have been drinking water containing arsenic at levels above the WHO provisional guideline value of 10 μg/L [22, 23]. In the evaluation conclusions, arsenic and other heavy metals exposure through drinking water is causally related to cancer in the lungs, kidney, bladder, and skin. There is also an increased risk of skin cancer and other skin lesions, such as hyperkeratosis and pigmentation changes. Ingestion of inorganic arsenic may induce peripheral vascular disease, which leads to black foot disease [24, 25].

It is therefore imperative to continually assess and monitor the concentration of heavy metals in water bodies in the environment due to anthropogenic activities, including gold mining, for evaluation of human exposure and for sustainable environment [26, 27].

This study investigated the extent of contamination by heavy metals of selected water bodies in the vicinity of gold mines and further compared the metal levels with those from the pristine sites to assess the possibility of mining activities causing toxicity (contamination) of the water bodies.

2. Materials and method

2.1 The study area

Samples were collected in eight regions of Ghana with the land cover ranging from 138 to 2950 km^2. The rivers that were sampled in the mining areas are Nyam river, Subri river, Birim river, and river Bonsa. The nature and the location of the rivers demonstrate the presence of metal contamination due to mining activities. The rivers from the pristine areas are Oda river, Bosomkese forest river, Ankasa river, Atewa forest river, Kalakpa river, Kakum river, and Mole river. The pristine rivers were used as background checks in order to assess the extent of metal contamination.

2.2 Sampling and sample collection

Water samples were collected from four selected rivers around the gold mining areas and seven rivers from the pristine areas. Sample collection was undertaken from January 2015 to January 2016. A total of 44 composite samples of water were collected from both mining and pristine areas. The rivers were sampled 100 m apart at four different points. 1.5 L plastic bottles that had been prewashed with detergent and 1:1 concentrated nitric acid/distilled water solution and eventually rinsed with only distilled water were used. The samples for metal analysis were acidified to a pH of 2 at site using concentrated HNO_3 before they were transported to the Chemistry Department laboratory of University of Cape Coast. The samples were kept in refrigerator at a temperature of 4°C for further analysis [28].

2.3 Digestion and analysis of water samples

Chemicals and reagents for analysis were acquired from the Central Analytical Facility of Queensland University of Science and Technology. 70% Nitric acid (HNO_3) was further distilled twice in Analab Sub-Boiling Distillation system. Water for the analysis was acquired from MilliQ water purification system (Millipore, Billerica, MA, USA). Water samples were analyzed in triplicates to check the efficiency of the analytical instrument. Centrifuge tubes were washed by rinsing three times in ultrapure water. They were then soaked in 3% analytical grade HCl and left on a hot plate for two days. The operating conditions for the instrument were the following parameters: Cell Gas flow rates: 5 ml/min; Carrier Gas Flow: 1.05 l/min; KED Voltage: 5 V; ICP RF Power: 1550 W; Octopole bias (V): −18, Octopole RF (V); 190: Spray Chamber t (C); 2: Sample depth (mm); 8.

The samples were digested by acidifying with 1 mL NHO_3. They were later centrifuged at 3500 rpm for 15 min. The samples were then filtered through 0.45 μm pore size cellulose acetate filters.

They were then analyzed with Agilent 8800 Triple Quadrupole Inductively Coupled Plasma Emission Spectrometer (ICP-QQQMS) in the Central Analytical Research Facility (CARF) laboratory of Queensland University of Technology, Australia. The same digestion procedure was applied to the Quality Control (QC) samples and the blank. The analytes were acquired using He mode, and those elements that do not suffer from polyatomic interferences were acquired in no gas mode.

Some physicochemical parameters such as pH, conductivity, and turbidity were also determined. The pH was determined alongside the temperature using a pre-calibrated JENWAY 3310 and JENWAY 3510 pH meter. Conductivity was measured using a pre-calibrated PHYWE 13701.93 and WAGTECH 4510 conductivity meter. The turbidity was measured with a Hachturbidimeter.

2.4 Recovery and reproducibility studies

Calibration solution was prepared by using Choice Analytical ICV-1 Solution and a Standard Agilent Technologies Multi Element Reference Standard 2A. The Agilent Standard was analyzed as unknown to monitor the accuracy of analytic process. The percent recovery was computed to range from 99.5% to 103.8% with the relative standard deviation ranging between 0.38 and 2.23. The recovery results indicate that the error associated with the determination of concentrations of the metals was negligible.

2.5 Data and statistical analysis

IBM SPSS Statistics version 22 and the Excel Analysis ToolPak were used to analyze the data from the study. Basic statistics such as mean and standard deviation were computed along the multivariate statistics. Relationships associated with the variables were tested using correlation analysis with statistical significance at $p < 0.05$. Hierarchical Cluster analysis (HCA) was also employed to provide a visual summary of the clustering process unsupervised pattern recognition technique. Factor analysis (FA) and principal component analysis (PCA) were computed to identify significant principal components in the data. The PCA was carried out by the Promax normalized rotation method for the results [29, 30]. PROMETHEE, a multicriteria outranking method, was employed to rank objects on the basis of range of variables and GAIA to add descriptive complement to the PROMETHEE rankings.

2.6 Human health risk assessment

The risk estimation was based on the United States Environmental Protection Agency (USEPA) risk assessment method for ingestion and dermal contact [29, 31].

The average daily dose (ADD) for the heavy metals (**Eq. 1**) was calculated using the following modified equations from USEPA protocol 1989 and 2004.

$$\mathrm{ADDing} = \frac{Cx \quad \times Ir \times Ef \times Ed}{\mathrm{Bwt} \times \mathrm{At} \times 365} \tag{1}$$

where Cx is the concentration of the metals in the drinking water (mg/L), Ir is the ingestion rate per unit time (L/day), Ed is the exposure duration (years), Ef is the exposure frequency (days/year), Bwt is the body weight of receptor (kg), and At is the average lifetime (years), which is equal to the life expectancy of a resident Ghanaian. In addition, ADDing is the quantity of heavy metals ingested per kilogram of body weight.

In this study, surface water ingestion is assumed to be the main pathway for risk assessment because the rivers are potential sources of drinking water. However, dermal contact is another important pathway, because residents sometimes swim in these rivers and thus may come into contact with the toxic metals through body contact.

Average daily dose for dermal contact was calculated using the formula in **Eq. 2** below:

$$\mathrm{ADDderm} = \frac{Cx \quad \times Sa \times Pc \times Et \times Ef \times Ed \times Cf}{\mathrm{Bwt} \times \mathrm{At} \times 365} \tag{2}$$

where Sa is the total skin surface area (cm^3), Cf is the volumetric conversion factor for water (1 L/1000 cm^3), Pc is the chemical-specific dermal permeability constant (cm/h).

The hazard for the metals was estimated as the ratio of the calculated dose to the reference dose (RfD) (mg/L/day) using **Eq. 3** below:

$$\mathrm{HQ} = \frac{ADD}{\mathrm{RfD}} \tag{3}$$

The chronic daily intake (CDI) of the metal was calculated using the **Eq. 4** below:

$$\mathrm{CDI} = C\,\frac{DIing}{\mathrm{Bwt}} \tag{4}$$

where C is the concentration of heavy metal in water, DI is the average daily intake rate (2 L).

The carcinogenic risks (CRs) of the metals were calculated using **Eq. 5 and 6** below for ingestion and dermal contact, respectively. The carcinogenic risk acceptable by USEPA ranges from $1x10^{-6}$ to $1x10^{-4}$.

$$\mathrm{CRing} = \frac{ADDing}{\mathrm{SFing}} \tag{5}$$

$$\mathrm{CRderm} = \frac{ADDderm}{\mathrm{SFing}} \tag{6}$$

where SF is the slope factor (mg/kg)/day. For As, Cd, and Cr, the slope factor values are 1.5, 6.1×10^2, and 5.0×10^2 (mg/kg)/day, respectively.

3. Results and discussion

3.1 Analysis of physical and chemical parameters

Even though people may not be affected directly by some of these parameters, elevated levels can cause unfavorable conditions and discomfort. For instance, drinking water with elevated pH will taste bitter [32]. Parameters such as electrical conductivity, pH, turbidity, and temperature as shown in (**Table 1**) were measured in this study. Water samples from the mining sites were acidic with pH values of some of the sites recording as low as 3.51. The pH of the pristine samples was, however, within the normal WHO's range of 6.5–8.5. The low pH values for mine samples might be responsible for the high metal levels measured.

The electrical conductivity values measured for the water samples were below the WHO normal range (400–600 μS/cm) [33]. The temperature values for the samples were below the recommended WHO's value of 29°C. Turbidity values for the mine samples were higher than those measured for the pristine samples due to activity of mining in those rivers. Other measured parameters such as salinity and total dissolved solids were relatively low. Low turbidity of the pristine samples indicates the absence of disease-causing organisms such as bacteria, viruses, and parasites that cause symptoms such as nausea, cramps, diarrhea, and associated headaches [34].

3.2 Concentration of heavy metals in water

The mean concentrations of the heavy metals obtained from ICP/MS instruments were presented in the attached **Table 2.** The mean concentrations were compared with the threshold/permissible values as shown in **Table 3**. The concentrations of Fe and Al especially from the mining sites were higher than the permissible values [35].

Sites	pH Range	Cond. (μS/cm) Range	Salinity Range	Turbidity Range	Temp. Range	TDS Range
AOBW	3.45–3.56	64.4–65.6	34.78–35.42	9–22	28.1–28.5	38.6–39.1
BAMW	5.34–5.86	2.98–3.19	1.64–1.72	3–21	28.6–28.9	6.06–6.35
EAMW	5.32–5.44	11.33–11.46	6.12–6.19	12–27	28.0–28.7	5.67–5.90
WTBW	5.10–5.41	9.54–10.05	5.15–5.41	5–15	28.3–28.6	2.70–2.71
EAW	6.20–6.09	4.46–4.89	2.41–2.64	1–6	28.0–28.2	1.96–1.99
WAW	6.14–6.41	0.52–0.54	0.28–0.29	0–4	28.6–28.9	0.31–0.32
BBW	6.38–6.48	1.65–1.66	0.89–0.90	3–9	27.4–27.9	0.98–0.99
AOD	6.24–6.33	2.11–2.19	1.14–1.18	5–7	28.2–28.4	1.64–1.66
VKPW	6.34–6.41	0.53–0.59	0.29–0.32	1–6	28.5–28.7	0.54–0.58
NM	6.06–6.99	0.66–0.73	0.36–0.39	2.-9	28.0–28.1	0.40–0.44
CK	6.55–6.77	0.41–0.47	0.22–0.25	0–1	28.0–28.3	0.70–0.76

Table 1.
Physical and chemical parameters for the water samples from the sites.

Sites	ICP/MS										
	Al	**V**	**Cr**	**Mn**	**Fe**	**Co**	**Ni**	**Cu**	**Zn**	**As**	**Pb**
						Mines					
AOBW	2.453 ± 0.83	0.080 ± 0.00	0.006 ± 0.00	1.827 ± 1.48	3.862 ± 1.55	0.008 ± 0.01	0.011 ± 0.00	0.016 ± 0.01	0.078 ± 0.02	0.226 ± 0.03	0.002 ± 0.00
BAMW	0.684 ± 0.24	0.006 ± 0.01	0.003 ± 0.01	2.667 ± 1.29	6.758 ± 1.76	0.006 ± 0.01	0.005 ± 0.00	0.004 ± 0.01	0.043 ± 0.02	0.010 ± 0.01	0.002 ± 0.00
EAMW	13.847 ± 4.57	0.077 ± 0.02	0.088 ± 0.03	1.213 ± 0.03	20.355 ± 5.60	0.044 ± 0.00	0.024 ± 0.00	0.092 ± 0.02	0.111 ± 0.01	0.007 ± .00	0.024 ± 0.00
WTBW	1.922 ± 0.65	0.003 ± 0.00	0.014 ± 0.01	0.230 ± 0.00	2.371 ± 1.16	0.003 ± 0.00	0.005 ± 0.00	0.245 ± 0.18	0.053 ± 0.03	0.006 ± 0.00	0.026 ± 0.05
						Pristine					
EAW	0.067 ± 0.01	—	—	0.064 ± 0.03	0.463 ± 0.02	—	—	—	0.044 ± 0.02	0.007 ± 0.00	—
WAW	0.111 ± 0.00	—	—	0.019 ± 0.02	0.715 ± 0.00	—	—	0.008 ± 0.02	0.037 ± 0.01	0.005 ± 0.00	—
BBW	0.142 ± 0.02	—	—	0.042 ± 0.01	0.929 ± 0.06	—	—	0.089 ± 0.03	0.044 ± 0.01	0.003 ± 0.00	0.010 ± 0.02
AODW	0.038 ± 0.03	0.010 ± 0.00	—	0.016 ± 0.02	0.157 ± 0.07	—	0.004 ± 0.00	0.006 ± 0.00	0.029 ± 0.01	—	—
VKW	0.030 ± 0.02	—	—	—	0.594 ± 0.18	—	—	0.004 ± 0.00	0.006 ± 0.01	—	—
NMW	0.047 ± 0.02	—	—	—	0.371 ± 0.28	—	—	—	0.021 ± 0.02	—	—
CKW	0.022 ± 0.01	—	—	0.010 ± 0.01	0.074 ± 0.01	—	—	—	0.006 ± 0.01	—	—

The dash (−) means below detection limit.

Table 2.
Mean concentrations (mg/L) of heavy metals in the rivers from pristine and mining areas (Hadzi et al., 2018).

Water Quality Guideline	As	Cr	Cu	Fe	Mn	Ni	Pb	Zn	Co
Drinking Water Quality									
EC(1998)	0.01	0.05	2	0.2	0.05	0.02	0.01	0.1	—
WHO (2004)	0.01	0.05	2	—	0.4	0.07	0.01	—	—
USEPA (2009)	0.01	0.1	1.3	0.3	0.05	—	0.015	5	0.11
USEPA (2006)	0.34	—	0.013	1	—	0.47	—	0.12	—

Table 3.
Maximum permitted heavy metal concentrations (mg/L) for drinking water quality and protection of freshwater aquatic life.

Metal concentrations from this study were safe for aquatic life. Hg and Cd were below detection limit. In general, higher concentrations of heavy metals were measured in mine sample with maximum concentrations of 13.847, 20.355, 2.667, 0.088, 0.245, 0.111, 0.226, and 0.026 mg/l for Al, Fe, Mn, Cr, Cu, Zn, As, and Pb, respectively. The concentrations of most metals in the pristine samples were either below their permissible limits or far below levels obtained from the mining sites, which suggests less anthropogenic activity in the pristine sites. Distribution of Al, Fe, and Zn is the same at the pristine and the mining sites. In assessing the heavy metal contaminations of the various sites, the levels were compared with previous studies from the same sampling sites and other natural rivers, and it was realized that the metal concentrations in this study are lower [36, 37]. A study conducted by Hadzi et al., in 2015 on the same rivers indicated a low metal input. However, similar low concentrations of Cd, Hg, As, Mn, Cu, and Zn in river Samre in the Wassa Amenfi West District in the Western region and Nangodi and Tinga drinking water sources in the Northern region of Ghana were reported. In a separate study in 2013, Cobbina et al., found relatively low concentrations of heavy metals in surface water and boreholes at Tinga in the Bole-Bamboi District of Ghana. According to Bowen [38], freshwater contains 0.1, 3.0, 3.0, and 15 mg/l of Cd, Cu, Pb, and Zn, respectively. However, the concentrations of metals reported at the pristine sites of this study are far less than those reported in freshwater bodies. Aladesanmi et al., in a similar study in Nigeria, 2014 [39], reported concentrations of Cd and As below detection limits and levels of Pb, Cr, Co, and Cu ranging from 0.003 to 0.009 mg/L.

3.3 Statistical analysis of data

Possible correlations and variability checks were conducted on the metal concentrations. The cluster analysis, as shown in **Figure 1 (attached),** indicates two main groups of metals. Cluster 1 comprised V, Co, Cr, Ni, Pb, Cu, Zn, and As with some association with Mn. Cluster 2 comprised Fe and Al with some association with Mn. The measurement of metals such as Pb, Co, Zn, Cu, As, and Cr indicates anthropogenic sources such as mining around the study sites. The PCA analysis identified two components that were significant with eigenvalues greater than 1 and were extracted accounting for total percent variance of 88.6% as shown in **Table 4.** Component 1 accounted for 74.1% of the total variance, and Component **2,** 14.5% of the total variance. This association of the metals into components as shown in **Figure 2** was confirmed by the correlation results in which As and Mn correlated

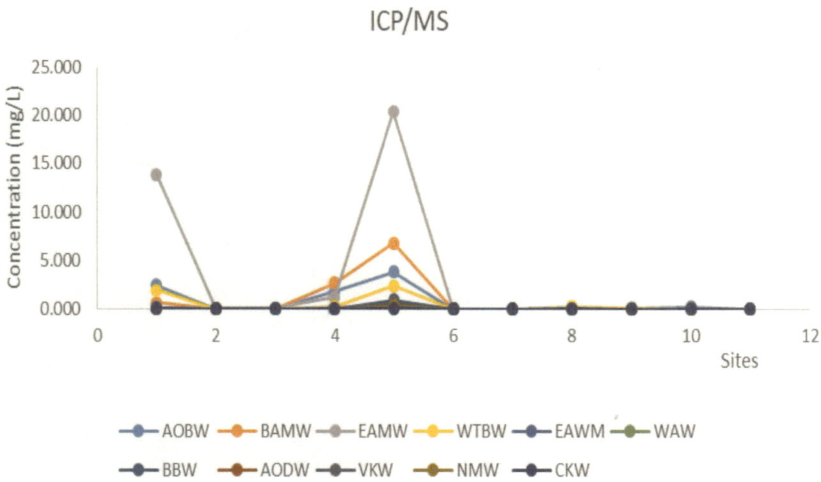

Figure 1.
A plot of concentration against sampling sites from ICP/MS results source: [12].

	PCA1	PCA2
Co	0.99	
Cu	0.98	
V	0.98	
Al	0.97	
Pb	0.97	
Cr	0.96	
Fe	0.95	
Zn	0.85	
Ni	0.77	
As		0.87
Mn		0.77
Eigenvalues	8.151	1.59
% total Variance	74.10	14.50
% cumulative variance	74.10	88.60

Table 4.
Factor loading for select heavy metals in water from mining and pristine sites.

weakly with all metals except Mn and Fe (0.76) as shown in **Table 5.** Manganese and As co-precipitate when Mn hydroxide and oxides in clay minerals act as nucleation sites for adsorption of As [40]. There was strong correlation between Pb and Cu, Co, V and Al. Lead was not detected in the pristine samples; therefore, the metal occurrence in the mining samples may be due to anthropogenic activities of mining.

Component 1, which explains majority of the total variance (74.1%), had strong loadings on Fe, Al, Pb, V, Cu, Zn, Co, Ni, and Cr. The presence of metals such as Pb, Cu, Co, Ni, Zn, and Cr suggests that mining might have contributed to metal

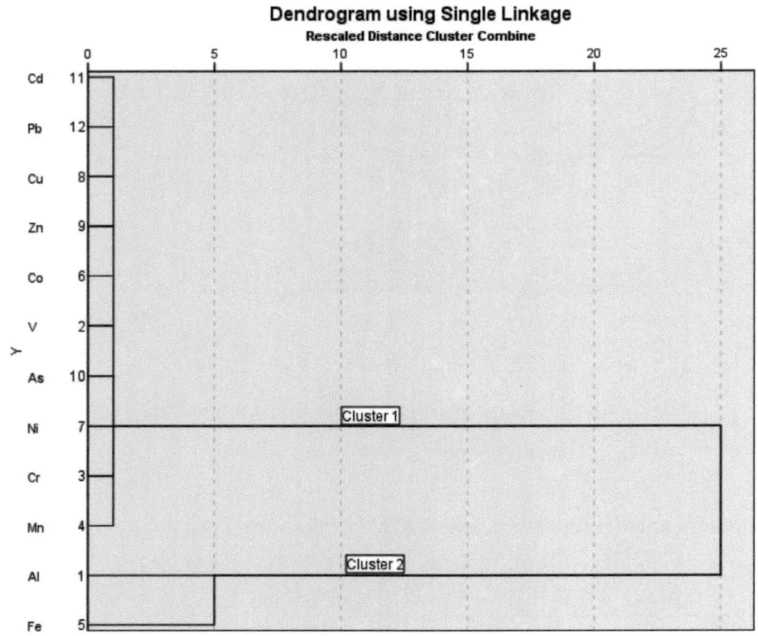

Figure 2.
Dendogram showing clustering of metals in rivers from pristine and mining sites.

Correlations											
	Al	V	Cr	Mn	Fe	Co	Ni	Cu	Zn	As	Pb
Al	1										
V	.921	1									
Cr	−.053	−.074	1								
Mn	.244	.429	−.063	1							
Fe	.740	.832	−.058	.755	1						
Co	.943	.953	−.063	.490	.847	1					
Ni	.498	.529	−.110	.437	.543	.558	1				
Cu	.965	.936	−.055	.259	.709	.957	.510	1			
Zn	.553	.418	−.041	.010	.331	.415	.272	.441	1		
As	.-.042	−.039	−.050	.348	.052	.123	.156	.073	−.088	1	
Pb	.967	.896	−.058	.159	.658	.929	.446	.984	.440	.040	1

Table 5.
Correlation matrix of select heavy metals in water samples from pristine and mining sites, n = 44.

contamination of the rivers [41]. Component 2 had strong loading on As and moderate loading on Mn suggesting that these two metals may be coming from different pollution sources. The ANOVA two-way computed indicates significant difference in metal concentrations since the probability associated with the p-value (0.005) is less than 0.05 (F = 2.89, Fcrit = 1.99). The p value (0.015) for the site study indicates significant

ANOVA						
Source of Variation	SS	*df*	*MS*	*F*	*P*-value	*F crit*
Rows	112.047	10	11.2047	2.372162	0.015297	1.937567
Columns	122.8127	9	13.64586	2.888984	2.888984	1.985595
Error	425.107	90	4.723411			

Table 6.
Two-way ANOVA showing differences between sites and metals.

differences in site concentrations (F = 2.37, Fcrit = 1.94) as shown in **Table 6**. These differences were confirmed by PCA, cluster analysis, and the correlation results. The study identifies anthropogenic activities as a major source of metal contamination of the rivers especially from the mining areas.

3.4 PROMETHEE and GAIA analysis of the heavy metals

Contamination of the rivers by heavy metals was ranked and recognized from site to site by simultaneously and systematically subjecting the concentrations to PROMETHEE and GAIA analysis. PROMETHEE II complete ranking of the sites (**Figure 3)** from least polluted to the highest polluted is shown as follows: CK ˃ AOD ˃ NM ˃ EA ˃ VKP ˃ WA ˃ BB ˃ WTB ˃ AOB ˃ BAM ˃ EAM.

The ranking shows that the pristine sites are less contaminated by the metals compared with the mining sites. The site with the least metal contamination is Kakum River (pristine site), and the highest contaminated river is the Birim River (mining site). GAIA, which is a pattern recognition tool, indicates that approximately 81.90% of the variance is explained by the first two principal components (PCs). The GAIA plot

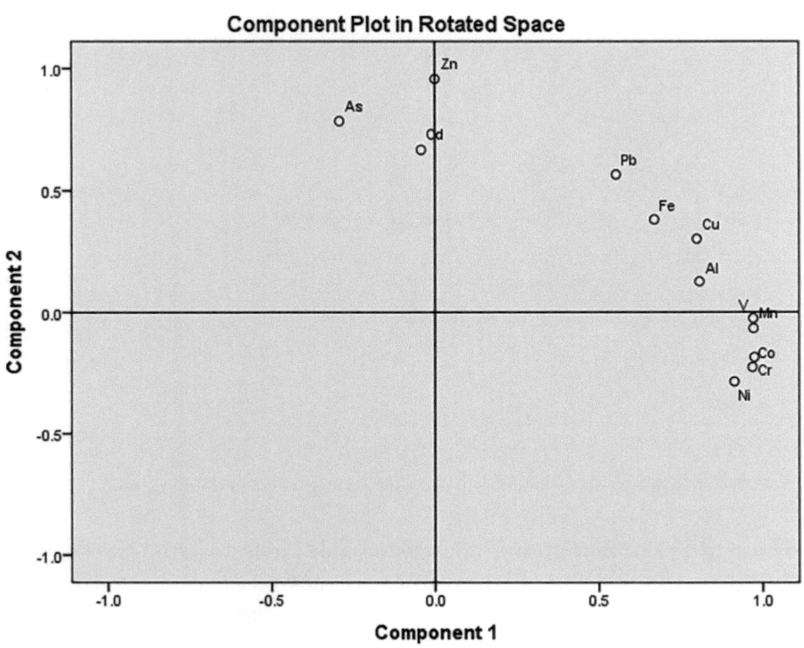

Figure 3.
Component plot showing metal loadings on components from pristine and mining sites.

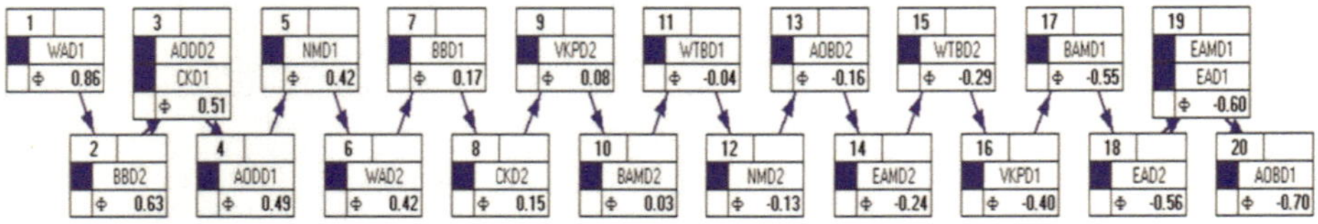

Figure 4.
PROMETHEE 2 outflow ranking of sampling sites based on heavy metals concentration in water samples from mine sites.

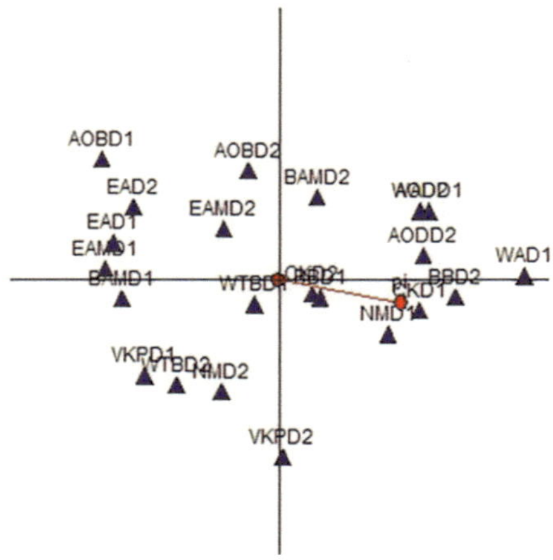

Figure 5.
GAIA plot of site distribution of metals in water samples from pristine and mine sites.

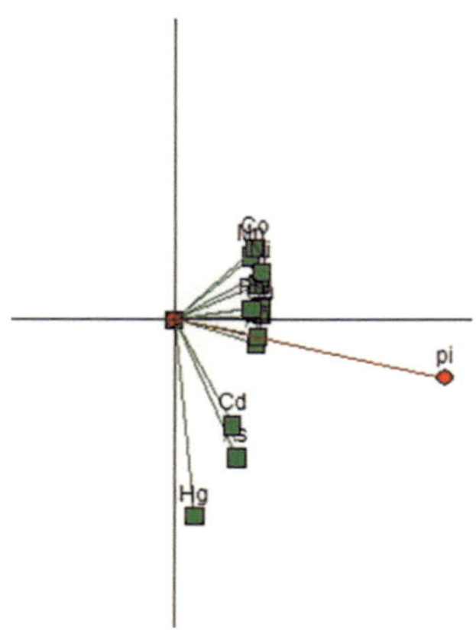

Figure 6.
GAIA plot showing heavy metals deviation from the decision axis (pristine and mining sites).

(**Figure 4**) identified similar groupings and trend as obtained from the PCA analysis. GAIA plot of the sampling sites (**Figures 5** and **6**) showed the decision axis (Pi) pointing toward the pristine sites. The PROMETHEE and GAIA analysis clearly indicates that the pristine sites are the least contaminated, while the mining sites are the most contaminated with the metals. The results also showed that anthropogenic activities such as mining may be impacting heavily on heavy metal contamination of the rivers.

Oral intake	Cr	Mn	Fe	Cu	Zn	As	Pb
ADD Range	2.5E-03 - 8.57E-05	0.076– 1.14E-04	0.582– 2.11E-03	0.007– 1.14E-05	0.003– 1.71E-04	0.006– 2.86E-05	0.001– 5.71E-05
CDI Range	0.0025– 8.571E-05	0.0762– 1.14E-04	0.5816– 2.114E-03	7.00E-03 - 1.14E-05	0.0032– 1.714E-04	6.50E-03 - 2.86E-05	7.00E-04 - 5.71E-05
CR Range	5.03E-06 - 1.71E-07					4.31E-03 - 1.91E-05	8.74E-05 - 6.72E-06
HQ Range	0.838– 0.0285	3.175– 0.005	0.831– 0.003	0.175– 0.0003	0.011– 0.0005	21.52– 0.095	0.743– 0.057
Dermal Contact							
ADD Range	2.12E-05 - 7.20E-07	6.40E-04 - 9.60E-08	4.89E-03 - 1.77E-05	5.88E-05 - 9.60E-08	1.44E-05 - 8.64E-7	5.42E-05 - 2.40E-07	6.24E-07 - 4.80E-08
CR Range	4.22E-08 - 1.44E-09					3.62E-05 - 1.60E-07	7.34E-08 - 5.65E-09
HQ Range	7.04E-03 - 2.40E-04	2.67E-02 - 4.00E-05	6.98E-03 - 2.54E-05	1.47E-03 - 2.40E-06	4.79E-05 - 2.90E-06	0.18– 8.00E-04	6.24E-04 - 4.80E-05
References							
RfDo, USEPA, 2004, 2013	0.003	0.024	0.7	0.04	0.3	0.0003	0.001
SF (DWSHA), 2012	500					1.5	8.5

Table 7.
Carcinogenic and non-carcinogenic assessment.

3.5 Carcinogenic risk assessment

Using the Central Tendency Exposure scenario (CTE) for child and adults, carcinogenic risks associated with ingestion and dermal contact with heavy metals (As, Cr, Ni, and Pb) were determined. For ingestion of water, the highest cancer risks for child and adult were measured from river EAM, a mining site for Cr as 3.45×10^{-1} and 3.70×10^{-1}, respectively. The highest cancer risks were measured for child and adult residents from river WA for Cr as 2.19×10^{-2} and 2.35×10^{-2}, respectively. Chromium posed the highest cancer risks in river EAM and WA for adult and child residents (**Table 7**). Chromium concentration from all the sites posed serious carcinogenic risk to both adult and child residents ranging from 9.39×10^{-2} to 1.35×10^{-1} and 8.77×10^{-2} to 1.26×10^{-1}, respectively. The carcinogenic risks for Ni, As, and Pb are within the USEPA risk limit (1.0×10^{-6} to 1.0×10^{-4}) [30, 31] except for As (3.35×10^{-3} and 3.12×10^{-3}) at site AOB and Pb (2.10×10^{-3} and 1.96×10^{-3}) at site EAM for resident adult and child, respectively. This implies that for As and Pb, there is a likelihood that up to 2–3 adults, out of 1000 and 1–3 children out of 1000 respectively if equally exposed continuously for 70 years would contract cancer. The carcinogenic risk via dermal contact (**Table 7**) or As, Ni, and Pb in the pristine and mining sites for adult and child is almost within the USEPA risk assessment guideline limit. However, the carcinogenic risks for Cr from all rivers in the mining sites were higher, ranging from 7.37×10^{-3} to 1.31×10^{-2} and 3.90×10^{-3} to 1.07×10^{-2} for child and adult residents, respectively. The risk values in this study are comparable with values obtained by other researchers [19, 32, 42]. The high carcinogenic risk values for As and Cr raise carcinogenic concerns for the local residents in the catchment areas. The method of risk estimation employed in this study provides ways to screen those pollutants that are of public health concern in order to prioritize research and policy interventions.

4. Conclusion

Rivers from pristine sites are less contaminated of heavy metals and are therefore safe for consumption. However, continual anthropogenic deposition of metals in the pristine rivers could accumulate with time and rise beyond acceptable limits resulting in human health risk. It was observed that the average concentrations of some of the toxic metals were low; however, direct consumption of water from these rivers could be harmful to residents since the concentrations of metals from the mining sites were far above the USEPA and WHO drinking water guideline limits. Though alternative sources of metal deposition could be accounting for high heavy metals presence in some of the rivers, anthropogenic activities, possibly mining, are suspected to be the major contributor. The first four most contaminated sites were all from the mining sites linking metal availability to mining activities.

Acknowledgements

This data reported in this paper were obtained at the Central Analytical Research Facility (CARF) operated by the Institute for Future Environments (Queensland University of Technology. Access to CARF is supported by generous funding from the Science and Engineering Faculty (QUT).

The authors are also grateful to CARF staff for their immeasurable support and training on the laboratory instruments and equipment during this work.

Conflicts of interest

The authors have declared that they have no conflict of interest.

Author details

George Yaw Hadzi
Department of Chemistry, School of Physical Sciences, College of Agricultural and Natural Sciences, University of Cape Coast, Cape Coast, Ghana

*Address all correspondence to: george.hadzi@ucc.edu.gh; geohadzi@yahoo.co.uk

References

[1] Aniefiok EI, Uwem UU, Usoro ME, Edet WN, Emmanuel JU, Akanino NE, et al. Heavy metals in Epiphytic Lichens and Mosses of Producing Communities of Ekel and Ibeno, Akwo Ibom State - Nigeria. American Journal of Environmental Protection. 2016;**4**(2):38-47

[2] Asamoah BD, Asare A, Okpati SW, Aidoo P. Heavy metal levels and their ecological risks in surface soils at Sunyani magazine in the bono region of Ghana. Scientific African. 2021;**13**:e00937

[3] Rai PK. Heavy metal phytoremediation from aquatic ecosystems with special reference to macrophytes. Critical Reviews in Environmental Science and Technology. 2009;**39**(9):697-753

[4] Okafor EC, Opuene K. Preliminary assessment of trace metals and polycyclic aromatic hydrocarbons in the sediments. International journal of Environmental Science and Technology. 2007;**4**(2):233-240

[5] Mohiuddin KM, Zakir HM, Otomo K, Sharmin S, Shikazono N. Geochemical distribution of trace metal pollutants in water and sediments of downstream ofan urban river. International journal of Environmental Science and Technology. 2010;**7**(1):17-28

[6] Nafea EM, Zyada MA. Biomonitoring of heavy metals pollution in Lake Burrullas, Northern Delta, Egypt. African Journal of Environmental Science and Technology. 2015;**9**(1):1-7

[7] Ali H, Khan E, Sajad MA. Phytoremediation of heavy metals—concepts and applications. Chemosphere. 2013;**91**(7):869-881

[8] Mudipalli A. Metals (micro nutrients or toxicants) & Global Health. The Indian Journal of Medical Research. 2008;**128**(4):331-334

[9] Harada M. Minamata disease: methylmercury poisoning in Japan caused by environmental pollution. Critical Reviews in Toxicology. 1995;**25**(1):1-24

[10] Debes F, Budtz-Jørgensen E, Weihe P, White RF, Grandjean P. Impact of prenatal methylmercury exposure on neurobehavioral function at age 14 years. Neurotoxicology and Teratology. 2006;**28**(5):536-547

[11] Proshad R, Kormoker T, Mursheed N, Islam MM, Bhuyan MI, Islam MS, et al. Heavy metal toxicity in agricultural soil due to rapid industrialization in Bangladesh: a review. International Journal of Advanced Geosciences. 2018;**6**(1):83-88

[12] Hadzi GY, Essumang DK, Ayoko GA. Assessment of contamination and health risk of heavy metals in selected water bodies around gold mining areas in Ghana. Environmental Monitoring and Assessment. 2018;**7**:1-7

[13] Miller JR, Hudson-Edwards KA, Lechler PJ, Preston D, Macklin MG. Heavy metal contamination of water, soil and produce within riverine communities of the Rıo Pilcomayo basin, Bolivia. Science of the Total Environment. 2004;**320**(2–3):189-209

[14] Windom HL, Byrd JT, Smith RG Jr, Huan F. Inadequacy of NASQAN data for assessing metal trends in the nation's rivers. Environmental Science & Technology. 1991;**25**(6): 1137-1142

[15] Acosta JA, Faz A, Martinez-Martinez S, Zornoza R, Carmona DM, Kabas S.

Multivariate statistical and GIS-based approach to evaluate heavy metals behaviour in mine sites for future reclamation. Journal of Geochemical Exploration. 2011;**109**:8-17

[16] Calas G. Mineral resources and sustainable development. Elements: An International Magazine of Mineralogy, Geochemistry, and Petrology. Vol. 3. no. 5. West Sussex, UK: John Wiley & Sons Ltd; 2011. pp. 301-306

[17] Hadzi GY, Essumang KD, Adjei KJ. Distribution and risk assessment of heavy metals in surface water from pristine environments and major mining areas in Ghana. Journal of Health and Pollution. 2015; **5**(9):86-99

[18] Asklund R, Eldvall B. Contamination of water resources in Tarkwa mining area of Ghana. Lund, Sweden. LUTVDG/TVTG – 5092 – SE: Department of Engineering Geology, Lund University; 2005

[19] Obiri S, Yeboah PO, Osae S, Adu-Kumi S, Cobbina SJ, Armah FA, et al. Human health risk assessment of artisanal miners exposed to toxic chemicals in water and sediments in the PresteaHuni Valley District of Ghana. International Journal of Environmental Research and Public Health. 2016; **13**(1):139

[20] Donato R. RAGE: A single receptor for several ligands and different cellular responses: the case of certain S100 proteins. Current Molecular Medicine. 2007;**7**(8):711-724

[21] Chon HT, Ahn JS, Jung MC. Environmental contamination of toxic heavy metals in the vicinity of some Au–Ag mines in Korea. Turku, Finland: Proceedings of the 4th Biennial SGA Meeting; 1997. pp. 891-894

[22] Okereafor U, Makhatha M, Mekuto L, Uche-Okereafor N, Sebola T, Mavumengwana V. Toxic metal implications on agricultural soils, plants, animals, aquatic life and human health. International Journal of Environmental Research and Public Health. 2020;**7**:2204

[23] Ravenscroft P, Brammer H, Richards K. Arsenic Pollution: A Global Synthesis. John Wiley & Sons; 2011

[24] Ahmad A, Bhattacharya P. Arsenic in drinking water: Is 10 μg/L a safe limit? Current Pollution Reports. 2019;**1**:1-3

[25] Caminero AG, Howe P, Hughes M, Kenyon E, Lewis DR, Moore M, et al. Environmental health criteria 224: Arsenic and arsenic compounds. 2nd ed. Geneva, Switzerland: World Health Organization; 2001. p. 512

[26] Ite AE, Udousoro II, Ibok UJ. Distribution of some atmospheric heavy metals in lichen and moss samples collected from Eket and Ibeno Local Government Areas of Akwa Ibom State, Nigeria. American Journal of Environmental protection. 2014;**2**(1): 22-31

[27] Etesin UM, Ite AE, Harry TA, Bassey CE, Nsi EW. Assessment of cadmium and lead distribution in the outcrop Rocks of Abakaliki Anticlinorium in the southern Benue Trough, Nigeria. Journal of Environment Pollution and Human Health. 2015;**3**(3): 62-69

[28] Naveedullah HMZ, Yu C, Shen H, Duan D, Shen C, Lou L, et al. Concentration and human health risk assessment of selected heavy metals in surface water of the Siling Reservoir watershed in Zhejiang Province, China. Polish Journal of Environmental Studies. 2014;**23**(3):801-811

[29] Bartholomew DJ, Steele F, Moustaki I, Galbraith J. Analysis of Multivariate Social Science data. 2nd ed. Boca Raton, FL: CRC Press; 2008. p. 384

[30] USEPA. United States Environmental Protection Agency. Risk Assessment Guidance for Superfund (Volume 1) Human Health Evaluation Manual Part A Interim Final. EPA/540/1–89/002. Washington, DC, USA: Office of Emergency and Remedial Response; 1989

[31] USEPA. Risk Assessment Guidance for Superfund Vol. 1 Human Health Evaluation Manual, Part E, Supplemental Guidance from Dermal Risk Assessment. Hawthorne, CA, Washington, DC, USA: Office of Emergency and Remedial Response. Academic Press; 1979

[32] Obiri S, Dodoo DK, Essumang DK, Armah FA. Cancer and non-cancer risk assessment from exposure to arsenic, copper, and cadmium in borehole, tap, and surface water in the Obuasi municipality Ghana. Human and Ecological Risk Assessment. 2010;**16**(3): 651-665

[33] WHO. Guidelines for drinking water quality. 3rd ed. Geneva: World Health Organization; 2004

[34] Akoto O, Adiyiah J. Chemical analysis of drinking water from some communities in the Brong Ahafo region. International Journal of Environmental Science and Technology. 2007;**4**(2): 211-214

[35] WHO. Aluminium. Geneva: World Health Organization, International Programme on Chemical Safety. 1997 (Environmental Health Critaria 194)

[36] Matthew N, Samuel Jerry C, Michael K. Assessment of endocrine disrupting trace metals in River Samre at Samreboi in the Wassa Amenfi west district of the western region of Ghana. Journal of Water Resource and Protection. 2013;**5**:983-992

[37] Cobbina SJ, Nkuah D, Tom-Dery D, Obiri S. Non-cancer risk assessment from exposure to mercury (Hg), cadmium (Cd), arsenic (As), copper (Cu) and lead (Pb) in boreholes and surface water in Tinga, in the Bole-Bamboi District, Ghana. Journal of Toxicology and Environmental Health Sciences. 2013;**5**(2):29-36

[38] Bowen HJ. Environmental Chemistry of the Elements. Academic Press; 1979

[39] Aladesanmi OT, Adeniyi IF, Adesiyan IM. Comparative assessment and source identification of heavy metals in selected fishpond water, sediment and fish tissues/organs in Osun State Nigeria. Journal of Health Pollution. 2014;7:42-53

[40] Takamatsu T, Kawashima M, Koyama M. The role of Mn2+−rich hydrous manganese oxide in the accumulation of arsenic in lake sediments. Water Research. 1985;**19**(8): 1029-1032

[41] Armah FA, Obiri S, Yawson DO, Onumah EE, Yengoh GT, Afrifa EK, et al. Anthropogenic sources and environmentally relevant concentrations of heavy metals in surface water of a mining district in Ghana: A multivariate statistical approach. Journal of Environmental Science and Health Part A. 2010;**45**(13):1804-1813

[42] Iqbal J, Shah MH. Health risk assessment of metals in surface water from freshwater source lakes, Pakistan. Human and Ecological Risk Assessment: An International Journal. 2013;**19**(6): 1530-1543

Chapter 11

Distribution of Potentially Toxic Elements in Water, Sediment and Soils in the Riparian Zones around a Kraft Pulp and Paper Mill in Western Kenya

Elijah S. Omutange, Julius O. Manyala, Anne N. Wekesa, Odipo Osano, Thomas Munyao, Alfred A. Otieno and Lazare Etiégni

Abstract

This chapter discusses the impact of effluent discharge on the distribution of potentially toxic elements (Pb, Cd, Cu and Zn) in water, sediments and soils near a Kraft mill in Western Kenya. Potentially Toxic Elements (PTEs) concentrations were determined at three sites: Water Intake Point, Effluent Discharge Point and Downstream Point. The mill liquid effluent parameters and gaseous emissions were also characterized. One-way ANOVA was used to analyze the spatial differences in PTEs concentrations. Principal component analysis determined the correlations between the proximity to the Kraft mill and the PTEs in soils, water and sediments. In riverine soils, Cd was 0.78 ± 0.01 mg/kg, while Pb was 94.38 ± 9.65 mg/kg. In sediments, the concentration was 16.81 ± 2.46 mg/kg for Zn, 6.16 ± 0.72 mg/kg for Cd and 75.28 ± 5.97 mg/kg for Pb. In water, Zn was 0.26 ± 0.038 mg/L, Cu was 0.75 ± 0.11 mg/L, Cd was 0.05 ± 0.004 mg/L and Pb was 1.26 ± 9.65 mg/L. The spatial distributions of PTEs in soils near the factory and across the river may have resulted from the factory's effluent discharge and gaseous emissions. These findings should help formulate more stringent industrial effluent management programs in Western Kenya.

Keywords: potentially toxic elements, gaseous emissions, industrial effluent, Kenya, pollution, pulp and paper

1. Introduction

Potentially toxic elements (PTEs) pollution is one of the most important environmental problems in many developed and developing countries [1–8]. PTEs are naturally occurring elements throughout the earth's crust and they exert key biochemical and physiological functions in plants and animals. At low concentrations as trace elements, they are important constituents of several enzymatic and biological functions and play important roles in various oxidation-reduction reactions in the human body.

At higher concentrations, however, PTEs become environmental contaminants, causing many ailments in humans through exposure to anthropogenic activities such as agriculture, mining and smelting operations [9–12]. Significant contribution to PTEs contamination may also come from natural phenomena i.e. metal corrosion, atmospheric deposition, soil erosion of metal ions and leaching of potentially toxic elements. Other natural phenomena include sediment re-suspension and metal evaporation from water resources to soil and groundwater volcanic eruptions and weathering [4, 13–15]. Industrial sources of PTEs include: metal processing in refineries, coal burning in power plants, petroleum combustion, nuclear power stations and high tension lines, plastics, textiles, microelectronics, wood preservation, leather tanning and paper manufacture plants [4, 14, 16–19].

Metal occurrence in the environment can increase significantly from industrial effluent leachates and accumulate near the surface layer of soil and at water column. Metal chemistry, biotic species or specific differences in water resource can affect patterns of metal exposure, uptake and metabolism. Analyzing sediments best assesses metal pollution in river, as they are less susceptible to flow conditions than water column and act as sink for metal pollutants. The metals attach to any of these phases (water, soil and sediments) in proportions that depend on the physico-chemical conditions such as conductance, pH and temperature prevailing in these phases.

The pulp and paper industry, in particular, has historically been a major contributor of pollutants into the environment, including PTEs [18, 20, 21]. There are five steps in the production of pulp and paper, namely, 1) preparation of raw materials, 2) pulping, 3) bleaching, 4) pulp handling and/or paper production, and 5) recovery of chemicals through the combustion of spent or black liquor [22, 23]. The paper manufacturing process consumes large volumes of fresh water, most of which becomes contaminated with recalcitrant compounds before effluent discharge.

The composition of each mill's effluent depends on the pulping process used; the grade of paper manufactured; the volume of fresh water consumed; the wastewater cleaning technique; and wastewater reuse programs. Paper mills effluents are complex mixtures with more than 200–300 different organic chemicals [24, 25] and approximately 400–500 inorganic compounds [26, 27]. Raw effluent from paper mills has characteristics of high BOD, COD, high levels of chlorinated ligno-sulphonic acids, total dissolved and suspended solids, chlorinated resin acids, chlorinated phenols and hydrocarbon. In addition, the effluent may contain PTEs such as Fe, Zn, Cu, Cd, Mn, Ni, Cr and Pb, that can be present in high concentrations well above permissible limits [20].

1.1 Source of PTEs in a Kraft pulp mill

The main chemical ingredients in a Kraft pulp mill are sodium hydroxide (NaOH) and sodium sulphide (Na_2S). The primary PTEs found in wood and wood products are Ca, Mg, K, Mn, Ba, Al, Ni, Cu, Zn, Cd, Pb. In addition, PTEs are generated through other make-up chemicals, such as sodium sulfate (Na_2SO_4); bleaching agents; water; the chemical recovery system where the cooking liquor is regenerated; Green Liquor Dregs (GLD); Lime Mud (LM); Slaker Grits (SG); Boiler Fly Ash (BFA) and waste

	GLD			SG		LM		BFA		
Element	[*]	[**]	[**]	[#]	[**]	[##]	[*#]	[**]	[*##]	[**##]
Pb	6.12	46.8	13	<3	34.1	6.79	<3	44.3	28.7	31
Cd	3.81	5.19	9.4	0.3	4.75	0.91	<0.3	4.7	2.9	3.3
Cu	229	80.9	102	<10	4.6	0.73	4.1	25.8	63.6	72
Cr	295	56.0	118	12.6	12.4	16.7	7.0	24.1	66.9	74
Ni	233	189	84	23.9	25.2	ni	4.0	97.4	32.4	33
Zn	3197	160	1000	9.9	15.0	ni	36	68.9	295.3	320
Hg	<0.05	ni	ni	<0.03	ni	<0.04	<0.03	ni	0.03	0.1
V	ni	ni	1.9	39.0	ni	ni	ni	ni	92.7	ni
Mo	0.29	ni	1.7	<1	ni	ni	2	ni	3.8	ni
As	<0.1	ni	0.3	<3	ni	0.38	2.7	ni	13.0	14

** [31]; #[32]; **[33]; ##[34]; #*[35]; *## [36]; **## [37].*

Table 1.
Content of potential toxic element [23].

paper recycling [28–30]. **Table 1** shows the chemicals composition of GLD, LM, SG, BFA determined in a few mills throughout the pulp and paper industry. These elements leave the mill system through wastewater, fugitive uncontrolled air emissions or flue gases from the recovery boiler [38, 39]. Apart from the fugitive emissions, the mill effluents (liquid and gaseous) undergo treatment in the wastewater purification plant or the air pollution control system depending on the local effluent discharge standards before final disposal into the environment.

There is increasing awareness of potentially toxic elements pollution from industrial discharges, especially their potential effects in a rural setting such as Western Kenya. The area houses several agro-industrial factories; including a Kraft pulp and paper mill, several sugar milling plants, tea factories and fish processing plants. Because of the various amounts of chemicals used in the manufacturing processes, the roles of these industrial discharges could be significant in metal concentration, fate and transport in the adjacent water bodies and soils. An extensive assessment of PTEs contamination in such a setting is therefore required. The objective of this study was to estimate PTEs (Pb, Cd, Zn and Cu) distribution in water, sediments and soils collected near a Kraft pulp and paper mill in Western Kenya.

2. Materials and methods

2.1 Study area

This study was carried out on a Kraft pulp and paper mill in Webuye (34–36° E, 0° 03′-1°15’ N) in the upper catchment of the Nzoia River (**Figure 1**). The Nzoia River basin is approximately 12,696 km^2 [40] and lies within the Lake Victoria basin in Kenya, East Africa. The river flows between 2700 and 1134 m above sea level. The region receives an average annual rainfall of 1350 mm. The average annual air temperature varies between 8°C and 28°C, with minimum temperatures between 8°C and 12°C and maximum temperatures of 24 and 28°C [41]. Food crops such as maize,

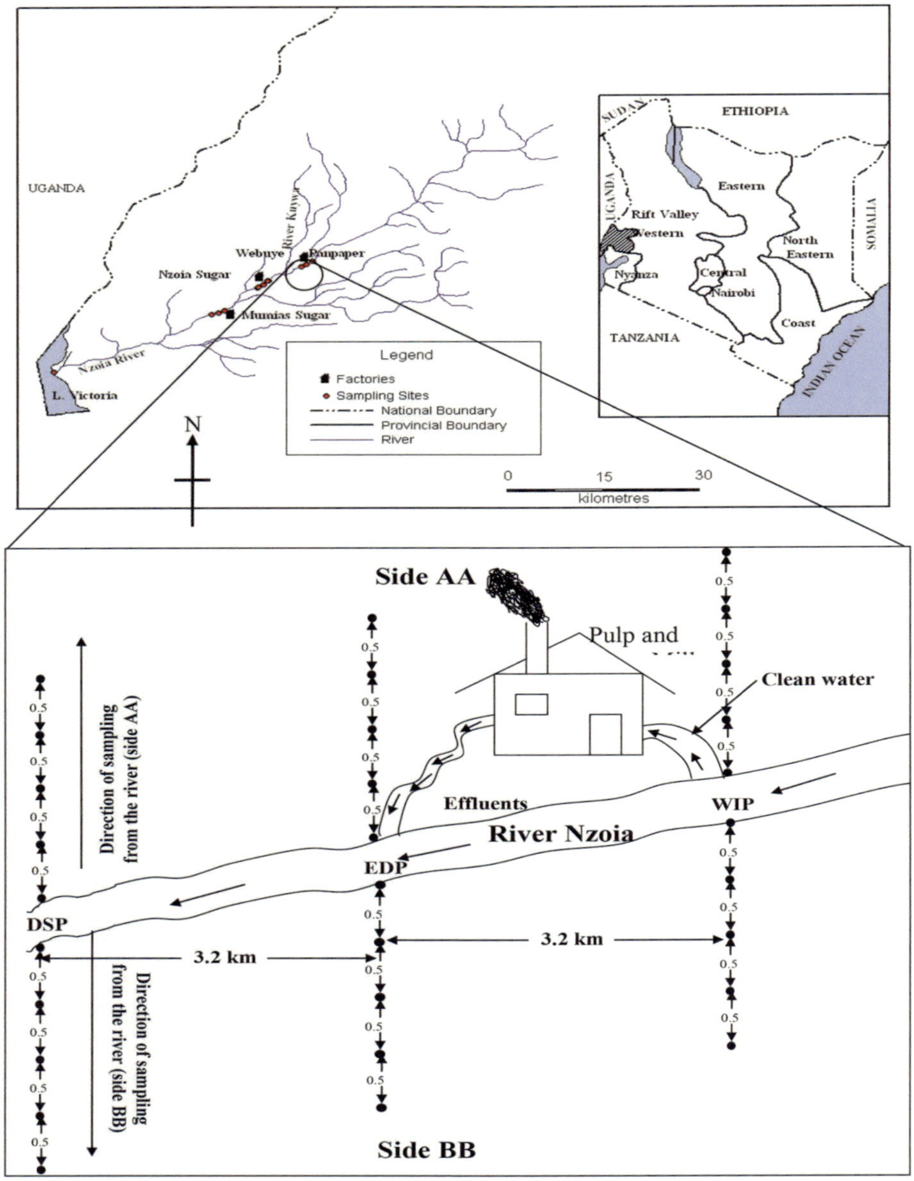

Figure 1.
Study area and study sites.

sweet potatoes and cassava, sorghum, millet and vegetables are grown on small-scale farms, usually extending up to the river banks. Livestock farming is practiced and River Nzoia provides water for both domestic and industrial use. The mill consumes about 40,000 m^3 of fresh water and discharges between 35,000 to 40,000 m^3 daily into the river at a dilution rate between 0.3 to 3.2%, depending on the seasonality of the river discharge. The mill's effluent takes 6 weeks to flow through a set of settling tanks (one primary and one secondary), two aerated lagoons, and two stabilization ponds before discharge into River Nzoia. Recent expansion programs within the mill have led to an overloaded wastewater treatment system, initially designed to treat only 25,000 m^3 of mill effluent per day.

2.2 Study sites

There were three sampling sites, namely, the 'Water intake point' (WIP), the 'Effluent deposition/discharge point' (EDP) and the 'Downstream point' (DSP). These sites were located along the profile of River Nzoia (**Figure 1**). The WIP was situated upstream of the factory discharge point and the river width at this location was 6.53 m. The EDP was approximately 3.2 km from the WIP and the river width was 5.92 m. The downstream point was 3.2 km from the EDP with a river width of 6.44 m. All the sampling sites located on the factory side were designated Side AA (**Figure 2**), while those on the opposite side were designated as Side BB.

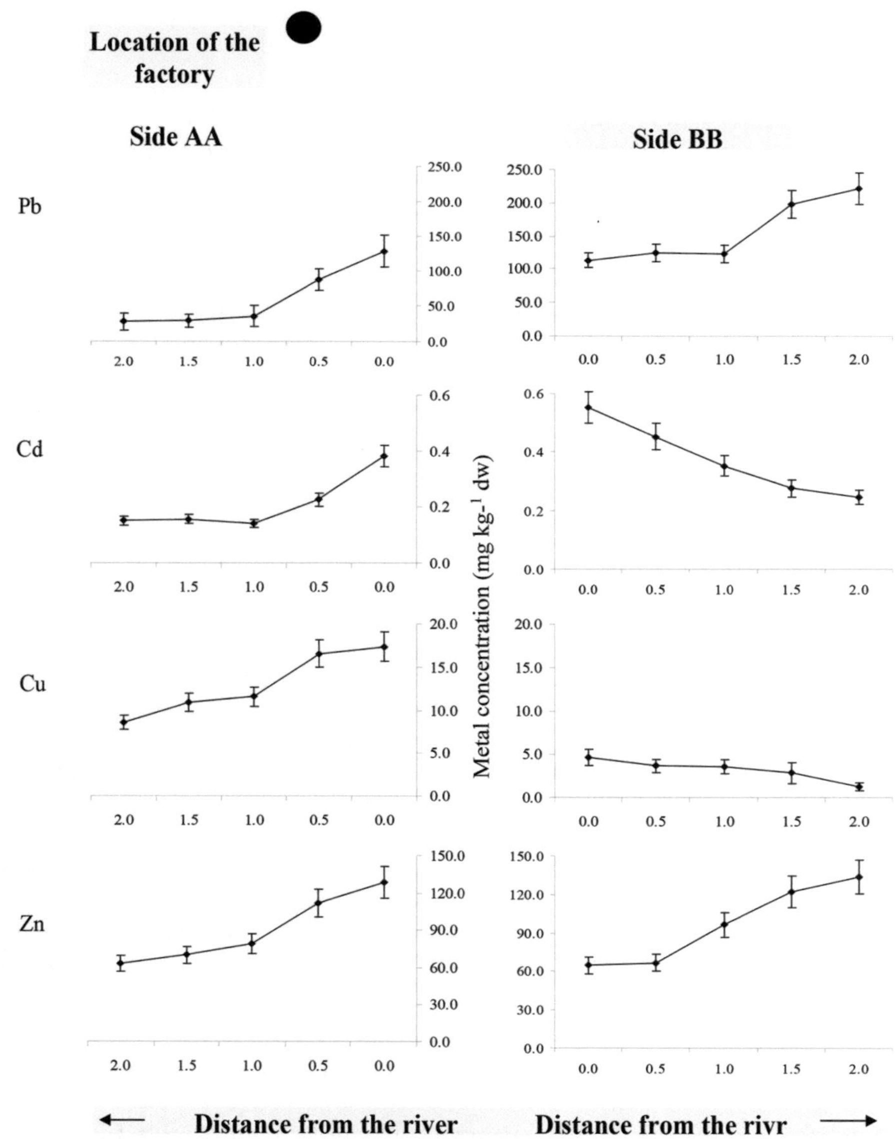

Figure 2.
Trends of heavy metal concentration mg/g in wetland soil at WIP in PanPaper.

2.3 Sampling and analysis of pulp and paper industrial wastes

Grab samples of 3 kg each of lime mud and boiler ash were collected from the recovery area of the pulp and paper mill in Webuye. These samples were spread on trays; air dried overnight, sieved using a No. 9 mesh sieve (2.00 mm) and stored in plastic bags. The pH of these samples was determined using the ASTM D 4972–01 Standard Test Method for soil pH (Electrometric method). The ASTM D 2216–98 Standard Test Method for Laboratory Determination of moisture was used to measure the moisture content. The concentration of PTEs was determined using an Inductively Coupled Plasma-Optical Emission Spectrometry (ICP-OES). The mill liquid effluent was tested for various parameters at different treatment stages and compared with national effluent discharge standards.

2.4 Sampling and analysis of river water

Grab river water samples were collected using half liter metal free *Van Dorn* bottle. These samples were collected at about 0.5 m below the water surface. Before collecting the samples, the *Van Dorn* bottle were washed in 2 L tap water, and rinsed three times in distilled water. The collected samples were then transferred to half liter polythene bottles pre-soaked in nitric and sulfuric acids solution at 1:1 volume ratio. The water samples were acidified to a pH 2 using concentrated nitric acid [42], and stored in an ice box before transportation to the laboratory for chemical analysis.

Temperature, pH and electrical conductivity (EC) were measured *in situ* using a calibrated JENWAY 3405 electrochemical analyzer (Barloword Scientific Ltd., Essex, UK), with a specific probe for each variable. The equipment was calibrated using de-ionized water before measurements. The measurements were conducted in triplicate for each site included in the study.

2.5 Sediments and soil sampling

Thirty-six sediment samples were collected from each sampling site. The sediments were collected at the bottom of the river using an *Ekmans* Grab Sampler. A polypropylene spatula was used to transfer the sediment sub-samples to acid rinsed polypropylene bottles and placed in an icebox for transportation to the laboratory for chemical analyses. Soil samples were collected using a soil auger along the river banks within the organo-mineral layer of 0–25 cm of the soil. The samples were then kept in black polythene bags, labeled and stored in an icebox before transportation to the laboratory for chemical analyses.

To determine PTEs concentration, soil samples were scooped at five equidistant points from the river bank, i.e. 0.0, 0.5, 1.0, 1.5 and 2.0 kilometers respectively away from the River Nzoia bank on both sides AA and BB. The 0.5 km distance was deemed sufficient to reflect lateral variations of potentially toxic elements concentrations in the soil.

2.6 Determination of leaching capability of PTEs in sediment and soils

A leachability tests were used to evaluate the possible elution of Pb, Cu and Zn from the sediment under selected treatment conditions. These tests took place because the paper mill periodically discharges the recovery boiler ash and lime mud into the river. The sediment and soil samples were filtered through a Ø 9.0 mm filter paper, air dried, crushed and then sieved through of 9.5 mm pore size. Distilled water was used as the leaching solution, adjusted to a pH 5 using a mixture of sulfuric and nitric acids (80, 20 by weight). Initially, 50 grams of sediments were added to the leaching solution followed by soil at a solid to liquid ratio of 1:20 (proportional to 1000 mL leaching solution). The mixture was agitated in a rotary agitator for 18 hours at 30 rpm and a constant temperature of 25°C. After agitation, the soil slurry was filtered through a 0.8 mm glass fiber filter. The liquid extract was digested using concentrated HNO_3 and analyzed for Cr, Cu, Pb and Zn using an ICP-OES. All the analyses were run in triplicate.

2.7 Analytical determination of PTEs in water

Water samples were digested using sulfuric and nitric acids before spectrophotometric analysis was conducted to minimize the interference of complex organic matter [43]. The samples were digested and concentrated on a hot plate from 100 mL to 25 mL for 3 hours. After digestion, the samples were allowed to cool to room temperature and then 2 mL of 30% hydrogen peroxide (H_2O_2) were added to oxidize any residual organic matter. Further cooling followed before the digested samples were filtered through a 0.45 μm nucleopore membrane filter over a vacuum pump. The filtrates were stored in 125 mL polyethylene sample bottles, at 4°C before analysis for Zn, Cu, Pb, and Cd in an Atomic Absorption Spectrophotometer (Model AA 10/20). In this study, at least two calibration standards were prepared for each metal before recording the measurements. The PTEs were determined at various spectrophotometric wavelengths and slit width (Pb = 17.0 nm; Zn = 13.9 nm; Cu = 324.8 nm; Cd = 228.8 nm and slit width for Pb and Zn = 1.0 nm and Cd and Cu = 0.5). The concentrations of PTEs in water were calculated and reported in mg L^{-1}.

2.8 Determination of PTEs in soil, sediment and industrial wastes

The study used analytical grade reagents and the stock solutions met the Merck certificate AA standards. In addition, all experiments used milli-Q water. Plastic and glassware were cleaned by soaking them in 14% (v/v) HNO_3 for 24 hours and rinsed with nano pure water. The sediments, soils samples and industrial wastes were crushed and homogenized using a Fritsch, Pulverisette 5, planetary mill (Fritsch GmbH Laborgerate, Idar-Oberstein, Germany) for 5 minutes at 400 rpm. About 0.20 g of soil and sediment samples and about 25 mL of water were weighed in Teflon (© poly-tetra-fluoretheen (PTFE), DuPont™) high pressure vessels. Then 4.0 mL concentrated nitric acid (65%), 1.0 mL concentrated hydrochloric acid (37%) and 1.0 mL ultrapure water were added to the samples. Six samples of each item were placed in the carousel of a *Paar* Microwave oven (Anton Paar GmbH – Graz –Austria). The samples were digested in a microwave oven (Anton Paarâ GmbH Kärntner Straße 322 A-8054 Graz/Austria) at a maximum temperature of 220°C and pressure of 75 bars for 15 minutes.

After cooling, the resultant clear solutions were poured into 50 mL volumetric flasks and diluted to the mark with ultra-pure demineralized water (Barnstead NanoPure, Thermo Fisher Scientific Inc., Barnstead International,

Iowa USA). Finally, the diluted solutions of the respective samples were transferred into acid cleaned polyethylene bottles. All elements were determined using an Inductively Coupled Plasma-Optical Emission Spectrometer (Perkin Elmer Optima 3000 XL, ICP-OES) with the PE calibration standards. The concentrations of PTEs in soil and sediments were calculated and reported in mg kg^{-1} dry weight. The quality of the analytical process was controlled by the analysis of IAEA MA-A-3/TM certified standard reference material of river sediment. Care was taken to ensure analytical results varied from certified values by less than 10%.

2.9 Estimation of atmospheric discharge of gaseous pollutants from the Kraft paper mill

Previous studies have indicated that atmospheric deposition is a major source of most PTEs entering agricultural land [44, 45]. Thus, emission (E) estimates of selected gases, particulates, ash and chemical elements were calculated using emission factors (E_f) from literature [31, 46]. The equation used was E = AixEf, where A_i = Activity or industrial production of the paper mill. The maximum production was 300 metric tonnes per day or 105,000 metric tonnes per year air-dried unbleached pulp. The study assumed that Kraft pulping was the only system at the mill and that the emissions came solely from the recovery boiler at the factory. In addition, we assumed that only an electrostatic precipitator was used to treat flue gases from the mill.

2.10 Data analysis

Statistical analyses were performed using STATISTICA 6.0 [47]. Data collected were first tested for homoscedasticity [48] before subjecting them to statistical analysis. Data on PTEs in water, soils and sediments were calculated as means ($\pm$ S.D) for each site. Spatial differences in metal concentrations were analyzed using One-way ANOVA. Duncan's Multiple Range Tests (DMRT) were used for Post-hoc discrimination between the means that differed significantly. The data for all the sites were subjected to Principal Component Analysis (PCA) to determine the correlation between the PTEs concentration in soil, water and sediments and their proximity to the Kraft mill [49].

3. Results

A summary of the liquid effluent from the paper mill at various stages of the treatment process are presented in **Table 2**. Most effluent parameters decreased or reduced as we moved from the primary clarifier to the final stabilization pond prior to discharge into the river and the difference was statistically significant ($p < 0.05$). Only turbidity and color of the treated effluent increased markedly. For example, the effluent color increased on average by almost 250% from 133 ^{O}H to 331 ^{O}H, while its turbidity rose by 155%, from 1574 NTU to 2431 NTU.

Table 3 summarizes emission estimates from the Kraft Paper mill in Western Kenya. The volume of total reduced sulfur (TRS) was still high during the study, and

Parameters	Primary clarifier overflow	Treated after last stabilization lagoon	Effluent discharge standards*
pH	8.5–9.3	6.9–7.5	6.5–8.5 (non-marine)
Alkalinity (mg/L)	330.0–346.4	70.0–88.15	—
Temp (OC)	39.0–39.6	19.0–21.2	± 3
TS (mg/L)	872.4–980.7	440.0–474.5	30
TDS (mg/L)	670.0–699.6	300.0–348.5	1200.00
TSS (mg/L)	212.5–291.5	94.6–133.0	30
Color (OH)	1280.5–1867.7	1600.0–3263.3	15
BOD5 (mg/L)	182.5–234.7	62.8–117.6	30
COD (mg/L)	536.0–591.54	296.7–401.5	50
Turbidity (NTU)	130.0–136.1	311.0–351.3	—
Conductivity (mScm-1)	1339.2–2109.3	790.0–891.3	—
Dissolved oxygen (mg/L)	0	0.40–0.30	—
Phosphorus (mg/L)	0.056–0.067	0.0050.0015	—
Nitrites (mg/L)	0.004–003	0.008–0.01	—
Nitrate (mg/L)	0.020.01	0.036–0.05	—
Cd (mg/L)	0.193–0.182	—	—
Co (mg/L)	0.041–0.035	—	—
Cr (mg/L)	0.568–0.63	—	—
Cu (mg/L)	0.105–0.146	—	—
Fe (mg/L)	0.211–0.235	—	
K (mg/L)	0.24–0.34	—	—
Mg (mg/L)	0.65–0.71	—	—
Mn (mg/L)	1.325–1.41	—	—
Na (mg/L)	0.083–0.92	—	—
P (mg/L)	0	—	—
Pb (mg/L)	0.08–0.09	—	—
Zn (mg/L)	0.04–0.46	—	—

Table 2.
*Raw and treated Kraft pulp and paper mill effluent by current treatment system (*source: Kenya effluent discharge standard. 3rd schedule).*

confirmed the characteristic strong foul smell in and around the mill. The concentration of Cu, Pb and Cr remained relatively high, while Zn, Ni and Cd were low over the course of the study period.

The river Nzoia's water physico-chemical parameters results within the three sampling sites (WIP, EDP and DSP) are presented in **Table 4**. Temperature, pH and electrical conductivity (E.C) in the three sites exhibited significant variations ($p < 0.05$) (Type III, One-Way ANOVA). Temperature, pH and EC were significantly higher at EDP as compared to other sites but there was not enough evidence of significant difference in pH between WIP and DSP sites.

Air Pollutants	Emission factor (kg/metric tonne)	Energy (MJ/ metric tonne)	Paper Production at PanPaper (tonne/year)	Emission (kg/year)
Particulates	1	33763.2	105,000	105,000
SO_2[a]	3.5	33763.2	105,000	367,500
CO[a]	5.5	33763.2	105,000	577,500
TRS[ab]	7.5	33763.2	105,000	787,500
NOX[ac]	1	33763.2	105,000	105,000
Hg[d]	1.0E-05	33763.2	105,000	1.06
Cd[d]	1.6E-04	33763.2	105,000	16.31
Cr[d]	6.7E-04	33763.2	105,000	70.19
Cu[d]	1.0E-03	33763.2	105,000	108.13
Ni[d]	4.0E-04	33763.2	105,000	42.19
Pb[d]	1.3E-03	33763.2	105,000	132.23
Zn[d]	4.0E-04	33763.2	105,000	42.19
Sn[d]	2.7E-04	33763.2	105,000	28.36
Co[d]	8.1E-05	33763.2	105,000	8.51
Ash[d]	1.1E-04	33763.2	105,000	11.70

[a]*Emission Factors expressed in unit weight of air –dried unbleached pulp (ADP) from Recovery boiler and direct evaporators –Controlled by Electrostatic Precipitators from US EPA, 1990.*
[b]*TRS (Total Reduced Sulfur) includes methyl mercaptan, dimethyl disulphide, and hydrogen sulphide.*
[d]*Emission factors from [42].*

Table 3.
Estimated Kraft pulp mill emissions.

	Sampling sites			One-Way ANOVA	
Parameters	**WIP**	**EDP**	**DSP**	**F**	**P**
Temperature °C	20.78 ± 0.14^a	23.83 ± 0.53^b	21.54 ± 0.25^a	20.735	<0.0005
pH	8.13 ± 0.11^a	8.43 ± 0.57^b	8.13 ± 0.07^a	4.871	0.014
Electrical conductivity (μS/cm)	109.6 ± 0.6^a	2720.5 ± 219.3^b	333.3 ± 21.5^a	129.403	<0.0005

**Mean values ± SEM followed by different superscripts 'a' or 'b' across the rows are significantly different at α = 0.05. Mean values across the rows with the same superscripts 'a' are not statistically different at α = 0.05. For example, the pH at DSP is not statistically different from the pH value at WIP because it carries the same superscript 'a'.*

Table 4.
**mean values ± SEM of the physico-chemical parameters along the sampling sites at PanPaper Mills, Webuye, Kenya during the study period.*

The results for lime mud and recovery boiler ash in **Table 5** showed their pH level, Cd and Zn concentrations were not significantly different, but the moisture content, Pb and Cu concentrations were significantly higher in lime mud compared to recovery boiler ash ($p < 0.05$). In terms of chemical compositions, calcite (calcium carbonate and calcium oxide) was probably the major component of lime mud, while recovery boiler ash was mainly composed of burkeite (sodium carbonate sulfate), aphthitalite (potassium sodium sulfate), and halite (sodium chloride).

Parameter	Pulp and paper industrial wastes	
	Lime mud	Recovery boiler ash
pH	11.9	11.5
Moisture content (% w/w)	1.11	0.25
Potentially Toxic Elements		
Pb (mg/kg)	32.11	20.66
Cd (mg/kg)	1.69	1.42
Cu (mg/kg)	8.73	3.36
Zn (mg/kg)	109.22	122.22
Chemical composition[1]	CaCO3	Na4SO4CO3
		K3Na(SO4)2
		NaCl

[1]*Determined by X-ray diffraction analysis (XRD).*

Table 5.
Chemical characteristics and metal concentration in the Kraft pulp and paper mill solid waste.

Metal	Soil	Sediments	P-value
Pb (mg/g)	0.250	0.451	0.002
Cd (mg/g)	0.008	0.009	0.432
Cu (mg/g)	0.036	0.037	0.124
Zn (mg/g)	0.060	0.192	0.019

Table 6.
Amount of leachable metals in soils and sediments near the pulp and paper industry.

	Solubility product constant (Kp)			
	Theoretical values	Experimental value		
Metal	LM 80 g/L	LM 110 g/L	RB 80 g/L	RB 110 g/L
As hydroxide				
Pb	2.5×10^{-16}[a]	1.95×10^{-12}	7.23×10^{-17}	3.11×10^{-16}
Cd	8.0×10^{-31}[a]	1.53×10^{-13}	6.44×10^{-21}	1.54×10^{-19}
Cu	2.0×10^{-19}[a]	8.45×10^{-12}	4.12×10^{-15}	1.77×10^{-14}
Zn	6.0×10^{-18}[a]	5.93×10^{-12}	3.24×10^{-16}	2.22×10^{-15}
As carbonate				
Pb	7.4×10^{-14}[a]	2.43×10^{-8}	6.50×10^{-8}	1.21×10^{-7}
Cd	—	—	—	—
Cu	2.3×10^{-10}[a]	1.22×10^{-7}	3.97×10^{-6}	5.93×10^{-6}
Zn	3.0×10^{-08}[b]	7.55×10^{-8}	2.96×10^{-7}	7.81×10^{-7}

No data reported for the precipitation of Cd as carbonate.
[a]*Values from [50].*
[b]*Values from [51].*

Table 7.
Solubility product constants between the theoretical values and experimental values.

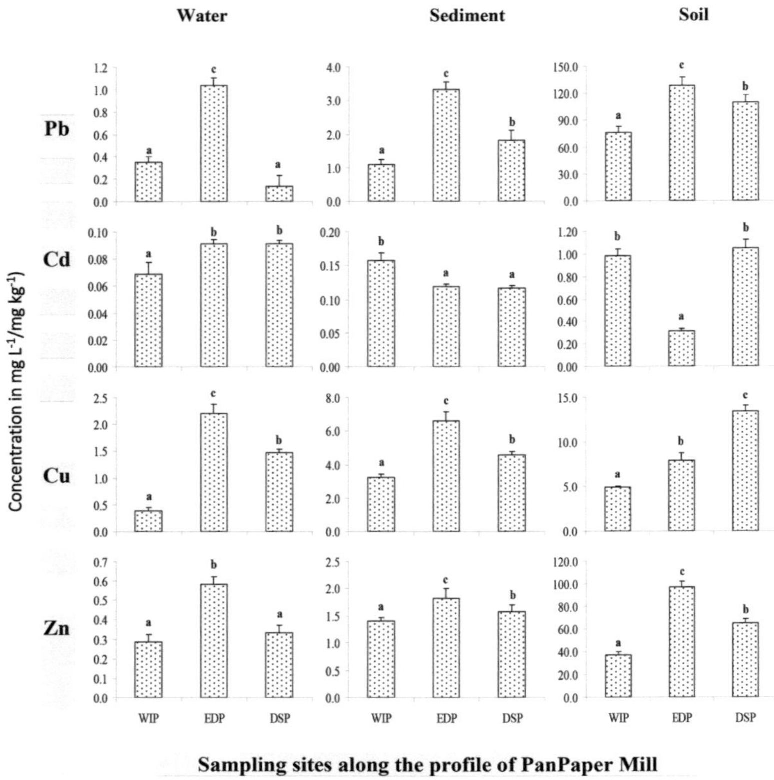

Figure 3.
Variations in concentration of potentially toxic elements in water, sediment and soils along the three sampling sites in PanPaper, Webuye during the study period. One way-ANOVA test for potentially toxic elements concentrations at WIP, EDP and DSP showed significant differences denoted by different letters 'a', 'b' and 'c'. For each sets where columns had the same letter 'b' or 'a', the difference was not statistically significant.

The leaching tests yielded only small amounts of Pb, Cd, Cu and Zn (**Table 6**). Statistical analysis confirmed that treatment with the different types and dosages of pulp and paper industrial wastes had an effect on the leaching behavior of these PTEs. However, treatments with 80 and 110 g L^{-1} of lime mud did not yield any statistical difference for Cu concentration ($p = 0.124$) and Cd ($p = 0.432$) (**Table 7**). Compared to the local standards, only the leaching of chromium and lead was within the regulatory limits.

The results of PTEs concentration in water at the WIP, EDP and DSP are presented in **Figure 3**. In this graph, the concentrations of PTEs in the soil were consolidated (8 values for each sampling point) and a mean concentration was calculated at WIP, EDP and DSP. In water, all the PTEs exhibited highest concentration at EDP but the difference in Cu concentration between ESP and DSP was not statistically significant. For soil samples, Pb, Cu and Zn remained higher at EDP, while Cd and Cu exhibited higher concentration at DSP, although the difference of Cd level between WIP and DSP was not statistically significant ($p > 0.05$).

PTEs distributions from the river, with respect to the paper mill are represented in **Figures 2**, **4**, and **5**. In **Figure 4**, at WIP, concentrations of Zn, Cu and Cd significantly ($p < 0.05$) reduced away from the river as one approached the factory (AA side), while Pb increased. On the BB side at WIP, Pb, Cd and Cu concentrations also decreased significantly ($p < 0.05$) away from the river, while Zn reduced (**Figure 4**).

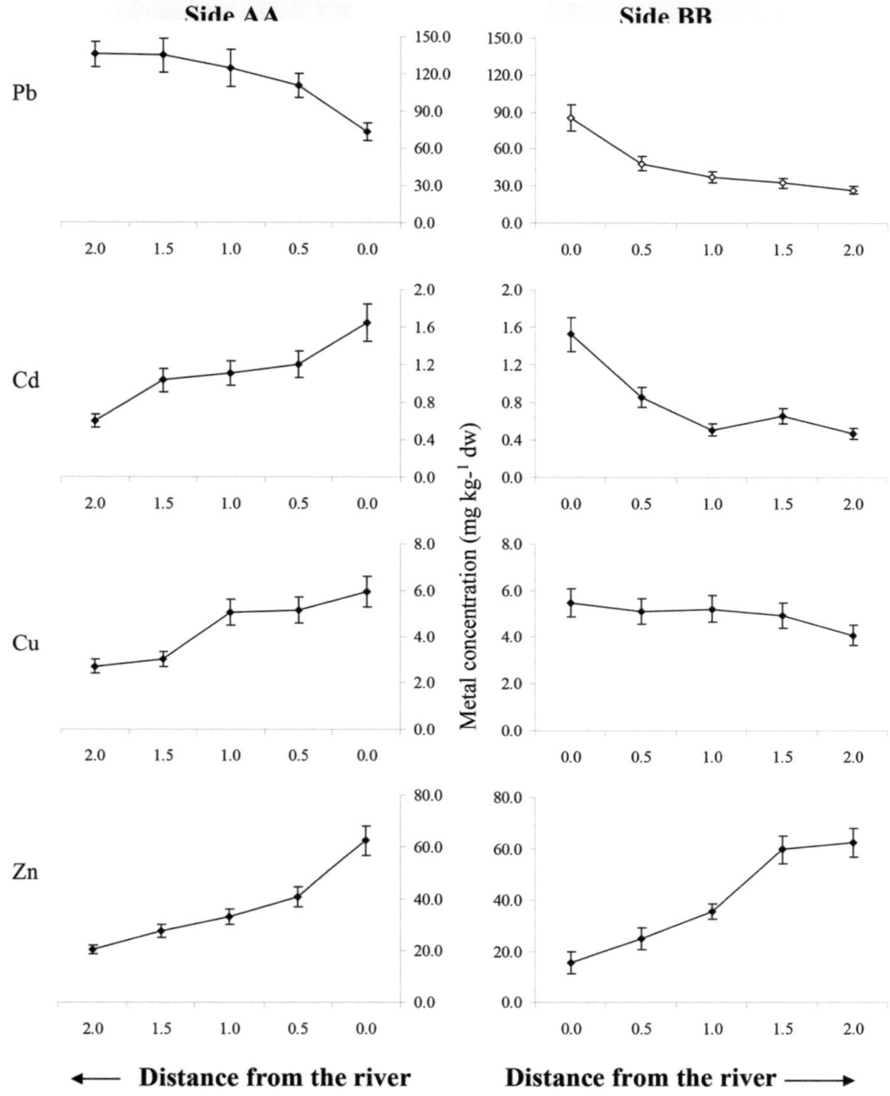

Figure 4.
Trends of heavy metal concentration mg/g in wetland soil at WIP in PanPaper.

In **Figure 2**, at EDP, all metals concentrations decreased ($p < 0.05$) away from the river as one moved closer to the factory (AA), while on the side opposite the factory away from the river, concentrations of Zn and Pb increased significantly as there was a marked reduction in Cu and Cd levels (BB side). Finally, at DSP (**Figure 5**), only Zn and Cu reduced away from the river bank to the factory, on the AA side, with an increase in Pb ($p < 0.05$). On the side BB, Pb, Cu, and Zn concentrations reduced away from the river ($p < 0.05$). Cd exhibited the same trend on both side of the river, with an initial increase followed by a slight and steady decrease as you moved away from the river.

The results of principal component analysis (PCA) on PTEs at 0 km, 0.5 km, 1 km, 1.5 km and 2.0 km on both sides (AA and BB) in the sediments, soil and water are

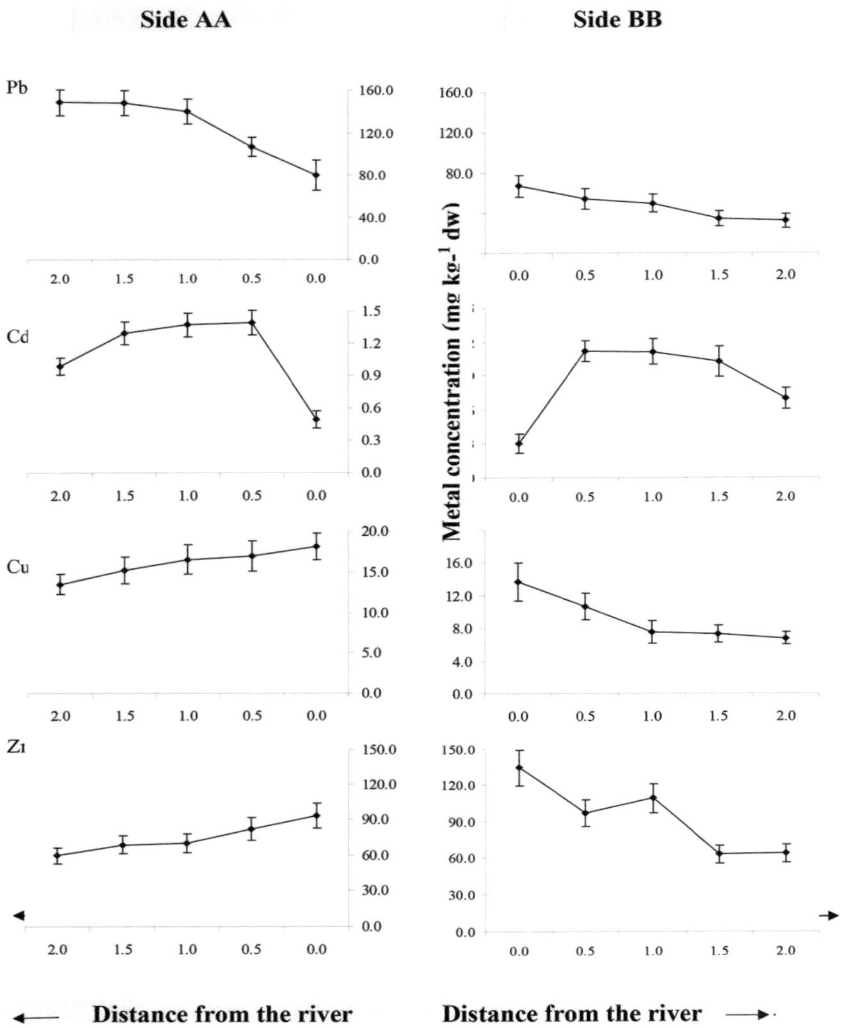

Figure 5.
Trends of heavy metal concentration mg/g in wetland soil at DSP in PanPaper.

presented in **Figures 6–10**. At 0 km from the river, two Principal Components (PC) could explain 96% of the variation in the data. There was a high positive score of Cd and Pb at EDP on AA side and high positive score of Zn and Cu at the DSP on both AA and BB sides. Generally, high score values were associated with the AA side at the EDP on the first PC (73.34%) but no clear association between PTEs concentration in soil, sediment and water. On the second PC (23.25%), there was a high positive score of Cd and Pb at the EDP but a high negative score of Zn and Cu at DSP at both AA and BB side (**Figure 8**).

In **Figure 7**, at 0.5 km from the river, two Principal Components could explain 93.26% of the variations in the data. There was a high positive score of Pb at the WIP on AA side and high positive score of Cd at EDP on BB side on PC1. A high positive score of Zn and Cu was recorded at DSP on BB side. Generally, high values were associated with the AA side at both EDP and DSP on the first PC (75.50%) but with no

DOI: http://dx.doi.org/10.5772/intechopen.102440

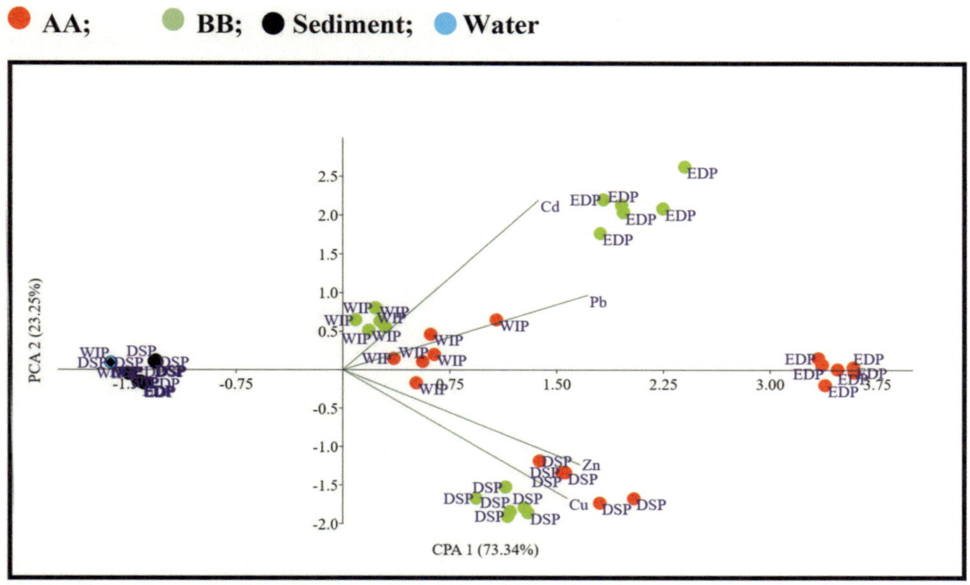

Figure 6.
Principal component analysis of potentially toxic elements (Cd, Pb, Zn and Cu) sampled in sediments, water and at 0 km (both sides 0 km of the buffer zone at Webuye, PanPaper mill. sample location at WIP, EDP, DSP.

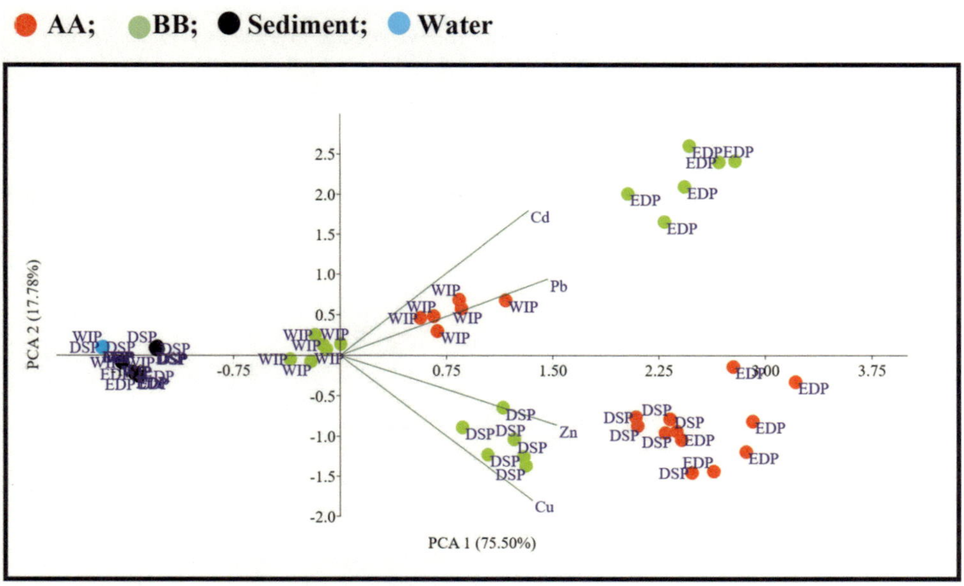

Figure 7.
Principal component analysis of potentially toxic elements (cadmium, Lead, zinc and copper) sampled in sediments, water and at 0.5 km (both sides of the buffer zone at Webuye, PanPaper mill. sample location at WIP, EDP, and DSP.

clear association between PTEs concentration in the soil sediments and water. PC2 explained 17.78% of the total variability, and there was a high positive score of Cd at EDP and Pb at WIP on BB and AA sides respectively. There was a high negative score of Zn and Cu at DSP on BB side (**Figure 7**).

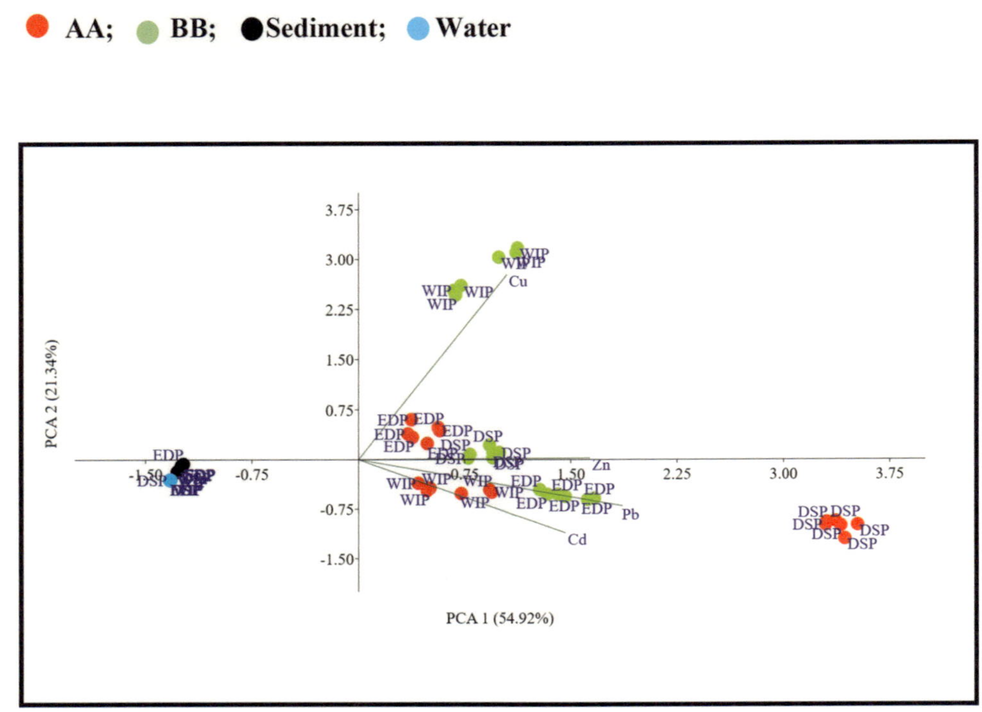

Figure 8.
Principal component analysis of potentially toxic elements (cadmium, Lead, zinc and copper) sampled in sediments, water and at 1 km (both sides of the buffer zone at Webuye pan-paper industry. Sample location at WIP, EDP and DSP.

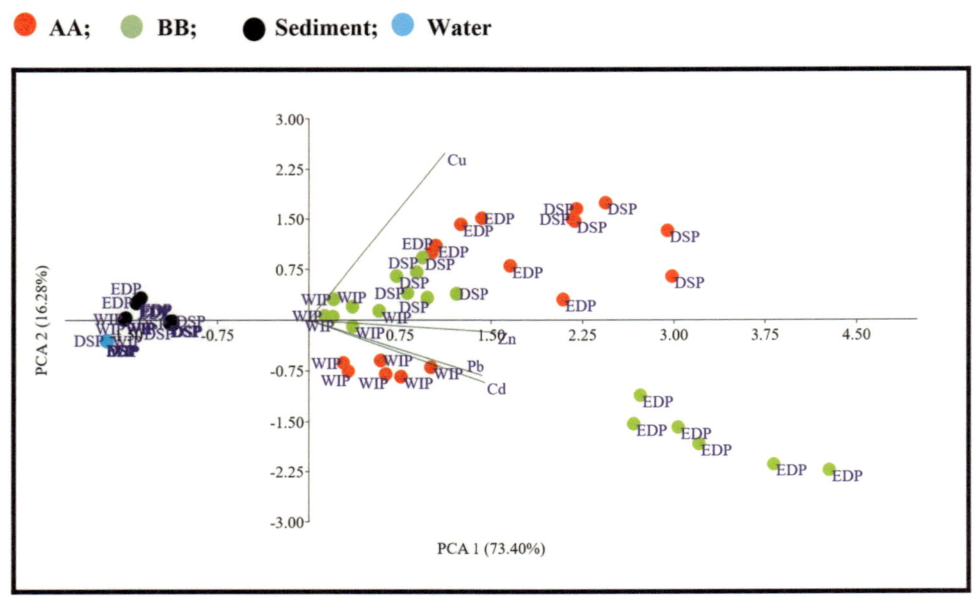

Figure 9.
Principal component analysis of potentially toxic elements (cadmium, Lead, zinc and copper) sampled in sediments, water and at 2 km (both sides of the buffer zone at Webuye pan-paper industry. Sample location at WIP, EDP, DSP.

In **Figure 8**, at 1 km from the river bank, there was a high loading of Cu on PC1 at WIP, Zn at DSP and Pb at EDP, all on the BB side, and high positive score of Cd at WIP on AA side. Generally, high scores were associated with AA side on PC1 (54.92%). On PC2 (21.34%), there was high positive loading of Cu at WIP on BB side and negative but weak loading of Cd at WIP on AA side. PC1 and PC2 explained up to 76% of the variability in the data.

In **Figures 9** and **11**, at 1.5 km and 2.0 km from the river bank, the concentrations of Pb and Cd had a high positive loading at WIP on the AA side. There was equally high positive score of both Cu and Zn, but they were not associated with any particular sampling points nor sides. At both distances (1.5 km and 2.0 km), there was a general strong influence of DSP and EDP on side AA and BB respectively on the PTEs loading from the river bank. The concentrations of PTEs followed a pattern similar to the one in **Figure 8** where Cu > Zn > Pb > Cd. However the concentration of Cu was higher on the side (AA) of the factory at DSP and EDP. While Zn and Cd levels were higher on the opposite side of the factory. The concentration of PTEs in sediment and water remain lower. The PCA1 and PCA2 could explain 73.4% of the observed variation in these measurements. WIP recorded the lowest concentrations of the PTEs at all the sampling points and on both sides (AA and BB) of the paper mill.

Figure 10 described the Principle Component Analysis of variation of HMs between PTEs concentrations at EDP with pH, Temperature and electrical conductivity as secondary variables. The results show that Zn had relatively high positive score at EDP in relation to the pH, temperature and electrical conductivity on the first axis (56.75%). Pb had a high negative loading at WIP. Cu had a high negative score at DSP but was not associated with any environmental variable. On the second axis, Cd had a high positive score at WIP but was also not associated with any environmental variable. The second axis could only explain 26.53% of the total variability.

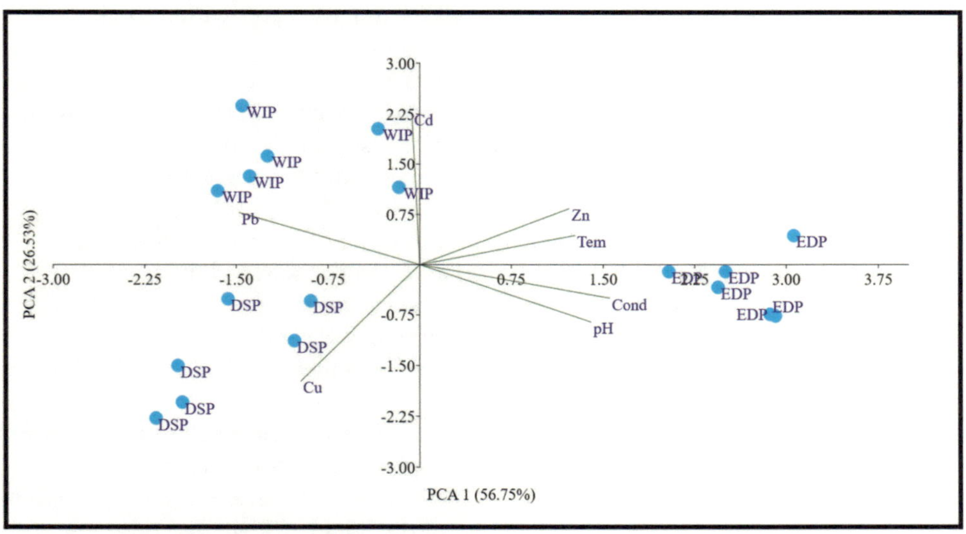

Figure 10.
Relationship between pH, EC and Temperature and Zn concentration at the EDP.

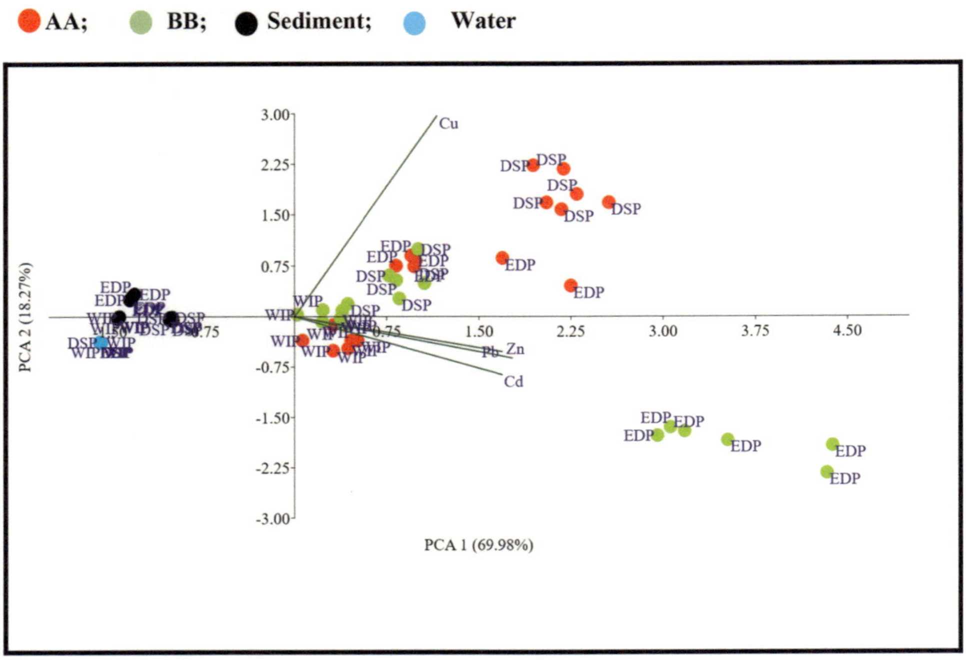

Figure 11.
Principal Component Analysis of potentially toxic elements (Cadmium, Lead, Zinc and Copper) sampled in sediments, water and at 2 km (both sides of the buffer zone at Webuye pan-paper industry. Sample location at WIP, EDP and DSP.

4. Discussion

The results of this study indicate that the factory had an impact on the river's physico-chemical parameters. For example, at EDP, pH, Temperature and EC were much higher than at WIP upstream and DSP downstream, with a slight distortion at 1 km from the river bank. Since there was no other point source between WIP and EDP, the effluent discharged from the pulp and paper mill at EDP probably accounted for much of the observed increased levels in the physico-chemical parameters. Higher pH at EDP sites was likely the result of alkaline industrial effluent discharges. In the factory, sodium hydroxide and sodium sulphide (Kraft process) used during the cooking process of wood pulps, [22, 23, 26], might have resulted in an increase in alkaline condition of the water body at the EDP. In addition, after Kraft cooking process at the mill, the wood pulp underwent bleaching through CEHP or CEHH bleaching sequences. It is likely that the strong caustic extraction in the bleaching sequence was not completely neutralized by the final hypochlorite stage, thereby leading to additional alkalinity into the river water.

In addition to pH, high conductivity at EDP was probably the result of the paper mill discharge into the river. Wastewater from pulp and paper mills is known to contain substantial amounts of dissolved ions and electrolytes. The discharge of such electrolytes ultimately increased the electrical conductivity of the liquid effluent and by extension the EC of the receiving river water. For instance, alkaline sulfates dissociate under high temperature to liberate the positively charged alkali metals (X^+, X^{2+} or X^{3+}) and the sulfate ions (SO_4^{2-}), which all increase the electrical conductivity of the water body [52].

The major components of lime mud and recovery boiler ash in the paper mill contain carbonate compounds [53]. Since the mill sometime releases lime mud and recovery ash into its effluent, this practice more likely exacerbated the alkaline conditions of its wastewater and the increase in pH and EC at the EDP when compared with WIP.

The leaching test results (**Table 6**) showed only marginal leachable concentrations of Cd and Cu. This can be explained by the fact the soils around the paper mill are of *Planesol* type, which are of fine texture intensely weathered and negatively charged. *Planesols* tend to attract positively charged ions such as Cd^{2+}, Cu^{2+}, Zn^{2+}, and Pb^{2+}, which will be strongly bound to the soils around the paper mill.

There were significant spatial variations in the potentially toxic elements concentrations in water, sediment and soils near the pulp and paper mill. Reports by Jumbe and Nandini, 2009 [54] had shown that average abundance of Pb in the Earth's crust is 13 ppm, but in natural soils the background level for Pb ranges from 2.6 to 25 ppm. In the same report, Cd in the Earth's crust is 0.16 ppm but in soils, it ranges from 0.1 to 0.5 ppm. The average abundance of Cu in the Earth's crust is 68 ppm while, in soils, it is between 9 and 33 ppm. The average abundance of Zn in Earth's crust is 76 ppm and in soils it ranges between 25 to 68 ppm. Our results show that Pb, Cd, Cu and Zn concentrations in the soils near the paper mill were approximately 5, 2, 0.5 and 2 times higher than the respective natural background levels of these metals in the earth's crust suggesting anthropogenic input of these metals. The high concentrations of Pb and Zn at EDP, suggest possible enrichment from the paper mill effluent as the source of the metals [18, 20, 29, 55–60]. The levels of the PTEs were not consistently high on the (AA) side of the mill, suggesting that the river bank was not an accurate reference point for the factory (AA) and opposite (BB) factory sides. In addition, high PTEs levels away from the river bank on the opposite (BB) side of mill indicate that river water was not the only source of the metals after effluent discharge.

Some of these PTEs may been deposited on soils through air emission from the paper factory, as suggested by Adoli *et al*, 2011 [61] while analyzing moss and topsoil to monitor metal emission from the same mill. Paper mills have been reported as potential hazardous substances emitters of metals. The presence of high Zn concentrations can be related to the use zinc dithionite as a bleaching agent of stone-ground wood pulps produced within the mill. Zn concentrations at EDP was probably influenced by pH, electrical conductivity, and temperature at the same sampling point. The reduced concentration of Cd as we moved from the EDP to DSP is due to absence of Cd compounds during pulp and paper processing. We can postulate that the factory effluents probably neutralized any geochemical Cd and higher Cd levels obtained at the soils at the downstream sites probably came from the recovery boiler system through the mill's gaseous emissions.

Most of the sulfates from industries are often associated with metal complexation causing insoluble inorganic substances [62], which have higher chances of removing Cd from the water. It is possible that the lack of such complexation agents from the paper mill probably resulted in increased Cd at DSP. Because the study area has higher rate of fertilizer use by farmers for sugarcane production, the net impact of fertilizer use on PTEs in the vicinity probably surpassed Cd concentration from the industrial effluents. However, it seems that it was efficiently reducing Cd, and therefore Cd enrichment of water was low. While it has generally been assumed that most of the metals are immobile in soils or sediments [63], there are factors that enhance their mobility resulting in differential metal distribution. These factors include the properties of the metals, soil texture, pH and competing cations in the soil and sediments solution [52]. Cd retention is reported to be greater in fine-textured soils with high CEC than in coarse-textured soils with lower CEC, while McBride, 1995 [63] noted that potentially toxic elements mobility was most closely associated with metal-organic complexation and soil pH, all these were likely to affect the distribution of Cd and Pb in water, soils and sediments. *Ultisol* soils around the paper mill are generally known to be poor in organic matter and therefore retained little of the Cd which was transported away from the EDP.

5. Conclusion

The concentration of Pb and Zn were consistently high near the effluent discharge points indicating enrichment through the Pulp and paper liquid effluents or gaseous emissions. Cu was consistently high from 1 km onward away from the river bank, while Cd was the lowest. Zn was highly influenced by pH, temperature and EC. The levels of PTEs in soils around the paper mill confirm anthropogenic enrichment and should therefore influence policy makers for industrial management in Western Kenya.

Acknowledgements

The authors would like to thank the Royal Netherlands Embassy in collaboration Victoria Institute for Research on Environment and Development (VIRED) International for funding this project. The LVEMP (Lake Victoria Environment Management Program) Project additional financial support is greatly appreciated. The authors also extend their gratitude to Mr. Lewela of Moi University, Chemical Analysis laboratory for his invaluable assistance in sample collection and analysis of PTEs.

Author details

Elijah S. Omutange[1], Julius O. Manyala[2], Anne N. Wekesa[3], Odipo Osano[4], Thomas Munyao[5], Alfred A. Otieno[6] and Lazare Etiégni[7]*

1 Department of Technology Education, Moi University, Eldoret, Kenya

2 School of Spatial Planning and Natural Resource Management, Jaramogi Oginga Odinga University of Science and Technology (JOOUST), Bondo, Kenya

3 School of Agricultural Sciences and Natural Resources, University of Kabianga, Kericho, Kenya

4 Department of Environmental Health, School of Environmental Studies, University of Eldoret, Eldoret, Kenya

5 Department of Earth Science, School of Environmental Studies, University of Eldoret, Eldoret, Kenya

6 Department of Fisheries and Aquatic Science, University of Eldoret, Eldoret, Kenya

7 Department of Forestry and Wood Science, University of Eldoret, Eldoret, Kenya

*Address all correspondence to: lazetiegni@amatala.org

References

[1] Ali H, Khan E, Ilahi I. Environmental chemistry and ecotoxicology of hazardous heavy metal: Environmental persistence, toxicity, and bioaccumulation. Journal of Chemistry. 2019;**2019**:1-14. DOI: 10.1155/2019/6730305

[2] Ciazela J, Siepak M, Wojtowicz P. Tracking heavy metal contamination in a complex river-oxbow lake system: Middle Odra Valley, Germany/Poland. Science of the Total Environment. 2018; **616**:996-1006

[3] Goher M, Abdo HM, Mangood HA, Hussein MM. Water quality and potential health risk assessment for consumption of Oreochromis niloticus from El-Bahr El-Pharaony drain, Egypt, Fresenius Environmental Bulletin, 2015. 24(11): p. 3590-3602

[4] Goher ME, Ali MHH, El-Sayed SM. Heavy metals contents in Nasser Lake and the Nile River, Egypt: An overview. The Egyptian Journal of Aquatic Research. 2019;**45**(4):301-312

[5] Mokarram M, Saber A, Sheykhi V. Effects of heavy metal contamination on river water quality due to release of industrial effluents. Journal of Cleaner Production. 2020;**277**:123380

[6] Nkwunonwo UC, Odika PO, Onyia NI. A review of the health implications of heavy metals in food chain in Nigeria. The Scientific World Journal. 2020;**2020**:1-11. DOI: 10.1155/2020/6594109

[7] Zhang Z, Wand JJ, Ali A, DeLaune DR. Heavy metal distribution and water quality characterization of water bodies in Louisiana's Lake Pontchartrain Basin, USA. Environmental Monitoring and Assessment. 2016;**188**(11):1-14

[8] Zhou Q, Yang N, Li Y, Ren B, Ding X, Bian H, et al. Total concentrations and sources of heavy metal pollution in global river and lake water bodies from 1972 to 2017. Global Ecology and Conservation. 2020;**22**:e00925

[9] Bigalke M, Ulrich A, Rehmus A, Keller A. Accumulation of cadmium and uranium in arable soils in Switzerland. Environmental Pollution. 2017;**221**:85-93

[10] Carnelo LGL, de Miguez SR, Marbán L. Heavy metals input with phosphate fertilizers used in Argentina. Science of the Total Environment. 1997;**204**(3):245-250

[11] El Hamiani O, El Khalil H, Sirguey C, Ouhammou A, Bitton G, Schwartz C, et al. Metal concentrations in plants from mining areas in South Morocco: Health risks assessment of consumption of edible and aromatic plants. CLEAN–Soil, Air, Water. 2015;**43**(3):399-407

[12] Rai KP, Lee SS, Zhang M, Tsang FY, Ki-Hyun K. Heavy metals in food crops: Health risks, fate, mechanisms, and management. Environment International. 2019;**125**:365-385

[13] Lee JS, Chon HT, Jung MC. Toxic risk assessment and environmental contamination of heavy metals around abandoned metal mine sites in Korea. In: Key Engineering Materials. Daejeon, Korea: Trans Tech Publications; 2005; **277-279**:542-547. DOI: 10.4028/www.scientific.net(KEM.277-279

[14] Nagajyoti PC, Lee KD, Sreekanth T. Heavy metals, occurrence and toxicity for plants: A review. Environmental Chemistry Letters. 2010;**8**(3):199-216

[15] Park JM, Lee SJ, Lee J-U, Chon TH, Jung CM. Microbial effects on geochemical behavior of arsenic in As-

contaminated sediments. Journal of Geochemical Exploration. 2006;**88**(1–3): 134-138

[16] Ferretti M, Cenni E, Bussotti F, Batistoni P. Vehicle-induced lead and cadmium contamination of roadside soil and plants in Italy. Chemistry and Ecology. 1995;**11**(4): 213-228

[17] Merian E. Introduction on environmental chemistry and global cycles of chromium, nickel, cobalt beryllium, arsenic, cadmium and selenium, and their derivatives. Toxicological & Environmental Chemistry. 1984;**8**(1):9-38

[18] Sharma P, Purchase D, Chandra R. Residual pollutants in treated pulp paper mill wastewater and their phytotoxicity and cytotoxicity in Allium cepa. Environmental Geochemistry and Health. 2021;**43**(5):2143-2164

[19] Tian HZ, Lu L, Hao JM, Gao JJ, Liu YK, Qiu PP, et al. A review of key hazardous trace elements in Chinese coals: Abundance, occurrence, behavior during coal combustion and their environmental impacts. Energy & Fuels. 2013;**27**(2):601-614

[20] Sharma P, Tripath S, Chandra R. Environmental impacts of pulp paper mill effluent: Potential source of chromosomal aberration and phytotoxicity. International Journal of Applied Environmental Sciences. 2020; **15**(1):77-92

[21] Thompson G, Swain J, Kay M, Forster FC. The treatment of pulp and paper mill effluent: A review. Bioresource Technology. 2001;**77**(3): 275-286

[22] Rintala JA, Puhakka JA. Anaerobic treatment in pulp-and paper-mill waste management: A review. Bioresource Technology. 1994;**47**(1):1-18

[23] Sthiannopkao S, Sreesai S. Utilization of pulp and paper industrial wastes to remove heavy metals from metal finishing wastewater. Journal of Environmental Management. 2009; **90**(11):3283-3289

[24] Fortin N, Fulthorpe RR, Allen GD, Greer WC. Molecular analysis of bacterial isolates and total community DNA from Kraft pulp mill effluent treatment systems. Canadian Journal of Microbiology. 1998;**44**(6):537-546

[25] Suntio LR, Shiu WY, Mackay D. A review of the nature and properties of chemicals present in pulp mill effluents. Chemosphere. 1988;**17**(7):1249-1290

[26] Čerňáková M, Golis E. Testing of the cytotoxic effects of sulfate pulp mill waste waters. Folia Microbiologica. 1994; **39**(4):307-314

[27] Muttray AF, Yu Z, Mohn WW. Population dynamics and metabolic activity of pseudomonas abietaniphila BKME-9 within pulp mill wastewater microbial communities assayed by competitive PCR and RT-PCR. FEMS Microbiology Ecology. 2001;**38**(1):21-31

[28] Nurmesniemi H, Pöykiö R, Keiski RL. A case study of waste management at the northern Finnish pulp and paper mill complex of Stora Enso Veitsiluoto Mills. Waste Management. 2007;**27**(12): 1939-1948

[29] Sharma P, Iqbal HM, Chandra R. Evaluation of pollution parameters and toxic elements in wastewater of pulp and paper industries in India: A case study. Case Studies in Chemical and Environmental Engineering. 2021;**4**: 100163

[30] Simão L, Hotza D, Raupp-Pereira F, Labrincha AJ, Montedo ORK. Wastes from pulp and paper mills-a review of generation and recycling alternatives. Cerâmica. 2018;**64**:443-453

[31] Jia Y, Hamberg R, Qureshi A, Mäkitalo M, Maurice C. Variation of green liquor dregs from different pulp and paper mills for use in mine waste remediation. Environmental Science and Pollution Research International. 2019; **26**(30):31284-31300. DOI: 10.1007/s11356-019-06180-0

[32] Watkins G, Pöykiö R, Nurmesniemi H, Dahl O. Earth construction and landfill disposal options for slaker grits. Research Journal of Applied Sciences, Engineering and Technology. 2010;**2**(8): 757-764

[33] Cabral F, Ribeiro MH, Hilário L, Machado L, Vasconcelos E. Use of pulp mill inorganic wastes as alternative liming materials. Bioresource Technology. 2008;**99**(17):8294-8298. DOI: 10.1016/j.biortech. 2008.03.001

[34] Mahmoudkhani M, Richards T, Theliander H. Recycling of solid residues to the forest: Experimental and theoretical study of the release of sodium from lime mud and green liquor dregs aggregates. Process Safety and Environmental Protection. 2004;**82**(3): 230-237

[35] Pöykiö R, Nurmesniemi H. Calcium carbonate waste from an integrated pulp and paper mill as a potential liming agent. Environmental Chemistry Letters. 2008;**6**(1):47-51

[36] Pöykiö R, Mäkelä M, Watkins G, Nurmesniemi H, Dahl O. Heavy metals leaching in bottom ash and fly ash fractions from industrial-scale BFB-boiler for environmental risks assessment. Transactions of Nonferrous Metals Society of China. 2016;**26**(1): 256-264

[37] Nurmesniemi H, Mäkelä M, Pöykiö R, Manskinen K, Dahl O. Comparison of the forest fertilizer properties of ash fractions from two power plants of pulp and paper mills incinerating biomass-based fuels. Fuel Processing Technology. 2012;**104**:1-6. DOI: 10.1016/j.fuproc. 2012.06.012

[38] Bordado JCM, Gomes JFP. Atmospheric emissions of Kraft pulp mills. Chemical Engineering and Processing: Process Intensification. 2002;**41**(8):667-671

[39] Dionne J, Walker TR. Air pollution impacts from a pulp and paper mill facility located in adjacent communities, Edmundston, New Brunswick, Canada and Madawaska, Maine, United States. Environmental Challenges. 2021;**5**: 100245

[40] Osano O, Nzyuko D, Mwakio T, Admiraal W. The fate of chloroacetanilide herbicides and their degradation products in the Nzoia Basin, Kenya. Ambio. 2003;**32**(6):424-427

[41] Government of Kenya. National Development Plan, 2007–2012. Nairobi, Kenya: Government Printing Office; 2008

[42] American Public Health Association, American Water Works Association, and Water Environment Federation. Standard Methods for the Examination of Water and Wastewater. 20th ed. Washington DC: American Public Health Association, American Water Works Association, and Water Environment Federation; 1999

[43] Food and Agricultural Organization. Manual of Methods in Aquatic Environment Research-pt. 1: Methods for Detection, Measurement and

Monitoring of Water Pollution. Rome (Italy): Food and Agricultural Organization; 1975

[44] Luo L, Ma Y, Zhang S, Wei D, Yong-Guan Z. An inventory of trace element inputs to agricultural soils in China. Journal of Environmental Management. 2009;**90**(8):2524-2530

[45] Nicholson FA, Smith RS, Alloway JB, Carlton-Smith C, Chambers JB. An inventory of heavy metals inputs to agricultural soils in England and Wales. Science of the Total Environment. 2003; **311**(1–3):205-219

[46] Tarnawski W. Emission factors for combustion of biomass fuels in the pulp and paper mills. Fibres and Textiles in Eastern Europe. 2004;**12**(3):91-95

[47] StatSoft Inc. Statistica for Windows. Tulsa, OK: StatSoft Inc.; 2001

[48] Splitstone DE, Ginevan ME. Statistical Tools for Environmental Quality Measurement. New York: CRC Press; 2003

[49] Borah P, Paul A, Bora P, Bhattacharyya P, Karak T, Mitra S. Assessment of heavy metal pollution in soils around a paper mill using metal fractionation and multivariate analysis. International journal of Environmental Science and Technology. 2017;**14**(12): 2695-2708

[50] Sawyer CN, McCarty PL, Parkin FG. Chemistry for Environmental Engineering and Science. 5th ed. New York: McGraw-Hill; 2003

[51] Day RA, Underwood AL. Quantitative Analysis. 4th ed. New Jersey; 1980

[52] Luoma SN, Rainbow PS. Metal Contamination in Aquatic Environments: Science and Lateral Management. Cambridge, New York: BioOne Complete; 2008

[53] Etiégni L, Campbell AG. Physical and chemical characteristics of wood ash. Bioresource Technology. 1991;**37**(2): 173-178

[54] Jumbe AS, Nandini N. Heavy metals analysis and sediment quality values in urban lakes. American Journal of Environmental Sciences. 2009;**5**(6):678

[55] Chen CW, Chih-Ming K, Chih-Feng C, Cheng-Di D. Distribution and accumulation of heavy metals in the sediments of Kaohsiung Harbor, Taiwan. Chemosphere. 2007;**66**(8):1431-1440

[56] Das K. Assessment of heavy metal contamination in soils irrigated with paper mill effluent in West Bengal, India. Annals of Plant and Soil Research. 2019;**21**(4):312-319

[57] Giri J, Srivastava A, Pachauri SP, Srivastava CP. Effluents from paper and pulp industries and their impact on soil properties and chemical composition of plants in Uttarakhand, India. Journal of Environment and Waste Management. 2014;**1**(1):26-32

[58] Mwamburi J. Variations in trace elements in bottom sediments of major rivers in Lake Victoria's basin, Kenya. Lakes & Reservoirs: Research & Management. 2003;**8**(1):5-13

[59] Mwamburi J, Oloo FN. The distribution and concentration levels of trace metals in water and sediments of Lake Victoria, Kenya. African Journal of Tropical Hydrobiology and Fisheries. 1996;**7**(1):37-48

[60] Rahman MM, Hassan MM. The amount of selected heavy metals in water, sediments and fish species from

the Rupsha River, Khulna, Bangladesh. Asian Journal of Fisheries and Aquatic Research. 2020;**6**:1-9

[61] Adoli FKL, Lalah JO, Okoth AO. Analysis of moss and topsoil to monitor metal emissions from a pulp and paper mill in western Kenya. Bulletin of Environmental Contamination and Toxicology. 2011;**86**(1):137-143

[62] Manahan SE. Toxicological Chemistry. Boca Raton: CRC Press Lewis Publishers; 1992

[63] McBride MB. Toxic metal accumulation from agricultural use of sludge: Are USEPA regulations protective? Journal of Environmental Quality. 1995;**24**(1):5-18

Chapter 12

Quantification of Heavy Metal Levels in Sediments of the "Palizada" River in a Protected Natural Area of Southeastern Mexico

Carlos Montalvo, Claudia A. Aguilar, Yunuen Canedo, Alejandro Ruiz, Brenda Zermeño, Ruby S. Gines and Rosa M. Cerón

Abstract

This chapter shows the results of the determination of the levels of Copper (Cu), Iron (Fe), Magnesium (Mg), Manganese (Mn), and Cadmium (Cd) in sediments of the "Palizada River"; evaluated by atomic absorption. The results show high levels of Fe, Mg, and Mn concerning previous studies and are directly related to agricultural and livestock activities in the area. The metal-metal correlation analysis show significant values, suggesting that the generation source is the same. Some of the pollutants produced by industrial or population centers; end up in natural water bodies; by different physical processes, such as precipitation, complex formation, and others. These contaminants can bioaccumulate in aquatic organisms and reach humans through the food chain. Certain heavy metals such as cadmium (Cd) produced by human activities have been found in the oceans and the tissues of different organisms; which constitutes a risk for consumption; Cd is classified as an element that predisposes to the presence of Cancer. Therefore, the study of sediments is fundamental for determining the degree of contamination of an ecosystem.

Keywords: heavy metals, contaminated sediments, Palizada river, southeast Mexican

1. Introduction

Currently, coastal ecosystems are exposed to a large load of pollutants from urban activities; this problem is because significant amounts of waste are dumped that generally has no treatment, which contributes to increasing the concentration of certain pollutants as heavy metals are of particular interest because they are highly persistent, toxic and can bioaccumulate in exposed organisms.

The sediments allow lies to determine, from the geochemical point of view, the content of trace metals, their possible origin (anthropogenic or natural), and the mobility that serves to infer the potential risks of toxicity for the biota. The study of sediments in an aquatic ecosystem allows a comprehensive estimation of pollution; metals associated with the organic fraction can form solutions and remain available to aquatic organisms [1].

An earlier study [2] shows the behavior of heavy metals; copper (Cu), nickel (Ni), chrome (Cr), lead (Pb), cadmium (Cd), and zinc (Zn) in sediments and organisms in Terminos Lagoon, Campeche, Mexico. The results for the sediments showed that the total levels of metals were higher on the west side of the lagoon, except for Cd and Pb, which had a high level on the east side. For the oyster (*Crassostrea Virginica*), metal levels increase in the rainy season except for Pb, which presents higher values during the northern season. The results show a tendency to increase in the levels of Cr and Pb and are related to human and industrial activities.

Petroleum hydrocarbons, heavy metals, and pesticides are the main contaminants in Mexican coastal [3]. These elements have generated harmful effects for the organisms, coastal ecosystems, and public health; however, certain metals are part of the biochemistry of living beings, are required by organisms, and are part of various metabolic processes. Some metals, such as Fe, Cu, Zn, and others, are essential elements. The problem arises when these essential elements and other nonessential elements such as Hg, Pb, and Cd are found at high levels in the organs or tissues of aquatic organisms.

Another studio [4] showed levels of heavy metals in water and sediment samples; the water samples were within the national standards for river water. However, the presence of all metals considered in the study was confirmed in the sediments, which indicates that the sediments are a reserve source of pollutants. In this same study, Cr, Pb, Fe, Ni, and Cu were very high in the sediments, mainly in summer and winter.

In Mexico, other studies [5] determined levels of Cu, Mn, Fe, Ni, Cd, Pb, and Zn in samples of oysters, water, and sediments from "San Andrés", lagoon to determine and understand the bioaccumulation mechanisms of the species and identify the potential risk for the consumption of oysters.

Several factors can predispose to the presence of heavy metals in certain studies [6], the bioavailable contents of Zn, Fe, Mn, Cu, Cr, Cd, Pb, and Ni in superficial sediments of the "Bocaripo" coastal lagoon were evaluated in two climatic periods. Showing the importance of the size and diameter of particles, the content of organic matter (OM), and the pH of the sediments.

Finally, the authors conclude that the bioavailable content of Cu may represent a potential risk of contamination and affect the mobility of other metals.

Studies carried out in the Bay of "Chetumal", Mexico [7] confirm that organic matter and other physicochemical parameters are essential in a global study of heavy metals; the authors determined the levels of Cd, Pb, Hg, and As, in surface sediments as well as organic matter and nutrients. The environmental disturbance was related to discharges of water from the "Deep River" and from the city. In spring, the highest levels were for Cd, Pb, Hg, and As. The parameters analyzed: PO4, NO3, CO3, and organic matter (OM), did not show significant variation concerning reference values of the official Mexican standards.

The study of contamination in sediments must include the analysis of the texture and determine the relationships with the levels of heavy metals; In this sense, in studies carried out in Chile [8], the granulometric and geochemical aspects of superficial and subsurface marine sediments were analyzed; with this value, the authors determined the environmental quality of the sediments. In this sense, areas with environmental deterioration were established, generated by enrichment or contamination of some metals such as Ni, Zn, Pb, Cd, Cu, Hg, and As. The main textural fractions: are silt, sand, and clay. In this same study, high values of OM were found. The authors conclude that this area is susceptible to contamination due to its sedimentological characteristics and geomorphology.

Gonzalez et al. [9] determined significant components such as organic and inorganic matter (IM), silica, and carbonates, as well as heavy metals (Al, Fe, Mn, Cu, Zn, Pb, Cr, and Ni), sedimentation, and texture studies. The authors conclude; that the environmental problems presented by the "Copper River", like most water bodies, are due to the contributions of residual waters (the majority without treatment). One of the sources of contamination has been the "Large Copper Mine", which continues to have an impact although it is not currently being exploited. A clear example is that sediments are a reservoir of heavy metals and other pollutants.

Several authors [10] have studied the spatial and seasonal distribution of heavy metals (Cd, Co, Cr, Cu, Fe, Mn, Ni, Pb, and Zn) in the dissolved fractions and particles of the Port of Mazatlán (a lagoon system coast in northwestern Mexico). The authors emphasize the effect of the anthropogenic contributions of heavy metals on the concentrations in the port and the adjacent seawater, finding that the effect of said agents in both fractions is limited to the immediate area of the discharge, with concentrations from outside of the port indistinguishable about the coastal waters. It is a verifiable fact that the sediments are temporary integrators of pollution [9–11] and are affected by different physical and chemical processes, which makes the concentration of the pollutant vary between climatic seasons.

Aguilar et al. [12] demonstrated in their studies that the climatic season has a great influence on metal levels; the contents of Cd, Cu, Zn, and Fe were analyzed in surface sediments of a "The Caleta" stream in the City of Carmen, Campeche. The experimental results indicated that there was no significant difference by time and sampling site, which was an expected behavior pattern since in the dry season there are no significant contributions of water.

Certain metals such as Pb, Cd, Hg, and Ni; are a significant risk to human health, are not classified as essential, and are often toxic at low concentrations. The Cd is considered a potent carcinogen according to the International Agency for Research on Cancer (IARC) [13]. These elements are found in the soil, air, and water and can reach man through the food or water he consumes.

The damage they cause is so severe that many species show no symptoms, but they accumulate large amounts in their internal tissues and organs [14–17], from this fact derives the importance of studying marine sediments and classifying pollution and the risk to which aquatic organisms are exposed; Likewise, sediments reflect the environmental quality of an ecosystem [18, 19].

This chapter shows the results of determining the levels of Cr, Cd, Fe, Cu, Zn, and their relationship to climatic seasons in the fluvial lagoon system of the "Palizada" river, which flows into the "Terminos" lagoon.

This area constitutes one of the Natural Protected Areas of Mexico and is the habitat of various species of flora and fauna endemic to Mexico.

2. Methodology

2.1 Description of the study area

Physical environment: Carmen City (**Figure 1**), is located southwest of the state of "Campeche", bordered to the north by the Gulf of Mexico and the municipality of "Champotón", to the south by the state of Tabasco, to the east by the municipalities of "Escarcega and Candelaria" and to the west by the municipality of "Palizada" [20].

The area of the "Terminos" lagoon includes Terminos lagoon and the adjacent fluvial -lagoon systems ("Palizada" from the east, "Chumpan-Balchacah", "Candelaria-Palau"), the lagoon complex of "Pom Atasta", "Puerto Rico", "Los Negros" and the estuary of "Sabancuy", the "Chacahito" lagoon, Carmen Island, and estuary Pargo. The study area is shown in **Figure 2**. The "Palizada" river delta-lagoon fluid system (SLFS) is in the southwestern portion of the Terminos lagoon hydrological basin between geographic coordinates 18°19′13″ and 18°29′04″ north latitude and 91°44′36″ and 91°51′31″ west longitude [20].

To carry out this study, ten points were included in the extraction of sediment samples, during two climatic seasons (2: rainy, 3: dry) in the year, which allowed obtaining significant information on the state of contamination of this important ecosystem (**Table 1**).

2.2 Selection of sampling sites

The selection of the ten sampling points was representative of the system, both in number and location. Given the influence of seasonal variability [2]. A standardized collection study was performed; that is the entire study area during a similar period at the same depth, of a similar substrate [21]. Samples (sediment and water) were taken according to standardized methods [22].

Figure 1.
The geographical location of the City of Carmen and "Terminos" lagoon.

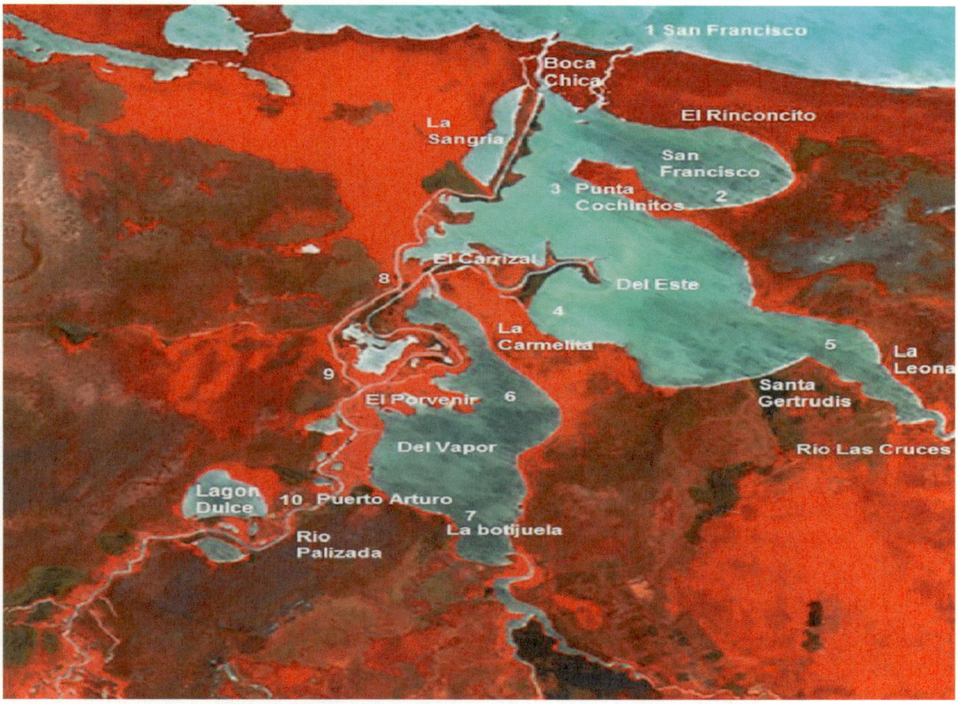

Figure 2.
The geographical location of the "Palizada" river.

Sampling sites	Name	Geographical coordinates
1	"Boca Chica"	18° 29′ 22″ North Latitude and 91° 47′ 37″ West Longitude.
2	"San Francisco"	18° 26′ 15″ North Latitude and 91° 45′ 27″ West Longitude.
3	"Punta Cochinos"	18° 26′ 14″ North Latitude and 91° 47′ 06″ West Longitude.
4	"Laguna del Este"	18° 23′ 39″ North Latitude and 91° 47′ 02″ West Longitude.
5	"Santa Gertrude"	18° 22′ 06″ North Latitude and 91° 44′ 17″ West Longitude.
6	"Laguna vapor"	18° 22′ 21″ North Latitude and 91° 49′ 32″ West Longitude.
7	"Botijuela"	18° 19′ 21″ North Latitude and 91° 50′ 06″ West Longitude.
8	Cañaveral	18° 24′ 59″ North Latitude and 91°50′ 16″ West Longitude.
9	"Porvenir"	18° 22′ 20″ North Latitude and 91° 51′ 46″ West Longitude.
10	"Puerto Arturo"	18° 20′ 27″ North Latitude and 91° 52′ 41″ West Longitude.

Table 1.
Sediment sampling points of the "Palizada" river.

2.3 Sample treatment

2.3.1 Treatment of glassware and preparation of standards

The glassware was treated with a 10% nitric acid (HNO_3) mixture, as suggested by US-EPA method 3050B [23], and dried at a temperature of 105°C for later use. Calibration standards for Cu, Pb, Cd, Fe, Mg, and Mn (**Table 2**), at 50 or 100 mL, were prepared. The calibration blank was a 5% solution of HNO_3.

Element	Wavelength (α)	Working range (μg/mL)
Copper	327.4	0.1–180
Cadmium	228.8	0–1.8
Lead	217	0.2–20
Iron	386	2–145
Magnesium	322.5	0–1.150
Manganese	219	0.2–20

Table 2.
Working ranges and wavelengths (α), used in the analysis of heavy metals.

2.3.2 Sediment sample treatment for heavy metal analysis

The sediment samples were dried at a temperature of 105°C for 24 hours, later they were treated in digestion with nitric acid (HNO_3), hydrochloric acid (HCl), and hydrogen peroxide (H_2O_2) [23]. The resulting solution is filtered, and deionized water is added to it up to 100 mL, as suggested by EPA method 3050B [23]. The equipment used for the analysis was an atomic absorption spectrum Mark Thermo Scientific, the model is ICE 3500.

2.3.3 Sediment texture and Organic Matter (OM)

Granulometry analyses were carried out using the hydrometer of the Bouyocos Technique [24], OM was analyzed by the method suggested in NMX-AA-034-SCFI-2001 [25]; The principle of this method is based on the quantitative measurement of OM contained in water by evaporation and calcination of the samples and is summarized by the following steps: the dry sediment sample, previously pulverized, is sieved to obtain a specific particle size. This method requires constant weight measurements of porcelain vessels (crucibles), and sample calcination at 550°C to measure organic matter by weight difference with the following Eq. (1):

$$MO = (GI - G) * 100 \tag{1}$$

Where:
OM: organic matter in mg/L
GI: is the weight of the crucible with the sample, before calcination (mg)
G: is the total weight of the crucible with the sample after calcination, in mg

2.4 Statistical analysis

To determine the variations by climatic season and by site, a one-way analysis of variance (ANOVA) was used. To establish the relationships between sediment texture, OM, and heavy metals, a Pearson correlation and multiple correlations (Software: Statistical 7.1) were used [26]. The normality of the data was verified using the Shapiro-Wilks statistical test (Software: Statistical 7.1).

3. Results

3.1 Heavy metals, texture, and organic matter in sediments

The results of heavy metals are shown in **Table 3** during the dry (1) and rainy (2) climatic seasons. These results confirm that Fe is one of the most abundant elements during the dry season (216.03–224.75 μg/g), suggesting evaporation phenomena and low sediment mobility. During the rainy season, the values decrease considerably (12.90–34.60 μg/g). During the dry season, the Mg (14.93–17.91 μg/g) did not show significant variations. In the rainy season (5.18–24.20 μg/g) the behavior pattern shows variations by the site. The Mn presents variations in both climatic seasons: dry season (6.89–13.44 μg/g), and rainy season (2.77–15.09 μg/g). Cu and Cd do not show a variable behavior. Although it is very difficult to differentiate between anthropogenic and natural contributions, the variations that occur during the climatic seasons; are an important element for contamination analysis.

The highest Cd values were in the rainy season at the "Laguna del Este" site with 2.34 μg/g. For Mg, the highest values occurred during the rainy season at the "El

Site	Climatic season	Metallic elements analyzed				
		Cu	Cd	Mg	Fe	Mn
1	1	0.19	1.01	17.24	216.06	9.82
2	1	0.19	1.05	16.97	222.20	13.44
3	1	0.19	1.13	16.62	215.40	11.70
4	1	0.19	1.20	17.74	224.75	12.90
5	1	0.19	1.28	14.93	222.73	6.89
6	1	0.19	1.48	17.70	219.17	8.07
7	1	0.19	1.52	14.69	219.67	8.85
8	1	0.19	1.53	17.52	222.54	9.54
9	1	0.19	1.55	17.72	219.62	7.15
10	1	0.19	1.60	17.91	217.34	9.00
1	2	1.13	2.25	5.18	33.31	7.58
2	2	1.23	2.30	5.97	18.05	15.09
3	2	1.18	2.30	7.81	20.63	7.12
4	2	1.10	2.34	9.36	20.63	5.25
5	2	1.10	2.20	21.05	12.90	2.77
6	2	1.18	2.26	10.10	34.60	6.90
7	2	1.21	2.26	22.16	19.89	5.18
8	2	1.17	2.25	6.26	28.71	8.01
9	2	1.13	2.19	24.20	19.52	6.56
10	2	1.11	2.26	10.29	29.45	5.40

Table 3.
Levels of heavy metals during the two climatic seasons: dry (1) and rainy (2) in μg/g.

Site %	Climatic season	% Sand	% Silt	% Clay	% OM
1	2	43.60	50.40	6.00	4.3971
2	2	71.20	26.20	2.60	3.857
3	2	56.80	32	11.20	3.8302
4	2	57.20	40.40	2.40	4.5454
5	2	31.60	68.80	3.60	2.944
6	2	37.60	58.80	3.60	2.414
7	2	44.06*	50.37*	5.57*	4.1561
8	2	7.60	80.80	11.60	4.1421
9	2	73.20	22.40	4.40	4.1655
10	2	18	73.60	8.40	2.2208
1	3	57.20	38.40	4.40	2.0269
2	3	47.60	40.80	11.60	3.7439
3	3	75.60	19.60	4.80	3.4965
4	3	33.60	55.20	11.20	3.3148
5	3	53.45*	40.3*	6.25*	3.1189
6	3	43.28*	50.07*	6.65*	2.9293
7	3	85.60	10.40	4.00	2.4688
8	3	15.20	76.40	8.40	2.244
9	3	77.60	20	2.40	2.4253
10	3	17.20	76	6.80	4.1819

**For the estimation of missing data, the block model approximation was used.*

Table 4.
Sediment texture and organic matter (OM) in %.

Porvenir" site with 24.20 μg/g. The results for Fe in the rainy season were higher at the "El Porvenir" site with 224.75 μg/g.

To analyze a contaminated site, it is important to establish correlations with the physical and chemical characteristics of the sediments, because different contaminants are related to organic matter and fine fractions. In this study, the texture of the sediments, organic matter, and their relationships with heavy metals (**Table 4**). The organic matter content is high and is related to the extensive area of mangroves, various species of macrophytes, and human settlements, which contributes to the variability of organic matter.

3.2 Statistical analysis

In **Table 5**, the ANOVA analysis of variance is shown, all the results are within the normal distribution ($p < 0.05$).

For Cu ($p = 0.5136, p > 0.05$) it shows that there are no statistically significant differences by site, however, for the climatic season, the statistical results show that there is a highly significant difference ($p = 0.000, p < 0.05$). The climatic season has a great influence on Cu abundance, as shown by a recent study [11, 27]. The maximum levels

Source	Sum of squares	Degrees of freedom	Mean square	F-statistic	*p*-statistic
(Cu)					
A = site	51.9116	9	5.76796	0.95	0.5136
B = climatic season	514.943	2	257.471	42.19	0.00000*
Residual	103.738	17	6.10222		
Total	687.832	28			
(Cd)					
A = site	0.584661	9	0.0649623	2.45	0.0531
B = climatic season	4.8115	2	2.40575	90.84	0.0000*
Residual	0.450235	17	0.0264844		
Total	5.90262	28			
(Fe)					
A = site	176.139	9	19.571	0.53	0.8342
B = climatic season	256511.0	2	128255.0	3463.58	0.00000*
Residual	629.505	17	37.0297		
Total	259823.0	28			
(Mg)					
A = site	145.696	9	16.1885	0.79	0.6275
B = climatic season	668.429	2	334.215	16.37	0.0001*
Residual	374.182	17	20.4225		
Total	1233.08	28			
(Mn)					
A = site	4833.04	9	537.004	0.99	0.4832
B = climatic season	190613.0	2	95306.6	175.54	0.0000*
Residual	9229.77	17	542.928		
Total	211579.0	28			

**Significant with a confidence value of 95%.*

Table 5.
Results of analysis of variance (ANOVA) for Cu, Cd, Fe, Mg, and Mn.

of Cu were recorded during the rainy season, which can be related to the very particular circulation mechanisms of the coastal zone at this time [2]. In previous studies [19, 28] similar levels of Cu were reported. Cu is associated with fine particles and OM [29] its high levels come from wastewater discharges and the hydrocarbon industry [30, 31]. In this study, no significant relationships between Cu, texture, and OM are shown (**Table 6**).

For Cd (**Table 5**) there is no significant difference for the site ($p = 0.0531$, $p > 0.05$). But there is a statistically significant difference in the climatic season ($p = 0.000, p < 0.05$), the greatest influence on the presence and distribution of heavy metals is due to the climatic season. **Table 7** shows the correlation results; none of the values show a significant relationship.

For Fe, only the climatic season influences the variation of the data (**Table 5**); **Table 8** shows the correlation analyses; there is no evidence of relationships between heavy metals and sediment texture. There is a significant relationship between organic matter and Fe, which can be found forming oxides.

Test stats	Sand	Slime	Clay	OM
Correlation coefficient (r)	0.1627	−0.20821	0.120697	0.418
p-value	0.493	0.3939	0.6122	0.066*

**Significant at 95% confidence level.*

Table 6.
Correlation values of texture and organic OM with Cu.

Test stats	Sand	Slime	Clay	OM
Correlation coefficient (r)	0.0376	−0.0780	0.1871	−0.466
p-value	0.8749	0.7436	0.4295	0.0383

Table 7.
Correlation values of texture and OM with Cd.

Test stats	Sand	Slime	Clay	OM
Correlation coefficient (r)	0.16394	0.2032	−0.1263	0.5173**
p-value	0.4898	0.3902	0.5956	0.0403

***Significant relationship ($r > 0.5$).*

Table 8.
Correlation values of texture and OM with Fe.

Test stats	Sand	Slime	Clay	OM
Correlation coefficient (r)	0.3101	−0.274	−0.3706	0.063
Value of the p statistic of the model	0.1832	0.2410	0.1077	0.7909

Table 9.
Values of the correlations of texture and organic matter with Mg.

Test stats	Sand	Slime	Clay	OM
Correlation coefficient (r)	0.01320	−0.0426	0.1607	0.4063
p-value	0.9559	0.8584	0.4984	0.0755

Table 10.
Correlation values of texture and OM with Mn.

Mg is a natural element of the sediments and forms part of the hardness of the waters, it is abundant in calcareous sediments and is a source of available minerals. The statistical results are shown in **Table 5** and confirm that only the climatic season influences the presence and availability.

The results of the relationships between sediment texture and OM with Mg are presented in **Table 9**. (No significant relationships are shown).

For Mn, the results show that only the climatic season influences the presence and distribution of the metal (**Table 5**), The relationships between sediment texture and organic matter are shown in **Table 10** (No significant relationships shown).

DOI: http://dx.doi.org/10.5772/intechopen.104657

	Cd	Fe	Mg	Mn	Cu
Cd	1				
Fe	−0.96	1			
Mg	−0.4352	0.4006	1		
Mn	−0.5225	0.4130	−0.133	1	
Cu	0.9475*	0.9964*	−0.434	−0.427	1

**Significant at 95% confidence.*

Table 11.
Values of metal-metal correlations.

A correlation analysis (**Table 11**) between metal-metal was performed to determine the degree of relationship and if they have sources of generation in common.

4. Conclusion

In this study, the levels of heavy metals showed significant variations only due to the climatic season; the dry season has a more significant influence on the increase in levels of all the metals included in this study. As discussed throughout the chapter, high temperatures in the area cause phenomena of evaporation, concentration, and precipitation of pollutants from the water column to the sediments. The organic matter levels have a high relationship with Fe. Likewise, this metal shows significant associations with Cd and Cu, which indicates that the generated source is the same. It is concluded that the contributions of Fe, Mg, and Mn come from the agricultural and livestock activities carried out along the "Palizada" river, where its high agricultural and livestock productivity is an essential characteristic of the municipality.

Likewise, the absence of effective programs for monitoring and controlling pollutants in the Mexican coastal environment, the growing industrialization, and urbanization have caused more than 30% of the rivers and lagoons in the Gulf of Mexico to be contaminated. The sediments represent a reservoir of heavy metals that contributes significantly to the bioavailability of aquatic organisms. On the other hand, the levels of organic matter in some sites exceed the criteria that have been established for marine areas where these levels should not be greater than 10%. These high levels of organic matter corroborate that the "Palizada" river directly influences anthropogenic discharges and rain slides.

Conflict of interest

The authors declare no conflict of interest.

Author details

Carlos Montalvo[1*], Claudia A. Aguilar[1], Yunuen Canedo[1], Alejandro Ruiz[1], Brenda Zermeño[2], Ruby S. Gines[3] and Rosa M. Cerón[1]

1 Autonomous University of Carmen, Campeche, Mexico

2 Faculty of Chemical Sciences, Autonomous University of San Luis Potosí, San Luis Potosí, Mexico

3 Technological Institute of Orizaba, Orizaba, Veracruz, Mexico

*Address all correspondence to: cmontalvo@pampano.unacar.mx

References

[1] Guiza González SY, Cedeño Ochoa CJ, González Quinteto J. Compilación y análisis de datos geoquímicos de metales traza en algunas zonas del río Magdalena. Ed. INGEOMINAS. Colombia, Bogotá; 2001

[2] Ponce G, Botello V. Aspectos geoquímicos y de contaminación por metales pesados en la Laguna de Términos, Campeche. Hidrobiológica. 1991;**1**:1-10

[3] Botello AV, Barrera G, Díaz G, Ponce G, Villanueva S, Wong I. Contaminación marina y costera. La pesca en Veracruz y sus perspectivas de desarrollo. Primera edición ed. Xalapa, Veracruz, México: Instituto Nacional de la Pesca. Universidad Veracruzana; 2002. p. 97

[4] Contreras PJB, Mendoza CL, Gómez A. Determinación de metales pesados en aguas y sedimentos del Río Haina. Ciencia y Sociedad, año/vol.29, número 001. Instituto Tecnológico de Santo Domingo; 2004. p. 38

[5] Vázquez FG, Sharma VK. Major, and trace elements in sediments of the Campeche Sound, Southeast Gulf of Mexico. Marine Pollution Bulletin. 2004;**48**:87-90

[6] Pérez M, Martínez G, Fermín I. Biodisponibilidad de metales traza en sedimentos superficiales del ecosistema lagunar costero Bocaripo-Chacopata. Península de Araya, estado Sucre: Instituto Oceanográfico de Venezuela; 2006

[7] González B, Carrión J, Yam G, Díaz L. Contaminación de aguas residuales municipales. México: Universidad de Quintana Roo: Departamento de Ciencias e Ingeniería; 2008

[8] Pineda M. Granulometría y Geoquímica de los Sedimentos Marinos en el área comprendida entre el seno Reloncaví y Golfo Corcovado, Chile. Ciencia y Tecnología del Mar. 2009;**32**:27-47

[9] González A, Marañón A, Pérez N, Espinosa M, Argota H. Granulometría, sedimentación, composición y contenido total de metales pesados en los sedimentos del río Cobre, de Santiago de Cuba. revista cubana de química. 2004;**16**:22-30

[10] Páez F, Osuna J. Metales pesados disueltos y particulados en el puerto de Mazatlán. Laboratorio de Oceanografía Química. Estación Mazatlán: Instituto de Ciencias del Mar y Limnología (UNAM); 1989

[11] Lima C, Olivares R, Columbia I, De la Rosa M, Gil C. Niveles de plomo, zinc, cadmio y cobre en el Río Almendares, Ciudad Habana, Cuba. Revista Internacional de Contaminación Ambiental. 2005;**21**:115-124

[12] Aguilar C, Montalvo C, Ramírez M, González C. Metales pesados en sedimentos del arroyo "La Caleta" de Ciudad del Carmen, Campeche, México: Estudio preliminar. Revista Latino americana de Recursos Naturales. 2006;**5**:232-237

[13] Boffeta P. Carcinogenicity of trace elements with reference to evaluations made by the: International agency for research on cancer. Scandinavian Journal of Work, Environment & Health. 1993;**19**:67-70

[14] Ploet DM, Fiits BE, Rice TM. Differential accumulation of heavy metals in muscle and liver of a marine fish, (*King Mackerel, Scomberomorus*

cavalla Cuvier) from the Northern Gulf of Mexico, USA. Bulletin of Environmental Contamination & Toxicology. 2007;**78**:134-137

[15] Montoya J, Alarcón E, Castañeda M, Lango F, Zamudio R. Heavy metals in muscle tissue of *Pterois volitans* from the Veracruz reef system national park, Mexico. International Journal of Environmental Research and Public Health. 2019;**16**:4611-4622

[16] Serviere E, Lluch S, Mazariego A, Balart E, Valencia H, Méndez L. Cadmium, lead, copper, zinc, and iron concentration patterns in three marine fish species from two different mining sites inside the Gulf of California, Mexico. International Journal of Environmental Research and Public Health. 2021;**18**:844-856

[17] Alvarado C, Ramírez J, Herrera E, Cortez D, Ramírez G. Bioaccumulation of Metals in Cultured Carp (*Cyprinus carpio*) from lake Chapala, Mexico. Biological Trace Element Research. 2020;**195**:226-238

[18] Azevedo H. Study of Heavy Metal Pollution in the Tributary Rivers of the Jacarepagua Lagoon. Rio de Janeiro, Brazil. Metals in Coastal Environments of Latin America. Springer, 1988. p. 58

[19] Santos J, Beltran R, Gómez A. Spatial variations of heavy metals contamination in sediments from Odiel river (Southwest Spain). Environmental International. 2003;**29**:67-77

[20] SEMARNAP. Programa de manejo del área de protección de flora y fauna Laguna de Términos, México. Mexico: Instituto Nacional de Ecología, Secretaría del Medio Ambiente, Recursos Naturales y Pesca; 1997

[21] Buchanan B, Kain M. Measurement of the physical and chemical environment. In: Holem NA, Intyre M, editors. Methods for the Study of Marine Benthos: International Biological Program Handbook. Oxford: Blackwell Scientific Publications; 1971

[22] Borchardt S, Burchert H, Hablizel L, Karbe R, Zeitner J. Trace metal concentrations in mussels: Comparison between estuarine, coastal, and offshore regions in the southeastern North Sea from 1983 to 1986. Marine Ecology Progress Services. 1988;**42**:17-31

[23] EPA. U. S. Method 3050B. Acid Digestion of Sediments Sludges and Soils. Research Triangle Park, NC: U.S. Environmental Protection Agency; 1996

[24] Bruder-Hubscher V, Lagarde F, Leroy MJF, Coughanowr C, Enguehard F. Application of a sequential extraction procedure to study the release of elements from municipal solid waste incineration bottom ash. Analytica Chemical Acta. 2002;**451**:285-295. DOI: 10.1016/S0003-2670(01)01403-9

[25] Aguilar C, Montalvo C, Rodríguez L, Cerón J, Cerón R. American oyster (*Crassostrea virginica*) and sediments as a coastal zone pollution monitor by heavy metals. International Journal of Environmental Science and Technology. 2012;**9**:579-586. DOI: 10.1007/s13762-012-0078-y

[26] Sokal R, Rohlf J. Biometry: The Principles and Practice of Statistics in Biological Research. New York: W. H. Freeman and Co.; 1997. p. 859

[27] Graham D, Daniel P, Hill M, Day W. A preliminary model of the circulation of Laguna de Terminus, Campeche, México. Vol. 8. An. Inst. Cienc, del Mar y Limmol. Universidad Nacional Autónoma de México; 1981. pp. 51-62

[28] Green R. Geoquímica de metales pesados y mineralogía de la fracción arcillosa de los sedimentos de cuatro puertos del golfo de California. Mazatlán: UNAM, México; 2000

[29] Leoni L, Santori Y. Heavy metals, and arsenic in sediments from the continental shelf of the northern Tyrrhenia/Eastern Ligurina seas. Marine Environmental Research. 1996;**41**:73-98

[30] González C, Méndez C, López G, Vázquez A. Evaluación de la contaminación en sedimentos del área portuaria y zona costera de Salina Cruz, Oaxaca, México. Interciencia. 2006;**31**:647-656

[31] Valdés J, Güines M, Castillo A, Ellen S. Cu, Pb, and Zn content in sediments and benthic organisms from San Jorge Bay (northern Chile): Accumulation and bio transference in subtidal coastal systems. International Journal of Marine Sciences. 2004;**4**: 304-309

Section 3

Remediation of Heavy Metals

Chapter 13

Leaching Technology for Precious Heavy Metal Recapture through (HCI + HNO_3) and (HCI + H_2SO_4) from E-Waste

Murugesan Manikkampatti Palanisamy, Akilamudhan Palaniappan, VenkataRatnam Myneni, Padmapriya Veerappan and Minar Mohamed Lebba

Abstract

The rapid growth of information technology and industrialization are the key components for the development of electronic equipment, and their inevitable role in human day-to-day life has an important stint in the generation of electronic waste (e-waste). This waste has far-reaching environmental and health consequences. One such e-waste printed circuit board (PCB) contains significant amounts of valuable heavy metals such as copper (Cu), lead (Pb), zinc (Zn), nickel (Ni), and others that can be extracted through various metallurgical routes. Recovery and recycle of heavy metal ions is a major challenge to prevent environmental contamination. The present study discusses the current e-waste scenario, health impacts and treatment methods in detail, and also presents experimental results of recovery of heavy metals from printed circuit boards (PCBs) by leaching using aqua regia (HCI + HNO_3 and HCI + H_2SO_4). Under varying conditions such as specified conditions of 80°C, 0.05 mm of thickness, 3 hrs of contacttime, 80rpm shaking speed, and concentration of PCB sample of 0.5 g ml^{-1}, it results in the composition of extracted heavy metal ions in such a way that 97.59% of copper, 96.59% of lead, 94.66% of tin, and 96.64% of zinc, respectively. The recovery of heavy metal ions from PCBs has an important leading contribution in electronic waste management and the result shows a higher rate.

Keywords: e-waste, printed circuit board, leaching, aqua regia, heavy metals, optimization

1. Introduction

The electronics industry is the largest and fastest-growing manufacturing sector in the world. The PCBs are waste sources from electronic machines such as television boards, CD players, and cell phones. Researchers have reported that in recent years, the average rate of PCB development has risen by 8.7% due to technological progress. The studies observed that the continuous increase in e-waste generation rates is due to the nation's population and technological growth. The studies predict that each individual would produce approximately 5173 kg of e-waste per year. The metallic composition consists primarily of 10–30% of copper (Cu) and other metals such as tin (Sn), zinc (Zn), lead (Pb), nickel (Ni), iron (Fe), silver (Ag), cadmium (Cd), gold (Au) and others, depending on the sources of printed circuit boards (PCB) [1–3]. A sample PCB is shown in **Figure 1**. Informal processing of e-waste in developing countries can lead to adverse effects on human health and environmental pollution. In 2016, 44.7 million metric tons of e-waste were generated globally [4, 5]. Health symptoms like headaches, dizziness, irritation in the eyes, nose, mouth, etc. are caused by exposure to Cu, which is present in landfills [6–8]. The methods that can be used to recover metals from PCBs are essentially physical, mechanical and chemical separations. Several studies on the feasibility of metal recovery from PCBs have been investigated in the last decade. Hydrometallurgical procedures, such as leaching, are very intentional in these studies.

1.1 Health hazards caused by informal disposal of e-waste

E-waste not only includes household and industrial electrical appliances but also includes their components such as batteries, capacitors, castings, etc. Recycling of such waste has been carried out both formally and informally in several countries like China, India, Ghana, Thailand, Vietnam, etc. [9]. Traditional recycling techniques are well developed techniques to ensure safe and efficient separation, but are highly expensive to install and run. So such techniques are not taken into consideration and cheap informal techniques are to be implemented. This may cause the release of several pollutants into the environment, which can lead to a variety of health problems [8–10]. The metals present in the PCBs are highly toxic and hazardous to living bodies. These metals follow media such as dust, air, water, and soil to reach the human frame. Exposure to metals such as lead (Pb) and cadmium (Cd) affects reproductive health, growth, and mental instability and damages human DNA [11–13]. Health symptoms like headaches, dizziness, irritation in the eyes, nose, mouth, etc. are caused by the exposure to copper (Cu) which is present in landfills [14–16]. The different e-waste sources, heavy metals, and effects are explained in **Figure 1**.

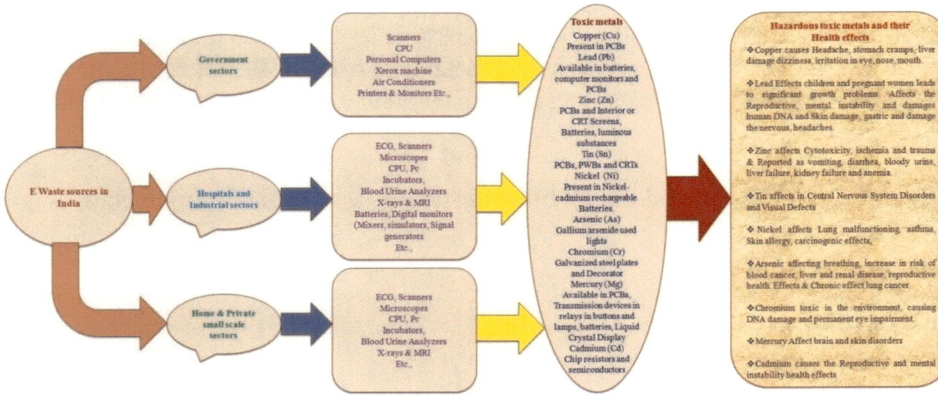

Figure 1.
Diagram of e-waste sources in various aspects and health effects.

Informal treatment strategies as well as innovative metal recovery approaches based on the material composition present in PCBs are accompanied by management and sustainable treatment techniques involving the reduction of waste circuit boards in developing countries. There are two techniques used to dispose of and extract heavy metals from PCBs. Incineration was used as a primary method through high-temperature PCB melting and is very hazardous as it releases strong toxic metal vapors such as polycyclic aromatic hydrocarbons and dioxins due to the emission of possible contaminants during incineration [17, 18], and the secondary method was followed at low temperature by hydrometallurgical techniques with the help of chemical reagents [19–21].

1.2 Pyrometallurgical method

Pyrometallurgical processing is the most common method used for the separation of heavy metals from PCBs. The smelting process consists of the melting of waste PCBs in a high-temperature furnace (up to 1500°C) and is primarily used for the recovery of copper from used waste circuit boards. The limitations of this process are relatively low performance, high energy consumption, and difficulty in distinguishing metallic and non-metallic components [19, 22]. The pyrometallurgy process involves the heating of e-waste at a high temperature to recover precious metals. This treatment leads to the release of dangerous gases into the air, which must be extracted from the air by the flue gas cleaning system [23]. The limitations of this process are:

- Pyrometallurgical techniques have a greater environmental impact due to the gaseous emissions from incineration.

- Plastic recovery is not possible due to the replacement of plastics by coke as a source of energy.

- Hazardous emissions such as dioxins are generated during the smelting of feed materials which have halogenated flame retardants and polyvinyl chloride (PVC), which lead to dioxins in the form of dioxins. Therefore, special installations (emission controls) are required to minimize environmental pollution.

- It is very difficult to separate all the metals.

- Low metals are recovered only by a fraction of metals.

- The process has high energy consumption.

1.3 Hydrometallurgical method

The hydrometallurgical method includes the application of aqueous solution leaching media, such as strong acid or base, oxidizing agents, and complexion agents, for the recovery of heavy metal separations. Previous studies have employed various leaching media such as strong acids (sulfuric acid, nitric acid), bases (sodium hydroxide, sodium hypochlorite), and complexing agents (cyanide, thiosulphate). This treatment approach has advantages over pyrometallurgical processes such as reduced pollution, radioactive contaminants, and moderately toxic chemicals for environmental effects. Therefore, these various recovery methods used for the treatment of used PCBs need to be reconsidered due to the enormous amount of flammable, toxic, and corrosive reagents used and the large volume of effluents and other solid waste produced [24]. In hydrometallurgical procedures, the following steps are widely used: leaching and extraction, purification, and concentration of liquefied solutions, as well as the recovery of heavy metals. Four operations are typically implemented in these recovery operations, as shown in previous studies. The hydrometallurgical procedure, such as leaching, has shown a great deal of strength in several studies. Several leaching reagents demonstrate major improvements in metal recovery. When treated with different acidic media, aq.HNO_3, aq.HCI, and aq.H_2SO_4, PCBs were cut to remove Cu^{2+} ions and the recovery percentage of Cu^{2+} was 97.5, 65, and 76.5%, respectively [20]. Only trace quantities of other metals can be extracted through this targeted extraction of copper. While using HCI as a leaching agent under specified conditions, the PCB sample size of 4×4 cm results in the separation of Cu, Zn, Sn, and Pb with a composition of 117.33, 28.97, 10.41, and 9.34 mg g^{-1}, respectively [23].

The amount of Zn and Pb leached was very small when compared to the typical PCB metal content. The recoveries for Cu, Pb, Zn, and Sn were 16, 2.0, 1, and 1%, respectively, when leaching was done in crushed PCBs (size between 0.43 and 3.33 mm) using sodium cyanide solution [25]. It has the least compositional value compared to the average weight of the total metals found in it. After 480 min, various metals leached from PCB waste, such as H_2SO_4 and H_2O_2, provide 76% Zn, 85% Cu, 82% Fe, 77% Al, and 70% Ni recovery [26]. Other valuable metals are retained in the leaching solution as residue. The effective treatment of PCBs will depend on choosing the suitable recovery method. The development of new technology for the recovery of toxic heavy metal ions from waste PCBs remains an important scientific endeavor. The literature study shows the more stable and effective metal ion recovery will be achieved by two-stage chemical leaching with adsorption from waste PCBs. However, a suitable carrier has to be selected for the selective recovery of heavy metals in an effective manner.

This chapter presents an overview of current e-waste scenario, its impacts and treatment methods. The experimental studies are carried out for the the extraction of copper (Cu), tin (Sn), zinc (Zn) and lead (Pb) from PCBs by leaching using aqua regia (a mixture of HCI and HNO_3 and HCI and H_2SO_4) for varying conditions of temperature, size of sample, contact time and shaking speed.

The main objectives of the present study are:

- To prepare the suitable leaching agents in chemicals and optimize the operational parameters like concentration, temperature, shaking speed, time of leaching, and bulk density for the recovery of metal ions such as copper, zinc, tin, and lead separate from PCBs.

- To determine the stability of the prepared leaching media by the recovery rate with the help of EDXs.

2. Materials and methods

2.1 Electronic waste collection

The waste PCBs are obtained from the e-waste disposal unit in India. The sample was initially cleaned manually to remove dust particles by the air blower. Later, other elements such as capacitors, resisters, integrated circuits, diodes, transistors, etc., were detached with the help of mechanical tools (saw metal cutter, sheet metal cutter, metal lathe cutting tool, cutting pliers, and materials separation toolkit). This separation is not as simple due to the difference in the physical characteristics of metals and non-metals. Hence, different separation methods, such as pneumatic separation, magnetic separation, filtering, eddy current separation, electrostatic separation, etc., are used to enrich metals and non-metals [10, 14, 23, 27].

The crushed PCBs obtained from the crusher are then pulverized and further exposed to milling operations for better size reduction using a ball mill, and particles of different mesh sizes are analyzed. The weight fraction of crushed PCBs obtained from the lower screens of jaw crushers with a capacity of 80 kg $hr.^{-1}$ and a clearance of 10 mm is much lower, making better ion recovery impossible. Thus, it is subjected to 5 mm of clearance in the same jaw crusher, yielding samples weighing 65, 53, 48, and 36 grams for sieves with mesh sizes of 0.3, 0.18, 0.05 mm, and pan, respectively, when screened using a rotary sieve shaker at a speed of 60 rpm with a power of 0.25 HP and a single-phase 80 volt supply. As the reduction in size increases the rate of recovery of metal ions [16], the resulting crushed samples are processed into powder form using a pulverizer with a disk diameter of 175 mm operated by a 3-phase motor at 1400 rpm in a 225–445 V supply (**Figure 2** and **Table 1**).

2.2 Chemical leaching experimentation with aqua regia

2.3 Leaching mediaPreparation

The leaching media is an important factor that should be considered while extracting heavy metals from PCBs. Various sorts of leaching agents show different leaching rates with respect to the type of metals present in PCBs. H_2SO_4, HCI, NaCl,

Figure 2.
Stepwise size reduction of PCBs under the various mechanical operations (jaw crusher, roll crusher, furnace and pulverized mills produced small sizes between 4 and 0.05 mm) and heavy metals presents before leaching by SEM with EDx analysis.

Mesh size		Weight fraction (grams)			
		Jaw crusher		Pulveriser	Ball mill
B.S.S	(mm)	Clearance 10 mm	Clearance 5 mm	Feed size 6 mm	Ball weight 500 g
4	4	155	118	45	27
7	2.3	125	92	57	35
25	0.6	95	76	69	58
52	0.3	52	65	87	64
85	0.18	30	53	60	78
300	0.05	22	48	85	82
pan	_	15	36	79	120

Table 1.
Analysis of PCB size reduction.

HNO_3, $Na_2S_2O_3$,etc. are commonly used leaching media for the extraction of heavy metals from PCBs. Aqua regia, which is a mixture of hydrochloric acid and nitric acid, is used as a leaching agent in this study. It is prepared by mixing HCI and HNO_3 in a 3:1 ratio under specified conditions of temperature, time, and surrounding conditions. Different leaching agents show different rates of recovery and metal extracted with respect to the nature of the leaching media, rate of ion exchange, degree of dissociation of ions and various parameters such as time, temperature, concentration etc. When remaining constraints are held constant, metal ions such as Cu^+, Zn^+, Pb^+, and Sn^+, among others, exhibit different ionic properties with acid medium. Even though the above studies result in significant metal recovery, they also possess demerits, such as the targeted extraction of a specific metal leads to the loss of several other valuable metals. By using aqua regia as a leaching reagent, heavy metals such as Cu, Sn, Pb, and Zn can be extracted from PCBs with a high recovery rate. Aqua regia preparation involves the mixing of strong acids. It produces heat and toxic gases, so it is necessary to follow safety protocols while preparing and handling this solution. In this experiment, HNO_3 is added to HCI contained in a beaker, which is placed in a water bath in order to reduce the fume generation. The two concentrated acids are mixed in the ratio of 3:1 (HCI:HNO_3); concentrated HCI is about 35% and concentrated HNO_3 is about 65%. So that volume ratio will be 4 parts concentrated HCI and 1 part concentrated HNO_3. The solution is kept away from organic contaminants because it leads to vigorous or violent reactions and a low temperature should be maintained.

2.4 Treatment of PCBs with aqua regia

All the experiments are carried out in a conical flask incorporated with a temperature-controlled shaker. Primary analysis was conducted by applying specific conditions to obtain a standard recovery rate. 5gm of PCB samples are allowed to react with 20 ml of leaching media inside the conical flask at 60°C and shaken in a mechanical shaker at a shaking speed of 120 rpm for 2 hrs. At the end of this effective contact time, the shaker is stopped and the solution in the conical flask is filtered using filter paper. After complete filtration, the filtrate is sent for SEM with EDX analysis to determine the composition of metals retained. The rate of leaching is affected by a number of factors, including sample size, concentration, temperature, shaking speed, and contact time. By varying these parameters, different values for the recovery rate and composition of heavy metals are obtained.

3. Results and discussion

3.1 Sample analysis of PCBs

It is important to conduct sample analysis before subjecting the crushed samples to the leaching process. The primary sample had followed three size reduction operations, namely crushing, pulverizing, and milling, and the weight fractions obtained at each operation are explained with their corresponding mesh size shown in **Figure 3**. The ultimate purpose of size reduction has been studied previously and data analyzed.

The graphical representation of size analysis shows that the fraction of sample obtained in the sieves with larger mesh sizes has been decreasing when subjected to a sequence of size reduction operations. However, the total weight obtained in the sieves is approximately conserved with a trace of negligible loss. From the sieve analysis data of each operation, the sample obtained from the ball mill has a fraction of weight in the pan that is less than 0.05 mm. Various studies used shredded samples with a size of less than 0.5 mm, which resulted in a high recovery rate of heavy metals [4, 5]. Present research comprises leaching particle sizes of 0.05 mm and 0.1 mm, which is the sample retained just above the pan (**Figures 4** and 5).

3.2 Optimization of various parameters for recovery of heavy metals

3.2.1 Effectof concentration

By varying the concentration, the leaching process shows a significant change in the recovery rate. The recovery rate increases with an increase in the concentration of the sample with respect to time. After attaining an equilibrium state, the rate of leaching becomes constant. 20 ml of aqua regia is used to leach heavy metals from 2, 4,

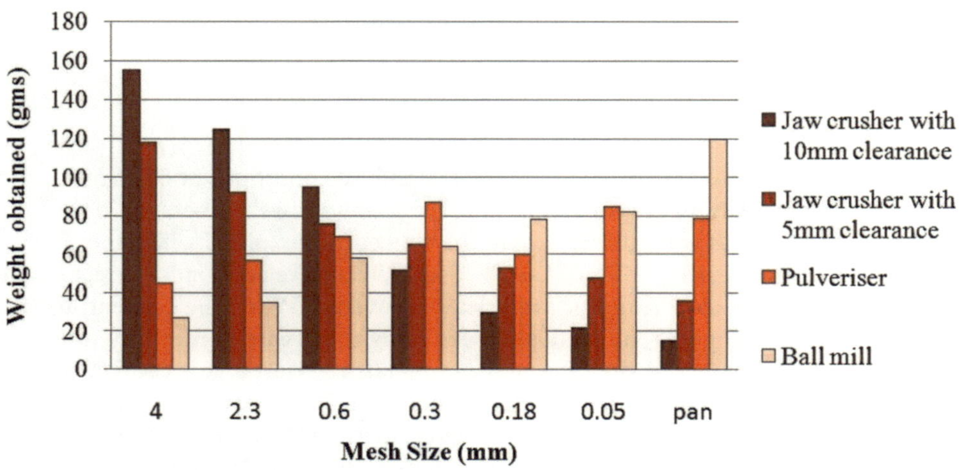

Figure 3.
Graphical representation of size reduction in different operation.

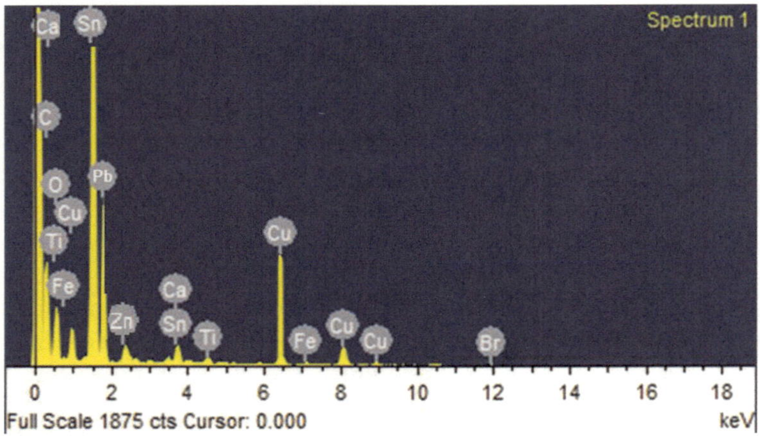

Figure 4.
Presents of metal components from PCBs by the EDXs.

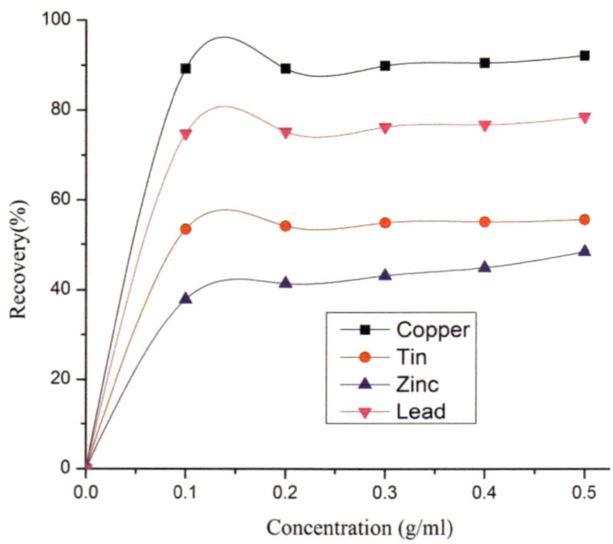

Figure 5.
Graphical representation of % recovery of metals with concentration.

6, 8, and 10 g of 0.3 mm sized PCB samples at standard conditions of 80°C of temperature and 200 rpm of speed for 2 hrs. The graph shows an increase in the recovery rate of metals with an increase in the concentration of PCB. The metals recovered in the decreasing order of Cu, Pb, Sn, and Zn were recovered. Copper is the most recovered metal, whereas zinc is the least recovered. When the concentrations were increased by 0.1 g ml^{-1}, all metals showed a slight increase in recovery rate.

As there is no decrease in the percentage of metals recovered, it is confirmed that the metallic distribution of powdered PCBs is uniform. When the concentration is 0.5 g/ml, the graph shows the maximum recovery with metallic composition as 92.06% of Cu, 55.42% of Sn, 48.27% of Zn, and 78.42% of Pb. Based on the previous studies of metal recoveries [28, 29].

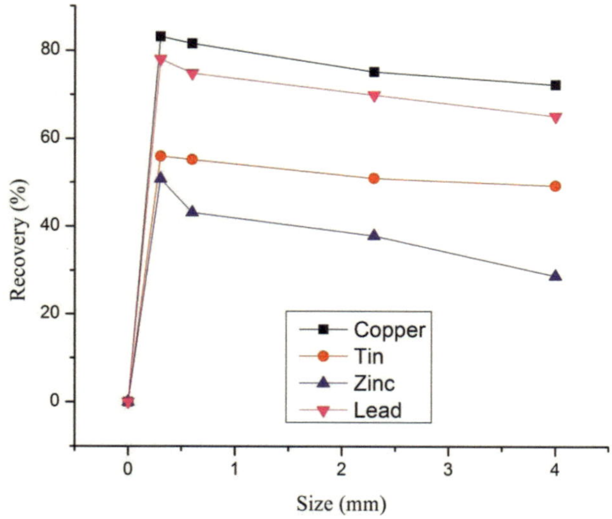

Figure 6.
Graphical representation of % recovery of metals with size.

3.2.2 Effect of size

The different sieve size particles are leached using aqua regia and the weight fraction of metallic components is analyzed. 5 gm of particles with sizes of 4, 2.3, 0.6, 0.3, and 0.05 mm are leached for hours at 80°C with a shaking speed of 200 rpm. **Figure 6** explains the relationship between size and recovery. It shows recovery increases with an increase in contacting surface.

The graph shows an appreciable increase in the percentage of metals recovered with a decrease in the size of the sample. The higher recovery rate is shown for the PCB sample at the lowest size, which is 0.05 mm. The uniformity of metallic distribution is also conserved here. Copper is the major component present in the leached sample and zinc is the minimum. It results in a percentage recovery of copper, tin, zinc, and lead of 83.49, 58.72, 57.75, and 78.42%, respectively.

3.2.3 Effect of temperatire

5 gm of PCB samples of size 0.3 mm are treated with aqua regia in a conical flask and shaken at a speed of 200 rpm for 2 hrs. Five samples of the same condition are maintained at varying temperatures of 40, 60, 80, 100, and 120°C. After completion of effective time, the leached PCB sample is analyzed. The data obtained is represented graphically and the relations between recovery and temperature are studied. The graph shows an appreciable increase in recovery rate with an increase in temperature for a particular point of temperature [30, 31]. When the sample is leached at a temperature above 80°C, there is no appreciable change in recovery, which is negligible. It indicates that the leaching depends on temperature only for a particular limit, and that after a certain point of temperature, leaching is independent of temperature. At 80°C, the result shows a metallic composition of 89.84% of Cu, 69.05% of Sn, 65.51% of Zn, and 82.45% of Pb (**Figure 7**).

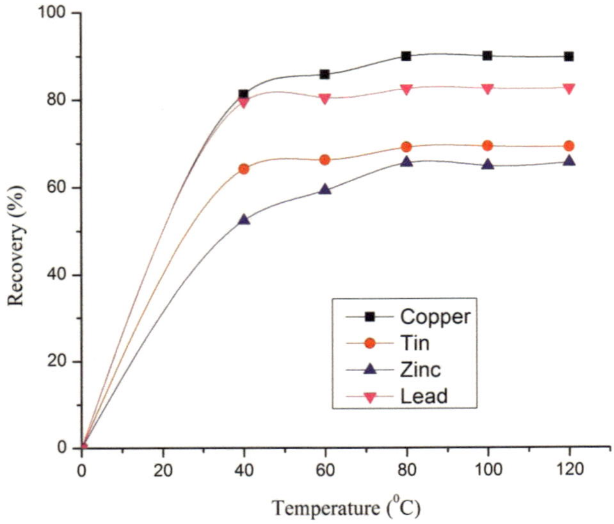

Figure 7.
Graphical representation of % recovery of metals with temperature.

3.2.4 Effect of time

To determine the effective leaching time, the sample is allowed to be leached for different intervals of time. The persistent condition is maintained as a 5gm sample of size 0.3 mm shaken with 20 ml of aqua regia in a conical flask and shaken at a speed of 200 rpm while the temperature is maintained at 80°C. Then it is allowed to be leached for 1, 2, 3, 4 and 5 hrs, respectively. The data collected is graphically represented in **Figure 8** below. It shows that the recovery percentage is almost constant when a sample is leached for more time after a certain period of time. The graph shows that the recovery of metals increases with an increase in time for a certain period, and after a particular point of time, the recovery becomes almost constant. That means all the

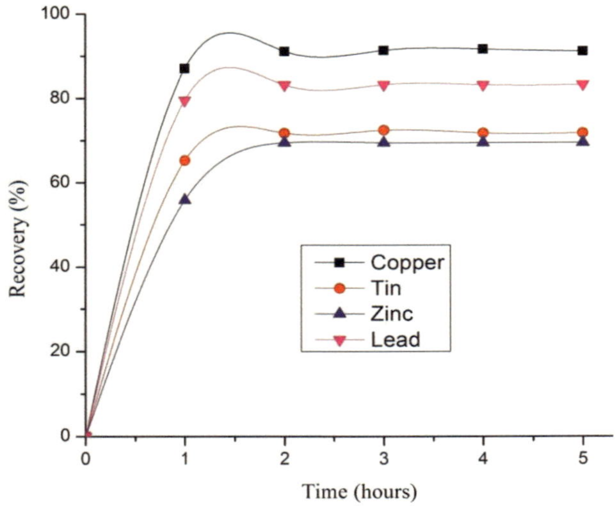

Figure 8.
Graphical representation of % recovery of metals with time.

metals in contact are leached from the sample within a particular time period, and there is no use in leaving the system under leaching condition after a certain period of time. The results show the maximum recovery when the sample is leached for 3 hrs. When the sample is leached for 3 hrs, the result shows the recovery of metals as 91.36, 69.43, 72.41, and 83.22% of Cu, Sn, Zn, and Pb, respectively.

3.3 Optimum conditional parameter studies

Once all the results for recovery with respect to various parameters are evaluated and studied as explained above, we get the optimum condition to obtain maximum recovery of metals. The optimum condition is the value of concentration, size, temperature, shaking speed, and time at which the maximum recovery is obtained. The results obtained at optimum conditions show that the recovery of heavy metals is as high as 99.9% of copper, 98.3% of lead, 96.8% of tin and 93.1% of zinc, respectively. In this study, specific conditions of 800°C, 0.05 mm of thickness, 3 hours of contacting time, 80 rpm shaking speed, and pulp density of PCB sample of 20gm L^{-1} were met in both stages, with a 3:1 ratio of first stage HCI and HNO3 and second stage HCI and H_2SO_4) prepared as a leaching agent. The experimental results were obtained under the above mentioned conditions and have been shown in (**Figure 9**) and (**Table 2**). Results found that the optimum recovery rate for stage I Cu was 89.5%,

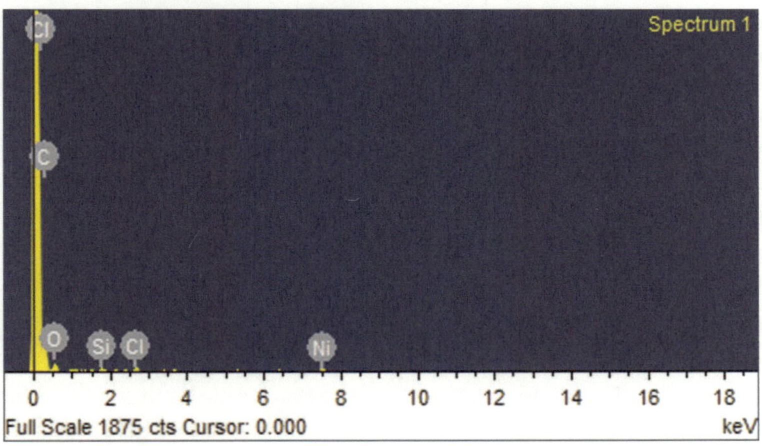

Figure 9.
EDXs spectrum analysis for metal ions obtained after leaching.

Metals	Initial C_O	Stage-I & II weight fraction				% Recovery	
		C_e-I (PCBs sample)	C_e-I Leached solution	C_e-II (PCBs sample)	C_e-II Leached solution	Stage-I	Stage-II
Cu	3.15	0.325	2.825	0.031	3.119	89.5%	99.0%
Sn	42.40	15.09	27.31	1.36	41.04	64.4%	96.8%
Pb	27.81	5.31	22.5	0.47	27.34	80.9%	98.3%
Zn	1.16	0.403	0.757	0.076	1.084	63.4%	93.1%
Others	27.81	12.4	15.41	1.168	26.642	52.6%	95.4%

Table 2.
Metallic composition of leached PCBs at optimum conditions by stage-I & II.

Sn 64.4%, Zn 63.4%, Pb 80.9%, and stage II Cu was 99.0%, Sn 96.8%, Zn 93.1%, and Pb 98.3%, respectively.

4. Conclusions and outlook

The study shows the dependency of the rate of recovery on the condition in an arbitrary manner. The recovery rate has a different approach with each parameter. The results show that the maximum percentage of metals recovered at 80°C, 0.05 mm thickness, 3 hours of contacting time, 80 rpm shaking speed, and PCB sample concentration of 0.5 g/ml^{-1}.Under this condition, the resultwas obtained with 97.59% of copper, 96.29% of lead, 94.66% of tin, and 96.64% of zinc, respectively. It is the most effective recovery condition for this experiment. However, targeted extraction of a particular metal can be made possible by varying a particular parameter only. In such a way, the percentage recovery of that particular metal can be increased with negligible loss. In such extractions, the other heavy metals are retained in the sample or less amounts of other metals are separated. Disposal of such residuals also causes environmental issues. Since this type of extraction is promoted in order to reduce the environmental problems caused by these kinds of heavy metals, the targeted extraction of a particular metal is not advisable, even if it is economically profitable. It was concluded that the combination of aqua regia (HCI and HNO_3 and HCI and H_2SO_4) leaching is an effective and economic way for the recovery of copper from leached solution. According to studies, modifying the dissolving of metal ions in the reagents increases the rate of leaching, but this raises the total cost and results in the introduction of additional chemicals into the atmosphere. As a result, attempts will be made in the future to resolve these issues. Only a few field trials have been performed, and more systematic studies are needed to decide the best conditions for using aqueregia as a leaching medium.

Acknowledgements

This study was carried out with utilization of the laboratory facilities in Excel Engineeeringcollege, Tamilnadu, India. The corresponding author would like acknowledge and thank to his parents and brother P. Selvarasu, PG Assist Zoology, Govt Higher Secondary school, Vellore for their kind support.

Highlights

- The heavy metals in PCBs were leached with two-step aqua regia (HCI and HNO_3 and HCI and H_2SO_4).

- Optimization of various parameters has been tested to enhance the recovery of heavy metals.

- The maximum recovery rates obtained were Cu 97.59%, Pb 96.89%, Sn 94.66%, and Zn 96.64%.

- The combination of aqua regia leaching and PCBs is an efficient and cost-effective method for recovering heavy metals from PCBs.

DOI: http://dx.doi.org/10.5772/intechopen.102347

Nomenclature

Cu	Copper (–)
EEE	Electrical and electronic equipment (–)
E–Waste	Electronic waste (–)
EDX	Energy-dispersive X-ray spectroscopy (–)
HCI	Hydrochloric acid (–)
PVC	Polyvinyil chloride (–)
Pb	Lead (–)
HNO_3	Nitric acid (–)
PCBs	Printed circuit boards (–)
SEM	Scanning electron microscopy (–)
H_2SO_4	Sulfuric acid (–)
Sn	Tin (–)
WEEE	Waste of electrical and electronic equipment (–)
Zn	Zinc (–)

Author details

Murugesan Manikkampatti Palanisamy[1]*, Akilamudhan Palaniappan[2], VenkataRatnam Myneni[3], Padmapriya Veerappan[4] and Minar Mohamed Lebba[2]

1 Department of Food Technology, Excel Engineering College, Namakkal, Tamil Nadu, India

2 Department of Chemical Engineering, Erode Sengunthar Engineering College, Erode, India

3 Department of Chemical Engineering, Mettu University, Metu, Ethiopia

4 Department of Applied Electronics, Erode Sengunthar Engineering College, Erode, India

*Address all correspondence to: engineermurugesh@gmail.com

References

[1] Bari F, Begum MN, Jamaludin B, Hussi K. Selective leaching for the recovery of copper from PCB. Proceedings of the Malaysian Metallurgical Conference. 2009;**2009**: 1-4

[2] Huang K, Guo J, Xu Z. Recycling of waste printed circuit boards : A review of current technologies and treatment status in China. Journal of Hazardous Materials. 2009;**164**:399-408

[3] Bandyopadhyay A. A regulatory approach for e-waste management: A cross-national review of current practice and policy with an assessment and policy recommendation for the Indian perspective. International Journal of Environment and Waste Management. 2008;**2**:139-186

[4] Rimantho D, Nasution S. The current status of e-waste Management Practices in DKI Jakarta. International Journal of Applied Environmental Sciences. 2016; **11**(6):1451-1468

[5] Cherukuri I, Sultana N, Podila SP. Status of e-waste in India—A review. IOSR Journal of Environmental Science, Toxicology and Food Technology. 2018; **12**(11):8-16

[6] Chatterjee P. Health costs of recycling. British Medical Journal. 2008; **337**:376-377

[7] Ghosh B, Ghosh MK, Parhi P, Mukherjee PS, Mishra BK. Waste printed circuit boards recycling: An extensive assessment of current status. Journal of Cleaner Production. 2015;**94**:5-19

[8] Frazzoli C, Ebere O, Dragone R, Mantovani A. Diagnostic health risk assessment of electronic waste on the general population in developing countries scenarios. Environmental Impact Assessment Review. 2010;**30**(6): 388-399

[9] Jian-jun D, Xue-feng WEN, Yue-min Z. Evaluating the treatment of E-waste—A case study of discarded refrigerators. Journal of China University of Mining and Technology. 2008;**18**:454-458

[10] Gupta S, Modi G, Saini R. A review on various electronic waste recycling techniques and hazards due to its improper handling. International Refereed Journal of Engineering and Science. 2014;**3**(5):05-17

[11] Zhang J, Jiang Y, Zhou J, Wu B, Liang Y, Peng Z, et al. Elevated body burdens of PBDEs, dioxins, and PCBs on thyroid hormone homeostasis at an electronic waste recycling site in China. Environmetal Science and Technology. 2010;**44**(10):3956-3962

[12] Xu X, Yang H, Chen A, Zhou Y, Wu K, Liu J, et al. Birth outcomes related to informal e-waste recycling in Guiyu, China. Reproductive Toxicology. 2012; **33**(1):94-98

[13] Huo X, Peng L, Xu X, Zheng L, Qiu B, Qi Z, et al. Research|children' s health elevated blood lead levels of children in Guiyu—An electronic waste recycling town in China. Environmental Health Perspectives. 2007;**115**(7): 1113-1117

[14] Poon CS. Management of CRT glass from discarded computer monitors and TV sets. Waste Management. 2008;**28**: 1449

[15] Zheng G, Xu X, Li B, Wu K, Yekeen TA, Huo X. Association between lung function in school children and

exposure to three transition metals from an e-waste recycling area. Journal of Exposure Science and Environmental Epidemiology, Nature Publishing Group. 2012;**23**(1):67-72

[16] Murugesan MP, Kandasamy K. Comparative studies on bentonite clay and peanut shell carbon recovering heavy metals from printed circuit boards. Journal of Ceramic Processing Research. 2020;**21**:75–85

[17] Needleman HL, Bellinger D. The health effects of low level exposure to lead. Annual Review of Public Health. 1991;**12**:111-140

[18] Cui J, Zhang L. Metallurgical recovery of metals from electronic waste: A review. Journal of Hazardous Materials. 2008;**158**:228-256

[19] Wang R, Xu Z. Recycling of non-metallic fractions from waste electrical and electronic equipment (WEEE): A review. Waste Management. 2014;**34**: 1455-1469

[20] Masavetas I, Moutsatsou A, Nikolaou E, Spanou S, Zoikis-Karathanasis A, Pavlatou EA. Production of copper powder from printed circuit boards by electrodeposition. Global NEST Journal. 2009;**11**(2):241-247

[21] Zhang K, Wu Y, Wang W, Li B, Zhang Y, Zuo T. Recycling indium from waste LCDs: A review. Resources, Conservation and Recycling. 2015;**104**: 276-290

[22] Ueberschaar M, Susanne V. Enabling the recycling of rare earth elements through product design and trend analyses of hard disk drives. (3RINCs) Journal of Material Cycles and Waste Management volume. 2015;**17**: 266-281

[23] Jadhav UU, Hocheng H. A review of recovery of metals from industrial waste. Journal of Achievements in Materials and Manufacturing Engineering. 2012; **54**:159–167

[24] Tuncuk A, Stazi V, Akcil A, Yazici EY, Deveci H. Aqueous metal recovery techniques from e-scrap : Hydrometallurgy in recycling. Minerals Engineering. 2012;**25**(1):28-37

[25] Montero R, Guevara A, De La Torre E. Recovery of gold, silver, copper and niobium from printed circuit boards using leaching column technique. Journal of Earth Science and Engineering. 2012;**2**:590-595

[26] Ficeriova J, Baláž P, Gock E. Leaching of gold, silver and accompanying metals from circuit boards (PCBs) waste Leaching of gold, silver and accompanying metals from circuit boards (PCBs) waste. Acta Montanistica Slovaca. 2011;**16**:128-131

[27] Xie F, Cai T, Ma Y, Li H, Li C, Huang Z, et al. Recovery of Cu and Fe from printed circuit board waste sludge by ultrasound: Evaluation of industrial application. Journal of Cleaner Production. 2009;**17**(16):1494-1498

[28] Canal Marques A, Cabrera JM, De Fraga Malfatti C. Printed circuit boards: A review on the perspective of sustainability. Journal of Environmental Management. 2013;**131**:298-306

[29] Chien YC, Paul Wang H, Lin KS, Huang YJ, Yang YW. Fate of bromine in pyrolysis of printed circuit board wastes. Chemosphere. 2000;**40**:383-387

[30] Hadi P, Xu M, Lin CSK, Hui CW, McKay G. Waste printed circuit board recycling techniques and product utilization. Journal of Hazardous Materials. 2015;**283**:234-243

[31] Chen M, Huang J, Ogunseitan OA, Zhu N, Wang Y. Comparative study on copper leaching from waste printed circuit boards by typical ionic liquid acids. Waste Management. 2015;**41**: 142-147

Chapter 14

Phytoextraction of Zn(II) and Cu(II) by *Canna indica*: Related Physiological Effects

Josefina Plaza Cazón, Matías Gonzalez and Marcela Ruscitti

Abstract

Phytoremediation is a technique for treatment areas with medium or low heavy metals concentrations. A pot experiment was carried out to determine the usefulness of *Canna indica* L. as phytoremediator species. The plants were treated with three increasing Zn(II) and Cu(II) solutions. 21 days later, dry weight, relative membrane conductivity, chlorophyll, carotene, malondialdehyde, soluble proteins, proline, and Zn(II) and Cu(II) contents were measured. Zn(II) and Cu (II) treatments caused a decline in the dry weight, chlorophyll, carotene, and soluble proteins content, whereas the relative conductivity, malondialdehyde, and proline content showed the opposite pattern. The bioaccumulation reached values approximately 48 and 15 times higher (5293 mg kg^{-1} and 1425 mg kg^{-1}), compared with the control, for Zn(II) and Cu(II), respectively. Our results suggest that this species can be used for the phytoremediation of polluted soils with moderate concentrations of Zn(II) and Cu(II).

Keywords: phytoremediation, *Canna indica*, copper, zinc, physiological response

1. Introduction

Heavy metal pollution of soil and water is a worldwide concern because of its harmful effect on human health. The constant accumulation of heavy metals in soil can pose a serious risk to living organisms including plants, animals, and microorganisms [1, 2]. To date, phytoremediation is confirmed to be the most environmentally friendly and cost-effective strategy. Types of phytoremediation include phytoextraction, phytovolatilization, phytostabilization, phytodegradation, and rhizosphere. The advantages of phytoremediation compared with traditional physical surface and chemical remediation methods are low cost and simplicity [3]. Phytoremediation is linked with the morphological, biochemical, and physiological effects on plant growth. During the phytoremediation process, some morphophysiological growth parameters have to be evaluated such as root growth, net biomass weight, leaf area, the net rate of photosynthesis, the effects on the plasma membrane of plants, reactive oxygen species (ROS) generation, hydrogen peroxide (H_2O_2) content, and malondialdehyde (MDA) level, linked to genotoxicity. Plants try to elude

IntechOpen

their harmful effects by adopting various defense mechanisms, which include antioxidant activation and other mechanisms of metal homeostasis. In response, plants have developed enzymatic and nonenzymatic antioxidant mechanisms and increased activities of catalase (CAT), superoxide dismutase (SOD), ascorbate peroxidase (APX), and peroxidase (POD) [4].

In literature, *C. indica* was investigated by different authors as phytoremediation species in pot, hydroponic, and wetlands systems [5–7]. Most of these works focused on the efficiency of the plant to accumulate heavy metals but did not evaluate the effect of these metals on the physiology of the plant.

This study aimed to evaluate the impact of Zn(II) and Cu(II) excess on the growth and metabolism of *C. indica* through the determination of physiological parameters and Zn(II) and Cu(II) bioaccumulation to establish the strategies used by the plant to overcome the stress and determine the correlation between metal accumulation and physiological parameters modification. Results obtained were compared with parameters measured and published in the scientific literature to provide information for future phytoremediation research.

2. Material and methods

2.1 Growth conditions

The test was carried out in a greenhouse with natural light, forced ventilation, and controlled temperature in La Plata city (Argentina) (34°54′45.5″ S–57°55′51.5″ W) from April to July (2019).

C. indica L. (achira) seeds were superficially disinfected with NaClO (10%) for 5 min, flushed with sterilized water, and placed in Petri dishes with filter paper moistened with water for their germination. Previously, they were subjected to a mechanical scarification treatment to break their dormancy.

Once germination had occurred, the seedlings were transferred to 0.5 L pots and then to 5 L pots with a substrate composed of soil and sand (2:1 v/v). After 45 days, when the plants were approximately 50 cm tall, metal solutions were applied by immersion for 24 h. Cu(II) was added in the form of $SO_4Cu{\cdot}5H_2O$ in three concentrations (500, 1000, and 1500 ppm) and Zn(II) in the form of $SO_4Zn{\cdot}7H_2O$ in three concentrations (1000, 2000, and 3000 ppm).

After 21 days of the application, plants were harvested to perform the different physiological and biochemical determinations.

2.2 Measurements performed

2.2.1 Biomass and leaf area

At harvest, the dry weight per plant (DW) was determined for all treatments by oven-drying them at 80°C until constant weight, distinguishing the shoot from roots.

2.2.2 Chlorophyll and carotene content

For all treatments, the contents of chlorophyll and carotene were determined from a 1 cm diameter leaf disk. Pigment content calculation was performed using Wellburn technique [8] with a Shimadzu UV 160-A spectrophotometer (Kyoto, Japan). The results were expressed in μg of chlorophyll cm^{-2} and μg of carotenoids cm^{-2}.

$$\text{Chlorophylla}\ (\mu g\ cm^{-2}) = 12 \times A_{643.8} - 3.11 \times A_{646} \quad (1)$$

$$\text{Chlorophyllb}(\mu g\ cm^{-2}) = 20.78 \times A_{646} - 488 \times A_{663.8} \quad (2)$$

$$\text{Totalchlorophyll}(a+b)(\mu g\ cm^{-2}) = 17.67 \times A_{646} + 7.12 \times A_{663.8} \quad (3)$$

$$\text{Carotenoids}\ (\mu g\ cm^{-2}) = \frac{(1000 \times A_{480} - 1.12\ \text{Ca} - 34.07\ \text{Cb})}{245} \quad (4)$$

where A is absorbance, Ca is chlorophyll a content, and Cb is chlorophyll b concent.

2.2.3 Soluble proteins content

The soluble protein content was measured from 100 mg of fresh leaves and root material, employing the Bradford method [9]. The protein content calculation was carried out using a standard curve prepared with different concentrations of bovine serum albumin (BSA) (SiFMa Chemical Co.).

2.2.4 Proline content

Proline determination was carried out taking 100 mg of fresh leaf and root material and homogenized with 2 ml of a 3% sulfosalicylic acid solution in water. The homogenate was centrifuged at 12,000g for 15 min, and 1 ml of the extract obtained was taken. Then 1 ml of the acidic ninhydrin reagent and 1 ml of glacial acetic acid were added to the extract in a 15 ml tube and put in a water bath at 100°C for an h. After this period, the reaction was stopped by rapidly cooling the tube. After, 2 mL of toluene was added to the above reaction mixture and vortexed for 15–20 s. The phases were allowed to separate and the aqueous phase containing the toluene-proline chromophore was taken. The absorbance at 520 nm was read using toluene as a blank. Proline content per unit of fresh weight was calculated according to:

$$\mu\text{mol proline} \cdot g^{-1}\ \text{FW} = \frac{\frac{\frac{\mu g\ \text{proline} \cdot ml^{-1}}{ml\ \text{toluene}}}{115.5\ \mu g \cdot \mu mol^{-1}}}{\frac{g\ FW}{5}} \quad (5)$$

where, FW is fresh weight.

2.2.5 Malondialdehyde content (MDA)

The amount of malondialdehyde (MDA) content in fresh tissues was determined by the reaction with thiobarbituric acid (TBA) described in the Heath and Packer method [10]. In total, 200 mg of fresh leaf tissue and 200 mg of fresh root tissue were ground with 1 ml of 0.1% trichloroacetic acid (TCA) and then centrifuged. The supernatant was reacted with 1 ml of the trichloroacetic acid (TCA), butylhydroxytoluene (BHT) and thiobarbituric acid (TBA) reagent (20% trichloroacetic acid (TCA), 0.37% thiobarbituric acid TBA and butylhydroxytoluene BHT 0.01 g), then the tubes were incubated for 30 min at 95°C. After this period, they were placed in an ice bath to rapidly stop the reaction, and then they were centrifuged at 10,000g for 10 min. Finally, the supernatant was separated, and the absorbance at 532 and 600 nm was read on a Shimadzu UV 160 UV/V spectrophotometer. The MDA concentration was calculated using an extinction coefficient of 155 mM^{-1} cm^{-1}:

$$\text{MDA equivalents}\left(\text{nmol ml}^{-1}\right) = \frac{A_{532} - A_{600}}{155,000} \tag{6}$$

where MDA is malondialdehyde content, A is the absorbance.

2.2.6 Relative conductivity (RC) of cell membranes

The determination of the relative conductivity (RC) of the cell membranes was made from 200 mg of fresh leaf material and 200 mg of fresh root material, from the different treatments, according to the Lutts method [11]. Immediately after sampling, the tissues were washed three times with redistilled water for 15 s, to remove the electrolytes adhering to the surface and those released by the wounds produced by the cut. Subsequently, each sample was immersed in a tube with 10 ml of double-distilled water where they remained for 4 h at room temperature. Following this, the electrical conductivity (dS m^{-1}) was determined using a Jenco model 3173 conductivity meter. Then, the tubes were capped and taken to an autoclave where they were kept for 20 min at a one-atmosphere pressure and 120°C, to affect the integrity of the membranes. Finally, the tubes were allowed to cool to room temperature, and the electrical conductivity of the medium was measured again. Based on the data obtained, the relative conductivity of cell membranes was estimated from the following formula:

$$RC\,(\%) = \left(\frac{L1}{L2}\right) \times 100 \tag{7}$$

where RC is the relative conductivity; $L1$ and $L2$ are the electrical conductivity readings before and after autoclaving, respectively.

2.2.7 Zn(II) and Cu(II) content in aerial part, root, and substrate

Plant tissues were digested in triplicate with concentrated perchloric and nitric acids in a 1:4 ratio (Merck, analytical grade), for the analyses of Cu(II) and Zn(II) (FAO & SIDA, 1983). Luoma method [12] was used to analyze the Cu(II) and Zn(II) labile fraction of sediments, being mineralized with hydrochloric acid (1 N, Merck analytical grade) by shaking for 24 h. Then, the absorbance was read using an atomic absorption spectrophotometer (Shimadzu AA6650F Atomic Absorption Spectrophotometer, Japan). The data obtained were employed for calculating the bioavailability, accumulation, translocations, and bioaccumulation indexes. All values were expressed on the dry weight of the respective sample [13].

$$\text{BAI} = \frac{\text{mg Zn(II)} \cdot \text{kg}^{-1}\text{in roots}}{\text{mg Zn(II)} \cdot \text{kg}^{-1}\text{in the substrate}} \tag{8}$$

$$\text{AI} = \frac{\text{mg Zn(II)} \cdot \text{kg}^{-1}\text{in aerial part}}{\text{mg Zn(II)} \cdot \text{kg}^{-1}\text{in the substrate}} \tag{9}$$

$$TI = \frac{\text{mg Zn(II)} \cdot \text{kg}^{-1}\text{in aerial part}}{\text{mg Zn(II)} \cdot \text{kg}^{-1}\text{in roots}} \quad (10)$$

$$BI = \frac{\text{mg Zn(II)} \cdot \text{kg}^{-1}\text{in the biomass}}{\text{mg Zn(II)} \cdot \text{kg}^{-1}\text{in the substrate}} \quad (11)$$

where BAI is bioavailability index and indicates if the metal is extracted and accumulated in the root; AI is accumulation index and indicates if the metal is extracted and accumulated in the aerial part; TI is translocation index and indicates if the metal is translocated to the aerial part; BI is bioaccumulation index and indicates if the metal is accumulated in the biomass.

2.2.8 Statistical analysis

The experimental design was fully randomized with a control (without addition of heavy metals solutions), two metals, and three concentrations for each one. The number of repetitions per treatment was $n = 5$. The data were subjected to analysis of variance (ANOVA) and the means compared by the 5% least significant difference test (LSD test) and the Pearson correlations using the software InfoStat version 2019.

3. Results

3.1 Growth, physiological and biochemical parameters

A negative effect on growth was found, expressed in a decrease in total biomass, as in **Figure 1A** and **B**. This result varied approximately 82 and 59% between the control (0 ppm) and the maximum concentration of Zn(II) (3000 ppm) and Cu(II) (1500 ppm), respectively. The dry weight of the root and the aerial part decreased by 82% for Zn(II), whereas 62 and 56% for Cu(II), respectively. A significant reduction was observed from the lowest concentration of Zn(II) (1000 ppm) while for Cu(II), this decrease was observed from the middle concentration (1000 ppm). The reduction of biomass, both shoot and root, shows the same pattern, as the metal concentration increases, the decrease of biomass becomes greater (**Figure 1A** and **B**).

Figure 1A and **B** represent chlorophyll and carotenes concentration in Zn (II) and Cu(II) systems, respectively. A significant decreased of chlorophyll and carotenes concentration was observed in Cu(II) treatment (1500 ppm) compared with the control (**Figure 1B**). This difference was approximately 47 and 16% for chlorophyll and carotenes content, respectively. However, chlorphyll and carotens concentration in Zn(II) systems (**Figure 1A**) were not affected.

Figure 3 shows the relativity conductivity (RC) percentage in roots and leaves in Zn(II) (A) and Cu(II) (B) systems. A gradual increase of relativity conductivity (RC) in roots with increasing Zn(II) and Cu(II) concentrations was noted. On the other hand, the RC in leaves biomass was not affected by Zn(II) and Cu(II) concentrations (**Figure 3A** and **B**).

Figure 4A and **B** represent malondialdehyde (MDA) content in the roots and leaves of *Canna indica* plants in Zn(II) (A) and Cu(II) (B) systems, respectively. As observed in **Figure 4A** and **B**, malondialdehyde (MDA) content in leaves had significant differences at maximum concentrations of Zn(II) and Cu(II) compared with the control. However, statistically significant increase of malondialdehyde (MDA) content was only detected in roots at 1500 ppm Cu(II) system (**Figure 4B**).

The soluble protein content in leaves and roots is shown in **Figure 5**. In general, it was determine there are not statistically significant differences of soluble protein content in roots for Zn(II) and Cu(II) systems, whereas the soluble protein content in leaves biomass decreased about 26% compared with the Cu(II) maximum concentration and the control (**Figure 5A** and **B**).

Figure 6 represents proline content in leaves and roots for Zn(II) and Cu(II) systems. The proline content in leaves increased with the increase of Zn(II) and Cu(II) concentrations, but statistically significant differences were determine only in the maximum concentrations for both metals compared with control system (**Figure 6A** and **B**).

3.2 Bioaccumulation and extraction of Zn(II) and Cu(II)

Figure 7A and **B** show the mean bioaccumulation values for Zn(II) and Cu(II) in shoot, roots, and total biomass of *Canna indica*, respectively. A higher bioaccumulation of Zn(II) and Cu(II) in the root than in the aerial part was observed. The results demonstrated that *C. indica* bioaccumulated 872.99 ± 694.68 mg Zn(II) kg^{-1} dry weight (DW) of total biomass (±SD), almost 77 times higher than the control (withouth heavy metal) (**Figure 7A**). The maximum concentration of Cu(II) in total biomass was 1432.15 ± 91.13 mg Cu(II) kg^{-1} DW (±SD) (**Figure 7B**).

On the other hand, the bioavailability (BAI), accumulation (AI), translocation (TI), and bioaccumulation (BI) indexes were calculated with the results mentioned above (**Table 1**). It was determined that BAI, AI, and BI indexes $>$ 1 for Zn(II) and Cu (II) system. These results mean *C. indica* plant was efficient in extracting Zn(II) and Cu(II) from the substrate. However, *C. indica* plant did not translocate Zn(II) and Cu (II) to the aerial part as TI index was $<$ 1 (**Table 1**).

3.3 Zn(II) and Cu(II) bioaccumulation correlated with physiological effects by Pearson stadistical method

Pearson coefficients (r) showed, for Zn(II), a significant negative correlation for shoot and root dry weight, whereas shoot malondialdehyde (MDA) and proline

Treatment	BAI (root/substrate)	AI (shoot/substrate)	TI (shoot/root)	BI (Biomass/substrate)
1000 ppm Zn(II)	5.409 ± 0.68	3.574 ± 0.32	0.663 ± 0.03	8.983 ± 1
2000 ppm Zn(II)	3.940 ± 0.12	1.922 ± 0.12	0.488 ± 0.04	5.862 ± 0.14
3000 ppm Zn(II)	14.283 ± 0.27	4.700 ± 0.27	0.329 ± 0.02	18.982 ± 0.31
500 ppm Cu(II)	3.966 ± 0.35	0.632 ± 0.02	0.160 ± 0.01	4.597 ± 0.37
1000 ppm Cu(II)	4.907 ± 0.74	0.706 ± 0.09	0.144 ± 0.01	5.613 ± 0.83
1500 ppm Cu(II)	2.540 ± 0.07	0.318 ± 0.02	0.125 ± 0.01	2.858 ± 0.09

Note: *(mean ± SD).*

Table 1.
BAI (bioavailability), AI (accumulation), TI (translocation), and BI (bioaccumulation) for Zn(II) and Cu(II) systems.

DOI: http://dx.doi.org/10.5772/intechopen.102450

Variable 1	Variable 2	Zn (II) (*r*)	*p*-value	Cu(II) (*r*)	*p*-value
Shoot Pearson correlation coefficient (r)					
Metal concentration	Shoot dry weight	−0.74	0.001*	−0.67	0.0048*
Metal concentration	Chlorophyll	−0.05	0.8524	−0.61	0.0113*
Metal concentration	Carotenes	0.33	0.2525	−0.36	0.175
Metal concentration	Relative conductivity	−0.23	0.3991	0.11	0.6825
Metal concentration	MDA content	0.53	0.0339*	0.32	0.2257
Metal concentration	Soluble proteins content	0.33	0.2068	−0.58	0.0195*
Metal concentration	Proline content	0.6	0.0144*	0.66	0.0053*
Root Pearson correlation coefficient					
Metal concentration	Root dry weight	−0.8	0.0002*	−0.78	0.0003*
Metal concentration	Relative conductivity	0.63	0.0086*	0.93	<0.0001*
Metal concentration	MDA content	−0.28	0.3254	0.44	0.1188
Metal concentration	Soluble proteins content	−0.1	0.6989	−0.22	0.4279
Metal concentration	Proline content	−0.03	0.9212	0.28	0.3267
Note: *Asterisks indicate significant differences* ($p < 0.05$)*, and* (r) *is Pearson correlation coefficient.*					

Table 2.
Zn(II) and Cu(II) bioaccumulation correlated with physiological effects by Pearson stastical method.

content and root-relative conductivity showed the opposite. For Cu(II), negative significant correlations were found for shoot dry weight, chlorophyll, and protein content while positive correlations were found for shoot proline content and root-relative conductivity. Positive correlations show an increase of both variables, whereas a negative correlation indicates a decrease in the second variable when the first variable increases (**Table 2**).

4. Discussion

4.1 Growth, physiological and biochemical parameters

Zinc is an essential trace element for normal plant growth. There are important enzymes that contain zinc, such as the enzyme alcohol dehydrogenase, carbonic anhydrase, ribonucleic acid (RNA) polymerase, and superoxide dismutase, a key enzyme in protection against oxidative stress. Zinc activates different enzymes responsible for the synthesis of certain proteins. It is involved in the formation of chlorophyll and some carbohydrates. It is essential in the formation of auxins, which help regulate stem development and elongation, in addition to being the precursor of tryptophan [14]. Copper also plays a key function in normal plant growth. For example, it participates in CO_2 assimilation and adenosine triphosphate (ATP) production [15]. It is the main constituent of diverse proteins such as plastocyanin of the photosynthetic system and cytochrome oxidase of the electron transport chain [16]. It plays a significant function in cell wall metabolism, signaling to the transcription protein trafficking apparatus, oxidative phosphorylation, iron armament, and biogenesis of molybdenum cofactor [17]. Both are essential micronutrients necessary for the correct growth and development of plants; however, in high concentrations, they turn out to be phytotoxic, generating various negative metabolism modifications.

The results of our experiment indicate that some physiological and biochemical parameters of *C. indica* were significantly different at high Zn(II) and Cu(II) concentrations (**Figures 1-6**). The biomass decreased (both aerial part and root) for both metals (**Figure 1**), but only Cu(II) treatments showed a decline in the content of chlorophyll and carotenes (**Figure 2**). Root-relative conductivity (RC) increased with the Zn(II) and Cu(II) increasing concentrations (**Figure 3**), and the same occurred for the malondialdehyde (MDA) content in shoots with both metals, whereas, in roots, only Cu(II) treatments showed an increase (**Figure 4**). The soluble proteins content increased in the roots of the plants treated with Zn(II) but decreased in shoots of Cu (II)-treated plants. (**Figure 5**). For proline shoot content, a decline was shown in the lowest concentrations of both metals but increased at the highest concentrations while, in roots, increased only in the lowest concentration of Zn(II) but then decreased again to the levels of control treatment, showing no significant difference (**Figure 6**).

The decrease observed in the biomass of *C. indica* is highly reported in this and other species for zinc [18–20] and copper [21, 22] toxicity as one of the most obvious symptoms of plants growing in these conditions.

The biomass reduction related to Zn(II) toxicity is a consequence of mitosis inhibition that causes growth alterations product of the inhibition of deoxyribonucleic acid (DNA) synthesis [23]. Also could be the result of the alteration in macronutrient absorption [24] or the micronutrient distribution in different parts of the plant [25] such as lower uptake of Fe^{+2} and Fe^{+3}; modification of the metabolic activity [26], inhibition of cellular division in the meristematic region, lengthening of root cells [27], reduction of cell viability, and death in the root tips [28].

Additionally, copper excess generates reactive oxygen species, which causes oxidative stress [29] that disrupts numerous metabolic pathways and modifies essential macromolecules [30]. Also, high copper concentrations cause negative modifications to DNA, photosynthesis, cell membrane integrity, enzyme activity, and respiration leading to general growth reduction [31]. Excess of copper in the roots can trigger alterations in the root system design that causes growth reduction, bronzing, necrosis, and nutritional inequities [32, 33].

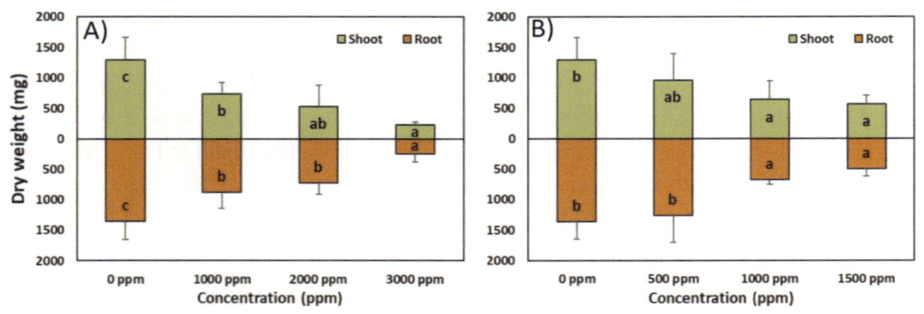

Figure 1.
Shoot and root dry weight (mg) of Canna indica *plants in Zn(II) (A) and Cu(II) (B) systems. Columns represent the mean* (n = 5)*, and vertical bars show the standard deviation (S.D.). Means followed by different letters (a-b-c) represent statistically significant differences* ($p < 0.05$)*, e.g., "a" is statistically different from "b" and "c", but not from "ab".*

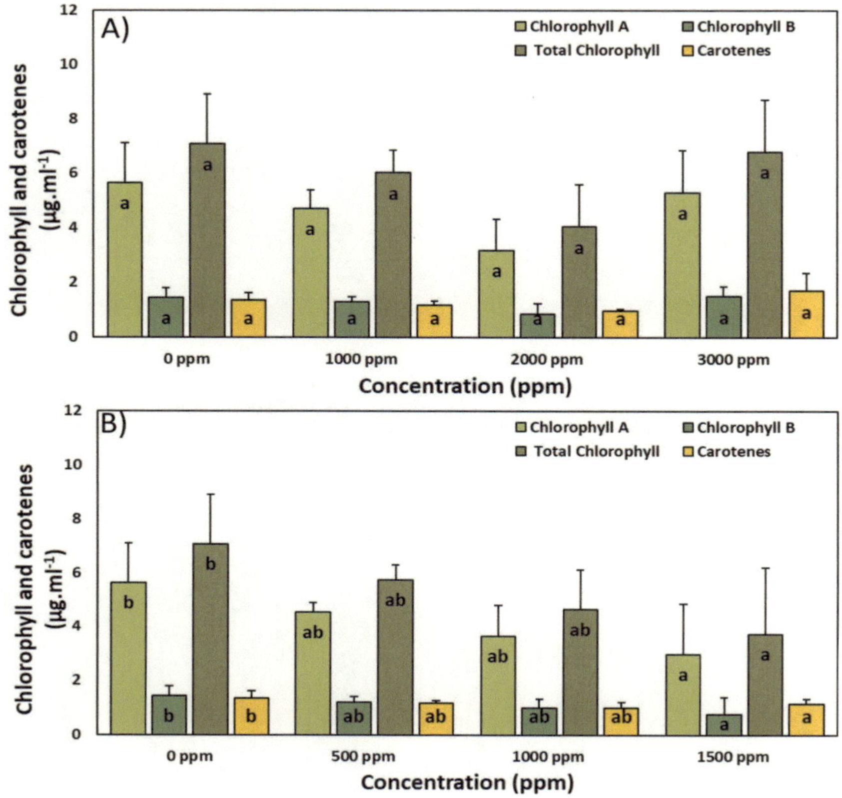

Figure 2.
Chlorophyll A, B, total and carotenes content of Canna indica *plant in Zn(II) (A) and Cu(II) (B) systems. Columns represent the mean (n = 5), and vertical bars show the standard deviation (S.D.). Means followed by different letters (a-b) represent statistically significant differences ($p < 0.05$), e.g., "a" is statistically different from "b", but not from "ab".*

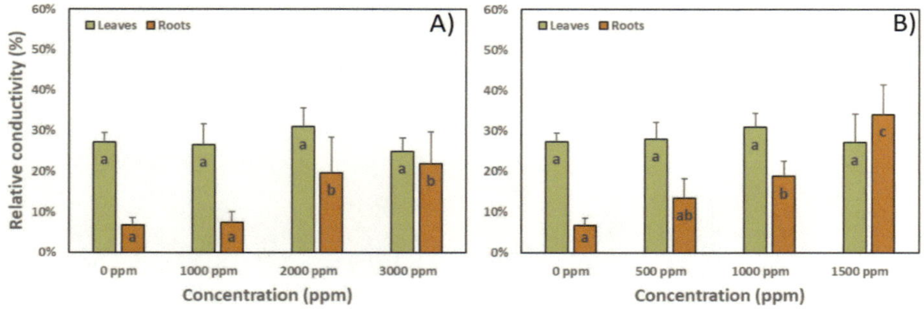

Figure 3.
Relative conductivity (RC) percentage (%) in roots and leaves biomass of Canna indica *plants in Zn(II) (A) and Cu(II) (B) systems. Columns represent the mean (n = 5), and vertical bars show the standard deviation (S.D.). Means followed by different letters (a-b-c) represent statistically significant differences ($p < 0.05$), e.g., "a" is statistically different from "b" and "c", but not from "ab".*

Zinc helps to maintain membrane integrity, preserving the structural orientation of macromolecules and protecting the transportation systems [18], but in high concentrations, triggers reactions that promote oxidative stress and the breakdown of membrane integrity [24]. Similar behavior happens with copper excess, causing the

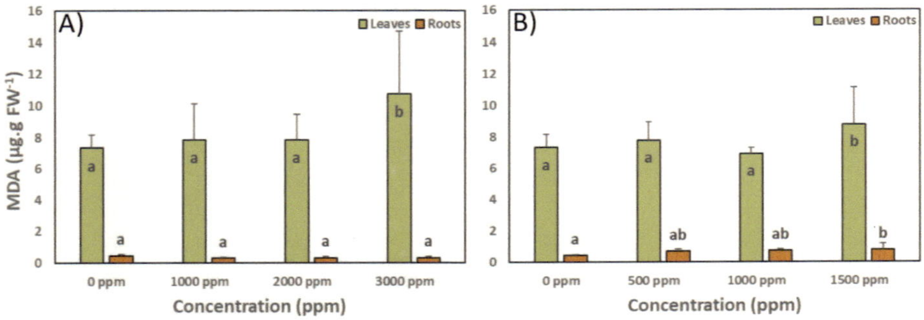

Figure 4.
Malondialdehyde (MDA) content in the roots and leaves of Canna indica *plant in Zn(II) (A) and Cu(II) (B) systems. Columns represent the mean (n=5), and vertical bars show the standard deviation (S.D.). Means followed by different letters (a-b) represent statistically significant differences (p < 0.05), e.g., "a" is statistically different from "b", but not from "ab".*

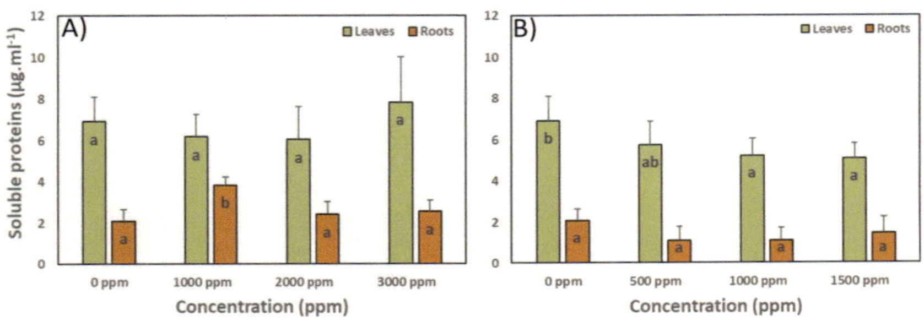

Figure 5.
Soluble protein content in the roots and leaves of Canna indica *plant in Zn(II) (A) and Cu(II) (B) systems. Columns represent the mean (n = 5), and vertical bars show the standard deviation (S.D.). Means followed by different letters (a-b) represent statistically significant differences (p < 0.05), e.g., "a" is statistically different from "b", but not from "ab".*

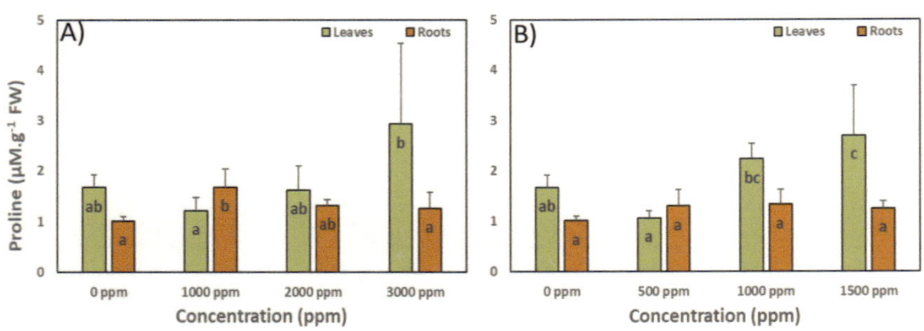

Figure 6.
Proline content in the roots and leaves of Canna indica *plants in Zn(II) (A) and Cu(II) (B) systems. Columns represent the mean (n = 5), and vertical bars show the standard deviation (S.D.). Means followed by different letters (a-b-c) represent statistically significant differences (p < 0.05), e.g., "b" is statistically different from "a" and "c", but not from "ab" and "bc".*

disruption of cell wall integrity and deposition of electron-dense material in the cytoplasmic membranes [34]. An increase in the relative conductivity (RC) of cellular membranes would indicate damage at the membrane level; higher values than 30%

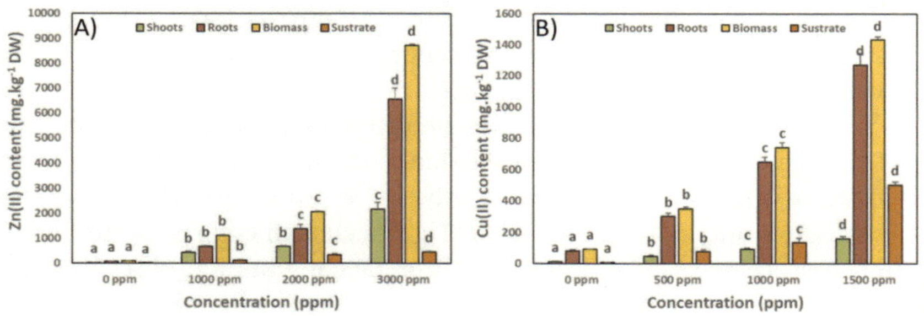

Figure 7.
(A) Zn(II) and (B) Cu(II) bioaccumulation in shoot, root, and total biomass of Canna indica *plants and heavy metal accumulation in substrate. Columns represent the mean* (n = 4)*, and vertical bars show the standard deviation (S.D.). Means followed by different letters (a-b-c-d) represent statistically significant differences* ($p < 0.05$)*, e.g., "a" is statistically different from "b", "c" and "d".*

indicate damage [35]. In this work, results show that RC significantly increased only in roots for both metals. However, the values obtained were relatively low, showing damage only in the highest concentrations. The degree of peroxidation of lipids and the degree of membrane damage are related and can be analyzed from the malondialdehyde (MDA) concentration and RC [36]. Increased levels of reactive oxygen species (ROS) caused by heavy metal stress could develop in damage to lipid membranes, proteins, pigments, and nucleic acids [37]. The malondialdehyde is a product of the lipid peroxidation of polyunsaturated fatty acids in cell membranes caused by oxidative stress and the production of ROS [35]. In this work, shoot MDA levels increased in the maximum concentration, in comparison to the control, for both metals, while in roots only copper treatments showed an increase in the maximum concentration. Also, this suggests that the antioxidant enzymes present in the roots of zinc treatments could have compensated the damage caused by ROS [38]. Similar results were found in different species such as *Salix fragilis* and *Salix aurita*, which showed an increase in the electrolytic leakage (similar parameter associated to relative conductivity) related to heavy metal concentrations [39], or *Canna orchioides*, which also showed an increase in the relative conductivity and MDA accumulation associated to this type of stress [40]. Metal-induced stress induces reactive oxygen species (ROS) generation, which can lead to lipid peroxidation, protein impairment, enzyme inactivation, and DNA damage [23]. Membrane disruption and lipid peroxidation are generally contemplated as dependable biomarkers of oxidative status in plants [24].

Another distinctive heavy metal toxicity symptom in plants is a reduction of the content of photosynthetic pigments [41]. They are directly related to photosynthesis and plant growth so, a decrease of the content of these pigments or damage done to chloroplasts results in lower CO_2 assimilation and a biomass decrease [42]. Carotenoids participate in antioxidant defense systems and impart a significant role in ROS sequestration [43], preventing the peroxidation of lipid membranes. [42]. Chloroplasts, mitochondria, and cellular membranes are some of the main sites that generate ROS. They are interconnected to the electron transport system, so when oxidative stress occurs, these sites are the first to be affected [44]. The decline in chlorophyll content in plants exposed to heavy metals stress is related to the inhibition of important enzymes, such as 6-aminolevulinic acid dehydratase (ALA-dehydratase) and protochlorophyllide reductase associated with chlorophyll biosynthesis, and the reduction of Mg^{+2} and Fe^{+2} supply. Zinc in phytotoxic concentrations may be equivalent to magnesium, causing processes of substitution of the central ion of the tetrapyrrolic chlorophyll ring, inhibiting its function and decreasing its concentration [45]. Similar effects are caused by excessive copper concentrations. Photosynthetic pigments decrease might be the result of displacement of magnesium required for chlorophyll biosynthesis or ultra-structural alteration of chloroplast under metal toxicity [46]. Also, this reduction might be due to the inhibited activities of various enzymes associated with chlorophyll biosynthesis [47]. A similar effect was observed in the present work but only with statistical significance in copper-treated *C. indica* plants where a decrease in chlorophyll and carotene contents was observed with the increment of this metal. This can be associated with the smaller biomass and the increment of oxidative stress indicated by the increase of MDA contents found in the highest concentrations of copper. Similar diminution in chlorophyll and carotenes caused by copper excess was found in different species such as *Citrus aurantium* [48], *Phragmites australis* [49], *Lemna minor* [50], and *Camellia sinensis* [51].

Shoot-soluble protein content of *C. indica* plants decreased with the increase of copper concentrations concerning the control, whereas the opposite was found in the roots of the lowest zinc treatment. Similar results were found in *L. minor* [52] and *Hordeum vulgare* [53] treated with high concentrations of heavy metals. The decrease in the level of soluble proteins is another symptom characteristic of the stress caused by metals [54]. Proteins not only can act as metal chelators; they can also act in the movement toward the interior of the cell, for compartmentalization in vacuoles, as well as the exterior by an ion flow [55]. Therefore, the increase of the protein content observed in the zinc-treated *C. indica* roots might be due to a nutritional boost caused by the lowest zinc concentration. Also, biosynthesis of various biomolecules is another way to tolerate zinc excess; this process includes the induction of metallochaperones, proteins of low molecular weight, or chelators such as nicotianamine, putrescine, spermine, mugineic acid, organic acids, glutathione, phytochelatins, and specific metallothioneins, such as proline and histidine [56]. A similar increment was found in different poplar clones [57] and was associated with antioxidant enzymes synthesis during oxidative stress induced by heavy metals. On the contrary, in this work, shoot-soluble protein content decreased in copper-treated *C. indica* plants. A similar reduction was found in *Brassica napus* growing on copper excess [58]. This decrease may be due to ROS generation. ROS are likely to target proteins that contain sulfur-containing amino acids and thiol groups [59]. Proteins can also be damaged in oxidative conditions by their reactions with lipid peroxidation products [60], and it can result in the deleterious effect of the normal protein form by disrupting the pathways and protein synthesis [61].

Proline is an amino acid that helps in activating many physiological and molecular responses in stress conditions. Its accumulation is a widespread response to heavy metal stress [62]. Shoot proline content con *C. indica* in this work showed a tendency to increase with the increment of both metal concentrations, whereas for roots only an increment in the first concentration of zinc treatment was observed. Proline accumulation increases the tolerance to heavy metals through several mechanisms, such as osmoregulation, stabilization of protein synthesis, and enzyme protection against denaturation [63]. It is suggested that proline accumulation is triggered by ROS, which allows their direct detoxification without the intervention of antioxidant enzymes [64]. Oxidative stress can lead to lipid peroxidation that produces a disruption at the cellular level, especially plasma membrane and leaking potassium from the plant cell; exogenous proline applications suppress the heavy metal induces [65]. Several authors found an increment in proline content in different species growing in excessive zinc [66–68] and copper [69–71] concentrations.

4.2 Bioaccumulation of Zn(II) and Cu(II)

Heavy metals are inorganic pollutants that cannot be degraded, so the principal strategy for plants should be to immobilize them in their rhizosphere, accumulate them in the roots, or translocate them to the aerial part [72]. They enter the root either by crossing the plasma membrane of the root endodermal cells, by entering the root apoplast through the space between cells, or with the aid of membrane transporter proteins. These transporters are present in membranes of different organelles such as tonoplasts, endoplasmic reticulums, mitochondria, or chloroplasts [73]. Inside the plant, they can be chelated by glutathione (GSH), phytochelatins (PCs), or metallothioneins (MTs), chelators that have thiol (—SH) groups, which gives them a high affinity for metal cations [74]. Also, this process may work synergistically with secondary stress-defensive antioxidative systems to combat metal-induced oxidative stress [75]. Metals in roots can be stored in vacuoles, cell walls or exported to the shoot via the xylem. Vacuoles are considered the main storage site for metals in plant cells, being a part of the tolerance mechanism [76].

In general, plants can contain, in their total biomass, Zn(II) in ranges from 30 to 100 mg kg^{-1} dry weight (DW); concentrations higher than 300 mg kg^{-1} DW are considered phytotoxic [77], but for other authors, this limit is set at 100 mg kg^{-1} DW [78]. For Cu(II), normal total biomass content ranges from 2 to 50 mg kg^{-1} DW, depending on the plant species. However, 5–20 mg kg^{-1} DW seems to be optimal, as toxicity symptoms appear above and deficiency symptoms below this critical range [79]. In the present work, *C. indica* accumulated values higher than the limits considered phytotoxic, reaching up to 8723.99 ± 694.68 mg kg^{-1} DW for Zn(II) ($\pm$SD) and 1432.15 ± 91.13 mg kg^{-1} DW for Cu(II) ($\pm$SD) in the total biomass in the maximum tested concentrations. Numerous authors showed the capacity of Zn(II) and Cu(II) accumulation of *C. indica* growing on different substrates [80–82].

Indexes are calculated to determine the phytoextraction efficiency, mainly being the bioaccumulation index (BI) and the translocation index (TI) [83]. An effective phytoextraction process requires the translocation of metals to easily harvestable parts. Plants with BI values less than 1 are unsuitable for phytoextraction. In this work, *C. indica* indexes suggest that this plant could act as a phytostabilizer because it showed low translocation to the aerial part but a high accumulation of both metals in the roots. Under this type of stress, the root suffers the first exposure, limiting transmission of heavy metals to other tissues [84]. Many studies found the same for the *Canna* genus for different heavy metals [85–87].

4.3 Correlation between physiological and biochemical parameters and Zn(II) and Cu(II) bioaccumulation: indicators for different applications

Some associations between physiological and biochemical parameters and the exposition of metals can be estimated by Pearson's correlation coefficient (r). In this work, *C. indica* plants showed a significant negative correlation for shoot ($r = -0.74$) and root dry weight ($r = -0.8$) in Zn(II) treatments and shoot dry weight ($r = -0.67$), chlorophyll ($r = -0.61$) and protein ($r = -0.58$) content in Cu(II) treatments showing that when the concentration of this metals increases, these parameters are affected negatively. The opposite occurred for shoot MDA ($r = 0.53$) and proline ($r = 0.6$) content and root-relative conductivity ($r = 0.63$) in Zn(II) treatments and shoot proline content ($r = 0.66$) and roots-relative conductivity ($r = 0.93$) in Cu(II) treatments. Proline accumulation in shoots, relative conductivity increment in roots, and the diminution of dry weight could be useful indicators of the strategies of this plant to overcome heavy metal stress and could be used to monitor the phytoremediation process.

The analysis of the correlation between metal accumulation and physiological parameters could be useful in different areas, such as variety selection, genetic improvement, environmental monitoring, or index construction as an indirect indicator of the phytoremediation process [88]. Various studies have demonstrated the correlation between metal accumulation and the antioxidant system. Antioxidant enzymes, such as superoxide dismutase (SOD), catalase (CAT), and peroxidase (POD), show an increased production to protect the plant from the damage caused by reactive oxygen species (ROS) under metals exposure [89]. Also, malondialdehyde (MDA) could act as an indicator of lipid peroxidation and is usually related to assessing oxidative damage [28]. Lipid peroxidation and oxidative damage cause alterations in metabolic processes [90] such as photosynthesis or protein productions leading to a decrease of photosynthetic pigments, less CO_2 assimilation, and diminution of biomass [91]. On the other hand, the accumulation of metabolites is another mechanism that plants use for stress tolerance. Proline is an amino acid that is involved in different stress mechanisms; it performs functions such as osmoregulation, stabilization of protein, and enzyme synthesis or even can chelate metal ions to help in the vacuolar sequestration [92]. These correlations are another way to demonstrate the tolerance mechanisms, and it helps to create comparations between species from the same genus or different cultivars to select the best for specific phytoremediation techniques becoming these, indicators of phytoremediation efficiency parallel to heavy metal accumulation [93].

Another use of these correlations is the construction of biomarkers. These represent the biological response to environmental disturbances or contamination, and they allow the detection of pollution at different contamination levels corresponding to concentrations difficult to achieve or when yield is not easy to form an integrative sample. There are three types of biomarkers: biomarkers of exposure: such as DNA breaks, stress proteins, and phytochelatins; biomarkers of effects such as morphological and physiological parameters; and biomarkers of susceptibility such as genetic mutations [94]. The use of such tools is currently increasing in the field of biomonitoring and bioremediation. Some biomarkers that have already been reported in plants are the following: oxidative stress by the production of reactive oxygen species [95], the reduction of macromorphological parameters such as plant height, stem diameter, and the number of leaves and negative modifications in chloroplasts with implications in photosynthesis [96]. These have been useful biomarkers for showing the adverse effects of metal exposition on the development, growth, and physiology of different plants exposed to this type of stress [97, 98].

5. Conclusion

Physiological and biochemical parameters are essential to understand the processes involved in the detoxification strategies employed by the plants during heavy metal stress. Some of them could be used as indirect indicators of the status of the phytoremediation process. In this work, *C. indica* plants could accumulate Zn(II) and Cu(II), mainly in roots. This affected some physiological and biochemical parameters due to the development of different physiological strategies, such as an increase of the antioxidant activity or the accumulation of proline, but these were not significant to produce high negative modifications in the physiological apparatus. Pearson analysis showed some negative correlations such as dry weight and chlorophyll, but also some positive correlations such as MDA, proline concentration, and relative conductivity, which could be useful to understand the strategies employed by *C. indica* plants to overcome heavy metal stress.

The plant could grow without great problems, accumulating high concentrations of both metals so it could be used in phytoremediation programs as a phytostabilization species, and parameters such as proline content, relative conductivity, and dry weight could be used to monitor the phytoremediation process.

Acknowledgements

This study was financial supported by the National Agency for Scientific and Technological Promotion of Argentina (PICT-2016-2535), National University of La Plata (UNLP), and National University of the Northwest of the Buenos Aires Province (UNNOBA). The authors like to thank Laura Wahnan (CONICET) and Cecilia Bernardelli (CONICET) for technical assistance.

Author details

Josefina Plaza Cazón[1]*, Matías Gonzalez[2] and Marcela Ruscitti[2,3]

1 CINDEFI (CCT-La Plata, CONICET, UNLP), La Plata, Argentina

2 INFIVE (CCT-La Plata, CONICET, UNLP), La Plata, Argentina

3 Department of Basic and Experimental Sciences, UNNOBA, Junín, Argentina

*Address all correspondence to: joplaca@hotmail.com

References

[1] Wei B, Yu J, Cao Z, Meng M, Yang L, Chen Q. The availability and accumulation of heavy metals in greenhouse soils associated with intensive fertilizer application. International Journal of Environmental Research and Public Health. 2020;**17**(15): 5359

[2] Liu Z, Tran KQ. A review on disposal and utilization of phytoremediation plants containing heavy metals. Ecotoxicology and Environmental Safety. 2021;**226**:112821

[3] Khan ZI, Ugulu I, Zafar A, Mehmood N, Bashir H, Ahmad K, et al. Biomonitoring of heavy metals accumulation in wild plants growing at soon valley, Khushab, Pakistan. Pakistan Journal of Botany. 2021;**53**(1):247-252

[4] Raza A, Charagh S, Najafi-Kakavand S, Siddiqui MH. The crucial role of jasmonates in enhancing heavy metals tolerance in plants. In: Jasmonates and Salicylates Signaling in Plants. Cham: Springer; 2021. pp. 159-183

[5] Subhashini V, Swamy AVVS. Phytoremediation of metal (Pb, Ni, Zn, Cd and Cr) contaminated soils using *Canna indica*. Current World Environment. 2014;**9**(3):780

[6] Cule N, Vilotic D, Nesic M, Veselinovic M, Drazic D, Mitovic S. Phytoremediation potential of *Canna indica* L. in water contaminated with lead. Fresenesius Environmental Bulletin. 2016;**25**(11):3728-3733

[7] Taufikurahman T, Pradisa MAS, Amalia SG, Hutahaean GEM. Phytoremediation of chromium (Cr) using *Typha angustifolia* L., *Canna indica* L. and *Hydrocotyle umbellata* L. in surface flow system of constructed wetland. In: IOP Conference Series: Earth and Environmental Science Vol. 308(1). Bristol, England: IOP Publishing; 2019. p. 012020

[8] Wellburn AR. The spectral determination of chlorophylls a and b, as well as total carotenoids, using various solvents with spectrophotometers of different resolution. Journal of Plant Physiology. 1994;**144**(3):307-313

[9] Bradford MM. A rapid and sensitive method for the quantitation of microgram quantities of protein utilizing the principle of protein-dye binding. Analytical Biochemistry. 1976;**72**(1–2): 248-254

[10] Heath RL, Packer L. Photoperoxidation in isolated chloroplasts: Kinetics and stoichiometry of fatty acid peroxidation. Archives of Biochemistry and Biophysics. 1968; **125**(1):189-198

[11] Lutts S, Kinet JM, Bouharmont J. NaCl-induced senescence in leaves of rice (*Oryza sativa* L.) cultivars differing in salinity resistance. Annals of Botany. 1996;**78**(3):389-398

[12] Luoma SN. Processes affecting metal concentrations in estuarine and coastal marine sediments. In: Furness RW, Rainbow PS, editors. Heavy Metals in the Marine Environment. USA: CRC Press; 1990

[13] Watson C, Pulford ID, Riddell-Black D. Screening of willow species for resistance to heavy metals: Comparison of performance in a hydroponics system and field trials. International Journal of Phytoremediation. 2003;**5**(4):351-365

[14] Castillo-González J, Ojeda-Barrios D, Hernández-Rodríguez A, González-

Franco AC, Robles-Hernández L, López-Ochoa GR. Zinc metalloenzymes in plants. Interciencia. 2018;**43**(4):242-248

[15] Marques DM, Júnior VV, da Silva AB, Mantovani JR, Magalhães PC, de Souza TC. Copper toxicity on photosynthetic responses and root morphology of *Hymenaea courbaril* L. (Caesalpinioideae). Water, Air, & Soil Pollution. 2018;**229**(5):1-14

[16] Zeng Q, Ling Q, Wu J, Yang Z, Liu R, Qi Y. Excess copper-induced changes in antioxidative enzyme activity, mineral nutrient uptake and translocation in sugarcane seedlings. Bulletin of Environmental Contamination and Toxicology. 2019; **103**(6):834-840

[17] Kumar V, Pandita S, Sidhu GPS, Sharma A, Khanna K, Kaur P, et al. Copper bioavailability, uptake, toxicity and tolerance in plants: A comprehensive review. Chemosphere. 2021;**262**:127810

[18] Gonzalez MA, Ruscitti MF, Plaza Cazón JDC, Arango MC. Bioaccumulation and physiological responses of Festuca arundinacea (Poaceae) to Zn (II) excess. Agronomía & Ambiente. 2021;**41**(1):13-21

[19] Rouphael Y, Colla G, Bernardo L, Kane D, Trevisan M, Lucini L. Zinc excess triggered polyamines accumulation in lettuce root metabolome, as compared to osmotic stress under high salinity. Frontiers in Plant Science. 2016;**7**:842

[20] Jayasri MA, Suthindhiran K. Effect of zinc and lead on the physiological and biochemical properties of aquatic plant Lemna minor: Its potential role in phytoremediation. Applied Water Science. 2017;**7**(3):1247-1253

[21] Cambrollé J, García JL, Figueroa ME, Cantos M. Evaluating wild grapevine tolerance to copper toxicity. Chemosphere. 2015;**120**:171-178

[22] Theriault G, Nkongolo K. Nickel and copper toxicity and plant response mechanisms in white birch (*Betula papyrifera*). Bulletin of Environmental Contamination and Toxicology. 2016; **97**(2):171-176

[23] Balafrej H, Bogusz D, Triqui ZEA, Guedira A, Bendaou N, Smouni A, et al. Zinc hyperaccumulation in plants: A review. Plants. 2020;**9**(5):562

[24] Kaya C, Ashraf M, Akram NA. Hydrogen sulfide regulates the levels of key metabolites and antioxidant defense system to counteract oxidative stress in pepper (*Capsicum annuum* L.) plants exposed to high zinc regime. Environmental Science and Pollution Research. 2018;**25**(13):12612-12618

[25] Sidhu GPS. Physiological, biochemical and molecular mechanisms of zinc uptake, toxicity and tolerance in plants. Journal of Global Biosciences. 2016;**5**(9):4603-4633

[26] Sidhu GPS, Bali AS, Singh HP, Batish DR, Kohli RK. Insights into the tolerance and phytoremediation potential of *Coronopus didymus* L. (Sm) grown under zinc stress. Chemosphere. 2020;**244**:125350

[27] Glińska S, Gapińska M, Michlewska S, Skiba E, Kubicki J. Analysis of *Triticum aestivum* seedling response to the excess of zinc. Protoplasma. 2016;**253**(2):367-377

[28] Li X, Yang Y, Jia L, Chen H, Wei X. Zinc-induced oxidative damage, antioxidant enzyme response and proline metabolism in roots and leaves of wheat plants. Ecotoxicology and Environmental Safety. 2013;**89**:150-157

[29] Huang WL, Wu FL, Huang HY, Huang WT, Deng CL, Yang LT, et al. Excess copper-induced alterations of protein profiles and related physiological parameters in citrus leaves. Plants. 2020; **9**(3):291

[30] Juang KW, Lo YC, Chen TH, Chen BC. Effects of copper on root morphology, cations accumulation, and oxidative stress of grapevine seedlings. Bulletin of Environmental Contamination and Toxicology. 2019; **102**(6):873-879

[31] Nazir F, Hussain A, Fariduddin Q. Hydrogen peroxide modulate photosynthesis and antioxidant systems in tomato (*Solanum lycopersicum* L.) plants under copper stress. Chemosphere. 2019;**230**:544-558

[32] De Conti L, Ceretta CA, Melo GW, Tiecher TL, Silva LO, Garlet LP, et al. Intercropping of young grapevines with native grasses for phytoremediation of Cu-contaminated soils. Chemosphere. 2019;**216**:147-156

[33] Marastoni L, Sandri M, Pii Y, Valentinuzzi F, Cesco S, Mimmo T. Morphological root responses and molecular regulation of cation transporters are differently affected by copper toxicity and cropping system depending on the grapevine rootstock genotype. Frontiers in Plant Science. 2019;**10**:946

[34] Minkina T, Rajput V, Fedorenko G, Fedorenko A, Mandzhieva S, Sushkova S, et al. Anatomical and ultrastructural responses of Hordeum sativum to the soil spiked by copper. Environmental Geochemistry and Health. 2020;**42**(1):45-58

[35] Ruscitti M, Arango M, Beltrano J. Improvement of copper stress tolerance in pepper plants (*Capsicum annuum* L.) by inoculation with arbuscular mycorrhizal fungi. Theoretical and Experimental Plant Physiology. 2017; **29**(1):37-49

[36] Wang Y, Yu Q, Li Y, Li J, Chen J, Liu Z, et al. Mechanisms of nitric oxide in the regulation of chilling stress tolerance in Camellia sinensis. Horticulturae. 2021;7(10):410

[37] Sosa-Torres ME, Saucedo-Vázquez JP, Kroneck PM. The magic of dioxygen, Sustaining Life on Planet Earth: Metallo enzymes Mastering Dioxygen and Other Chewy Gases. Metal Ions in Life Sciences. In: Kroneck P, Sosa Torres M, editors. Vol. 15. Switzerland: Springer, Cham; 2015. pp. 1-12. DOI: 10.1007/978-3-319-12415-5_1

[38] Jan S, Parray JA. Approaches to Heavy Metal Tolerance in Plants. Singapore: Springer; 2016. pp. 1-18

[39] Evlard A, Sergeant K, Printz B, Guignard C, Renaut J, Campanella B, et al. A multiple-level study of metal tolerance in *Salix fragilis* and *Salix aurita* clones. Journal of Proteomics. 2014;**101**: 113-129

[40] Pan X, Zhao Q, Zhao T. Plant growth, antioxidative enzyme, and cadmium tolerance responses to cadmium stress in Canna orchioides. Horticultural Plant Journal. 2021;7(3): 256-266

[41] Saleem MH, Ali S, Rehman M, Rizwan M, Kamran M, Mohamed IA, et al. Individual and combined application of EDTA and citric acid assisted phytoextraction of copper using jute (*Corchorus capsularis* L.) seedlings. Environmental Technology & Innovation. 2020;**19**:100895

[42] Laidinen GF, Kaznina NM, Batova YV, Titov AF. The resistance of

Phleum pratense and Elytrigia repens to high concentrations of zinc. Biology Bulletin. 2018;**45**(5):454-460

[43] Arif N, Yadav V, Singh S, Kushwaha BK, Singh S, Tripathi DK, et al. Assessment of antioxidant potential of plants in response to heavy metals. In: Plant Responses to Xenobiotics. Singapore: Springer; 2016. pp. 97-125

[44] Gill SS, Tuteja N. Reactive oxygen species and antioxidant machinery in abiotic stress tolerance in crop plants. Plant Physiology and Biochemistry. 2010;**48**(12):909-930

[45] Bechaieb R, Fredj AB, Akacha AB, Gérard H. Interactions of copper (II) and zinc (II) with chlorophyll: Insights from density functional theory studies. New Journal of Chemistry. 2016;**40**(5): 4543-4549

[46] Rahman Z, Singh VP. The relative impact of toxic heavy metals (THMs) (arsenic (As), cadmium (Cd), chromium (Cr) (VI), mercury (Hg), and lead (Pb) on the total environment: An overview. Environmental Monitoring and Assessment. 2019;**191**(7):1-21

[47] Sánchez-Pardo B, Fernández-Pascual M, Zornoza P. Copper microlocalisation and changes in leaf morphology, chloroplast ultrastructure and antioxidative response in white lupin and soybean grown in copper excess. Journal of Plant Research. 2014;**127**(1):119-129

[48] Giannakoula A, Therios I, Chatzissavvidis C. Effect of lead and copper on photosynthetic apparatus in citrus (*Citrus aurantium* L.) plants. The role of antioxidants in oxidative damage as a response to heavy metal stress. Plants. 2021;**10**(1):155

[49] Wu J, Hu J, Wang L, Zhao L, Ma F. Responses of *Phragmites australis* to copper stress: A combined analysis of plant morphology, physiology and proteomics. Plant Biology. 2021;**23**(2): 351-362

[50] Bodnar IS, Cheban EV. Combined action of gamma radiation and exposure to copper ions on *Lemna minor* L. International Journal of Radiation Biology. 2021:1-10

[51] Dey S, Mazumder PB, Paul SB. Effect of copper on growth and chlorophyll content in tea plants (*Camellia sinensis* (L.) O. Kuntze). International Journal of Research in Applied, Natural and Social Sciences. 2014;**2**(5):223-230

[52] Hou W, Chen X, Song G, Wang Q, Chang CC. Effects of copper and cadmium on heavy metal polluted waterbody restoration by duckweed (*Lemna minor*). Plant Physiology and Biochemistry. 2007;**45**(1):62-69

[53] Guo TR, Zhang GP, Zhang YH. Physiological changes in barley plants under combined toxicity of aluminum, copper and cadmium. Colloids and Surfaces B: Biointerfaces. 2007;**57**(2): 182-188

[54] Radić S, Babić M, Škobić D, Roje V, Pevalek-Kozlina B. Ecotoxicological effects of aluminum and zinc on growth and antioxidants in *Lemna minor* L. Ecotoxicology and Environmental safety. 2010;**73**(3):336-342

[55] Akladious SA, Mohamed HI. Physiological role of exogenous nitric oxide in improving performance, yield and some biochemical aspects of sunflower plant under zinc stress. Acta Biologica Hungarica. 2017;**68**(1):101-114

[56] Emamverdian A, Ding Y, Mokhberdoran F, Xie Y. Heavy metal stress and some mechanisms of plant

defense response. The Scientific World Journal. 2015;**2015**:756120

[57] Trudic B, Kebert M, Popovic BM, Štajner D, Orlovic S, Galovic V. The level of oxidative stress in poplars due to heavy metal pollution in soil. Baltic Forestry. 2012;**18**:214-227

[58] Habiba U, Ali S, Farid M, Shakoor MB, Rizwan M, Ibrahim M, et al. EDTA enhanced plant growth, antioxidant defense system, and phytoextraction of copper by *Brassica napus* L. Environmental Science and Pollution Research. 2015;**22**(2): 1534-1544

[59] Hancock J, Desikan R, Harrison J, Bright J, Hooley R, Neill S. Doing the unexpected: Proteins involved in hydrogen peroxide perception. Journal of Experimental Botany. 2006;**57**(8): 1711-1718

[60] Pitzschke A, Forzani C, Hirt H. Reactive oxygen species signaling in plants. Antioxidants & Redox Signaling. 2006;**8**(9–10):1757-1764

[61] Goswami S, Das S. Copper phytoremediation potential of *Calandula officinalis* L. and the role of antioxidant enzymes in metal tolerance. Ecotoxicology and Environmental Safety. 2016;**126**:211-218

[62] Szepesi Á, Szőllősi R. Mechanism of proline biosynthesis and role of proline metabolism enzymes under environmental stress in plants. In: Plant Metabolites and Regulation Under Environmental Stress. USA: Academic Press; 2018. pp. 337-353

[63] Siddique A, Kandpal G, Kumar P. Proline accumulation and its defensive role under diverse stress condition in plants: An overview. Journal of Pure and Applied Microbiology. 2018;**12**(3): 1655-1659

[64] Ku HM, Tan CW, Su YS, Chiu CY, Chen CT, Jan FJ. The effect of water deficit and excess copper on proline metabolism in Nicotiana benthamiana. Biologia Plantarum. 2012;**56**(2):337-343

[65] Hasanuzzaman M, Alam M, Rahman A, Hasanuzzaman M, Nahar K, Fujita M. Exogenous proline and glycine betaine mediated upregulation of antioxidant defense and glyoxalase systems provides better protection against salt-induced oxidative stress in two rice (*Oryza sativa* L.) varieties. BioMed Research International. 2014; **2014**:757219

[66] Al Khateeb W, Al-Qwasemeh H. Cadmium, copper and zinc toxicity effects on growth, proline content and genetic stability of *Solanum nigrum* L., a crop wild relative for tomato; comparative study. Physiology and Molecular Biology of Plants. 2014;**20**(1): 31-39

[67] Redondo-Gómez S. Abiotic and biotic stress tolerance in plants. In: Molecular Stress Physiology of Plants. India: Springer; 2013. pp. 1-20

[68] Vijayarengan P, Mahalakshmi G. Changes in the photosynthetic responses and biochemical constituents of tomato (*Lycopersicon esculentum* Mill.) under zinc stress. World Applied Sciences Journal. 2013;**24**(5):654-659

[69] Szabados L, Savouré A. Proline: A multifunctional amino acid. Trends in Plant Science. 2010;**15**(2):89-97

[70] Dar MI, Naikoo MI, Rehman F, Naushin F, Khan FA. Proline accumulation in plants: Roles in stress tolerance and plant development. In: Osmolytes and Plants Acclimation to

Changing Environment: Emerging Omics Technologies. New Delhi: Springer; 2016. pp. 155-166

[71] Pena LB, Méndez AA, Matayoshi CL, Zawoznik MS, Gallego SM. Early response of wheat seminal roots growing under copper excess. Plant Physiology and Biochemistry. 2015;**87**:115-123

[72] Saxena G, Purchase D, Mulla SI, Saratale GD, Bharagava RN. Phytoremediation of heavy metal-contaminated sites: Eco-environmental concerns, field studies, sustainability issues, and future prospects. Reviews of Environmental Contamination and Toxicology. 2019;**249**:71-131

[73] Komal T, Mustafa M, Ali Z, Kazi AG. Heavy metal uptake and transport in plants. In: Heavy Metal Contamination of Soils. Champions: Springer; 2015. pp. 181-194

[74] Machado-Estrada B, Calderón J, Moreno-Sánchez R, Rodríguez-Zavala JS. Accumulation of arsenic, lead, copper, and zinc, and synthesis of phytochelatins by indigenous plants of a mining impacted area. Environmental Science and Pollution Research. 2013;**20**(6):3946-3955

[75] Berni R, Luyckx M, Xu X, Legay S, Sergeant K, Hausman JF, et al. Reactive oxygen species and heavy metal stress in plants: Impact on the cell wall and secondary metabolism. Environmental and Experimental Botany. 2019;**161**:98-106

[76] Guo J, Xu W, Ma M. The assembly of metals chelation by thiols and vacuolar compartmentalization conferred increased tolerance to and accumulation of cadmium and arsenic in transgenic *Arabidopsis thaliana*. Journal of Hazardous Materials. 2012;**199**:309-313

[77] Bharagava RN, Saxena G, Mulla SI. Introduction to industrial wastes containing organic and inorganic pollutants and bioremediation approaches for environmental management. In: Saxena G, Bharagava R, editors. Bioremediation of Industrial Waste for Environmental Safety. Springer: Singapore; 2020. pp. 1-18. DOI: 10.1007/978-981-13-1891-7

[78] Hajiboland R, Amirazad F. Growth, photosynthesis and antioxidant defense system in Zn-deficient red cabbage plants. Plant, Soil and Environment. 2010;**56**(5):209-217

[79] Reichman SM. The Responses of Plants to Metal Toxicity: A Review Forusing on Copper, Manganese & Zinc. Melbourne: Australian Minerals & Energy Environment Foundation; 2002. pp. 22-26

[80] Sricoth T, Meeinkuirt W, Saengwilai P, Pichtel J, Taeprayoon P. Aquatic plants for phytostabilization of cadmium and zinc in hydroponic experiments. Environmental Science and Pollution Research. 2018;**25**(15): 14964-14976

[81] Premarathne HKMMP, Udagedara T, Young SM. Assessment of Phytoremediation to Treat Selected Metals in TextileWastewater. In: International Research Conference of UWU. Sri Lanka; 2019

[82] El Faiz A, Duponnois R, Winterton P, Ouhammou A, Meddich A, Boularbah A, et al. Effect of different amendments on growing of *Canna indica* L. inoculated with AMF on mining substrate. International Journal of Phytoremediation. 2015;**17**(5):503-513

[83] Antoniadis V, Levizou E, Shaheen SM, Ok YS, Sebastian A, Baum C, et al. Trace elements in the soil-plant interface: Phytoavailability, translocation, and phytoremediation–A

review. Earth-Science Reviews. 2017;**171**: 621-645

[84] Usman K, Al-Ghouti MA, Abu-Dieyeh MH. The assessment of cadmium, chromium, copper, and nickel tolerance and bioaccumulation by shrub plant Tetraena qataranse. Scientific Reports. 2019;**9**(1):1-11

[85] Bose S, Jain A, Rai V, Ramanathan AL. Chemical fractionation and translocation of heavy metals in *Canna indica* L. grown on industrial waste amended soil. Journal of Hazardous Materials. 2008;**160**(1):187-193

[86] Nevena C, Ljubinko J, Dragana D, Milorad V, Suzana M, Marija N. Potential use of *Canna indica* L. for Phytoremediation of Heavy Metals. Republic of Macedomia: Balwois-Ohrid; 2012. pp. 1-8

[87] Cule N, Lucic A, Nesic M, Veselinovic M, Mitrovic S, Sredojevic Z, et al. Accumulation of chromium and nickel by *Canna indica* and decorative macrophytes grown in floating treatment wetland. Fresenius Environmental Bulletin. 2021;**30**(6b):7881-7890

[88] Krayem M, El Khatib S, Hassan Y, Deluchat V, Labrousse P. In search for potential biomarkers of copper stress in aquatic plants. Aquatic Toxicology. 2021; **239**:105952

[89] Liang L, Liu W, Sun Y, Huo X, Li S, Zhou Q. Phytoremediation of heavy metal contaminated saline soils using halophytes: Current progress and future perspectives. Environmental Reviews. 2017;**25**(3):269-281

[90] Rasheed F, Zafar Z, Waseem ZA, Rafay M, Abdullah M, Salam MMA, et al. Phytoaccumulation of Zn, Pb, and Cd in Conocarpus lancifolius irrigated with wastewater: Does physiological response influence heavy metal uptake? International Journal of Phytoremediation. 2020;**22**(3):287-294

[91] Chtouki M, Naciri R, Soulaimani A, Zeroual Y, El Gharous M, Oukarroum A. Effect of cadmium and phosphorus interaction on tomato: Chlorophyll a fluorescence, plant growth, and cadmium translocation. Water, Air, & Soil Pollution. 2021;**232**(3):1-11

[92] Ghosh UK, Islam MN, Siddiqui MN, Cao X, Khan MAR. Proline, a multifaceted signalling molecule in plant responses to abiotic stress: Understanding the physiological mechanisms. Plant Biology. 2021;**24**(2): 227-239

[93] Yan A, Wang Y, Tan SN, Mohd Yusof ML, Ghosh S, Chen Z. Phytoremediation: A promising approach for revegetation of heavy metal-polluted land. Frontiers in Plant Science. 2020;**11**:359

[94] Hook SE, Gallagher EP, Batley GE. The role of biomarkers in the assessment of aquatic ecosystem health. Integrated Environmental Assessment and Management. 2014;**10**(3):327-341

[95] Ercal N, Gurer-Orhan H, Aykin-Burns N. Toxic metals and oxidative stress part I: Mechanisms involved in metal-induced oxidative damage. Current Topics in Medicinal Chemistry. 2001;**1**(6):529-539

[96] Tovar-Sánchez E, Cervantes-Ramírez T, Castañeda-Bautista J, Gómez-Arroyo S, Ortiz-Hernández L, Sánchez-Salinas E, et al. Response of *Zea mays* to multimetal contaminated soils: A multibiomarker approach. Ecotoxicology. 2018;**27**(8):1161-1177

[97] Goncalves AC Jr, Schwantes D, de Sousa RFB, da Silva TRB, Guimaraes VF,

Campagnolo MA, et al. Phytoremediation capacity, growth and physiological responses of Crambe abyssinica Hochst on soil contaminated with Cd and Pb. Journal of Environmental Management. 2020;**262**: 110342

[98] Santoyo-Martínez M, Mussali-Galante P, Hernández-Plata I, Valencia-Cuevas L, Flores-Morales A, Ortiz-Hernández L, et al. Heavy metal bioaccumulation and morphological changes in Vachellia campechiana (Fabaceae) reveal its potential for phytoextraction of Cr, Cu, and Pb in mine tailings. Environmental Science and Pollution Research. 2020;**27**(10): 11260-11276